T0205950

Urban Ecology

This is the urban century. For the first time the majority of people live in towns and cities. Understanding how people influence, and are influenced by, the 'green' component of these environments is therefore of enormous significance.

Providing an overview of the essentials of urban ecology, the book begins by covering the vital background concepts of the urbanisation process and the effect that it can have on ecosystem functions and services. Later sections are devoted to examining how species respond to urbanisation, the many facets of human–ecology interactions, and the issues surrounding urban planning and the provision of urban green spaces.

Drawing on examples from urban settlements around the world, it highlights the progress to date in this burgeoning field, as well as the challenges that lie ahead.

KEVIN J. GASTON is Professor of Biodiversity and Conservation in the Department of Animal and Plant Sciences, University of Sheffield. He has 20 years of research experience in environmental science, conducting studies both in the UK and overseas on a wide range of topics, including conservation prioritisation, ecosystem services, invasive species, people–wildlife interactions, and urban ecology.

Ecological Reviews

Ecological Reviews publishes books at the cutting edge of modern ecology, providing a forum for volumes that discuss topics that are focal points of current activity and likely long-term importance to the progress of the field. The series is an invaluable source of ideas and inspiration for ecologists at all levels from graduate students to more-established researchers and professionals. The series has been developed jointly by the British Ecological Society and Cambridge University Press and encompasses the Society's Symposia as appropriate.

Biotic Interactions in the Tropics: Their Role in the Maintenance of Species Diversity
Edited by David F. R. P. Burslem, Michelle A. Pinard and Sue E. Hartley

Biological Diversity and Function in Soils
Edited by Richard D. Bardgett, Michael B. Usher and David W. Hopkins

Island Colonization: The Origin and Development of Island Communities
By Ian Thornton
Edited by Tim New

Scaling Biodiversity
Edited by David Storch, Pablo A. Marquet and James H. Brown

Body Size: The Structure and Function of Aquatic Ecosystems
Edited by Alan Hildrew, Dave Raffaelli and Ronni Edmonds-Brown

Speciation and Patterns of Diversity
Edited by Roger Butlin, Jon Bridle and Dolph Schluter

Ecology of Industrial Pollution
Edited by Lesley C. Batty and Kevin B. Hallberg

Ecosystem Ecology: A New Synthesis
Edited by David G. Raffaelli and Christopher L. J. Frid

Urban Ecology

Edited by

KEVIN J. GASTON
University of Sheffield

CAMBRIDGE
UNIVERSITY PRESS

University Printing House, Cambridge CB2 8BS, United Kingdom

Cambridge University Press is part of the University of Cambridge.

It furthers the University's mission by disseminating knowledge in the pursuit of education, learning and research at the highest international levels of excellence.

www.cambridge.org
Information on this title: www.cambridge.org/9780521743495

First published 2010
3rd printing 2013

A catalogue record for this publication is available from the British Library

Library of Congress Cataloguing in Publication data

Urban ecology / [edited by] Kevin J. Gaston.
    p.   cm. – (Ecological reviews)
    ISBN 978-0-521-76097-3 (Hardback) – ISBN 978-0-521-74349-5 (pbk.)
    1. Urban ecology (Sociology)   2. City planning–Economic aspects.   3. Sustainable development.   I. Gaston, Kevin J.   II. Title.   III. Series.
    HT241.U7243 2010
    307.76–dc22

                                                                    2010031323

ISBN 978-0-521-76097-3 Hardback
ISBN 978-0-521-74349-5 Paperback

Contents

Contributors

JO BARTON
Centre for Environment & Society,
Department of Biological Sciences,
University of Essex, Colchester,
Essex CO4 3SQ, UK

ADAM BATES
School of Geography,
Earth and Environmental Sciences,
Birmingham University,
Birmingham B15 2TT, UK

CHRISTOPHER BOONE
School of Human Evolution
and Social Change,
Arizona State University,
Tempe, AZ 85287, USA

ZOE G. DAVIES
Biodiversity and Macroecology Group,
Department of Animal and Plant
Sciences,
University of Sheffield,
Sheffield S10 2TN, UK

JILL L. EDMONDSON
Biodiversity and Macroecology Group,
Department of Animal and Plant
Sciences, University of Sheffield,
Sheffield S10 2TN, UK

KARL L. EVANS
Department of Animal and
Plant Sciences,
University of Sheffield,
Sheffield S10 2TN, UK

RICHARD A. FULLER
The Ecology Centre,
The University of Queensland,
St Lucia, Qld 4072, Australia

KEVIN J. GASTON
Biodiversity and Macroecology
Group,
Department of Animal and Plant
Sciences,
University of Sheffield,
Sheffield S10 2TN, UK

JAMES HALE
School of Geography,
Earth and Environmental Sciences,
Birmingham University,
Birmingham B15 2TT, UK

SHARON L. HARLAN
School of Human Evolution
and Social Change,
Arizona State University,
Tempe, AZ 85287, USA

KATHERINE N. IRVINE
Institute of Energy and Sustainable
Development,
De Montfort University,
The Gateway, Leicester LE1 9BH, UK

PHILIP JAMES
School of Environment and Life
Sciences,
University of Salford, Salford,
Greater Manchester M5 4WT, UK

ANN P. KINZIG
School of Life Sciences,
Arizona State University,
Tempe, AZ 85287, USA

STEFAN KLOTZ
Helmholtz Centre for Environmental
Research – UFZ,
Department of Community Ecology,
Theodor-Lieser-Str. 4, 06120 Halle,
Germany

INGOLF KÜHN
Helmholtz Centre for Environmental
Research – UFZ,
Department of Community Ecology,
Theodor-Lieser-Str. 4, 06120 Halle,
Germany

SUSANNAH B. LERMAN
Graduate Program in Organismic
and Evolutionary Biology,
University of Massachusetts-Amherst,
Amherst, MA 01003, USA

GARY W. LUCK
Institute for Land, Water and Society,
Charles Sturt University,

PO Box 789, Albury,
NSW 2640, Australia

MICHAEL L. McKINNEY
Department of Earth & Planetary
Sciences,
University of Tennessee,
Knoxville,
TN 37916, USA

PAUL OPDAM
Department of Land Use Planning &
Alterra Landscape Center,
Wageningen University Center,
PO Box 47, 6700 AA Wageningen,
The Netherlands

JULES PRETTY
Centre for Environment & Society,
Department of Biological Sciences,
University of Essex, Colchester,
Essex CO4 3SQ, UK

JON SADLER
School of Geography,
Earth and Environmental Sciences,
Birmingham University,
Birmingham B15 2TT, UK

EYAL SHOCHAT
Global Institute of Sustainability,
Arizona State University,
Tempe, AZ 85287, USA

LISA T. SMALLBONE
Institute for Land, Water and
Society,
Charles Sturt University,
PO Box 789, Albury,
NSW 2640 Australia

ROBBERT SNEP
Department of Land Use Planning &
Alterra Landscape Center,
Wageningen University Center,
PO Box 47, 6700 AA Wageningen,
The Netherlands

PAIGE S. WARREN
Department of Natural Resources
Conservation,
University of Massachusetts-Amherst,
Amherst, MA 01003, USA

Preface

Recent years have witnessed an extraordinary explosion of interest in the field of urban ecology. Not only has the academic ecological community decided that it is both an inherently interesting and practically significant topic of study, but there has also been a growing realisation in other quarters that an improved understanding of the ecology of urban areas may be vital for major issues such as human health and wellbeing.

This volume is intended to provide a broad overview of the field of urban ecology, particularly for those wanting to find out what it is essentially about, either as part of their general ecological education or perhaps with a view to begin working in this area. Some of the chapter topics could be equally relevant to almost any ecosystem (although the chapter contents would vary), others particularly reflect the intimate links between people and ecological pattern and process in urban areas. The book opens with chapters addressing what urban ecology is (Chapter 1) and some of the essentials of the urbanisation process (Chapter 2) and of ecosystem function in urban areas (Chapter 3). It moves on to consider how individual species (Chapter 4), species richness (Chapter 5) and alien species (Chapter 6) respond to urbanisation. The next three chapters tackle various facets of human–ecology interactions in urban environments, respectively those between people and wildlife (Chapter 7), between human social organisation and ecology (Chapter 8), and between human health and wellbeing and ecology (Chapter 9). Two chapters then follow that focus on urban green spaces and urban planning (Chapters 10 and 11). The book concludes with consideration of the future of urban systems and urban ecology (Chapter 12).

I am grateful to the British Ecological Society for the invitation to put this volume together, and to all of the authors for producing and revising their chapters amidst many competing pressures on their time. Other contributors and a number of additional people kindly provided helpful comments on individual chapters. Thanks are also due to P. Johnson for all her assistance.

CHAPTER ONE

Urban ecology

KEVIN J. GASTON

What is urban ecology?

Real ecologists study wild and natural places. At least, that has been the view which has dominated the field of ecology for much of its existence. Some lone voices argued that this was a narrow, unrealistic and unhelpful perspective, but their arguments were easily ignored by the majority as they focused on their supposedly pristine field sites. This was always something of an irony, as surely there was never any real doubt that less wild and natural areas equally functioned as ecosystems, containing the same essential components (e.g. water, soil, microorganisms, plants, animals), and being subject to much the same processes (e.g. carbon, nutrient and water cycles). Thus, towards the close of the twentieth century and the dawning of the twenty-first, this viewpoint changed radically. Growing numbers of studies focused on the patterns and processes structuring systems that had been heavily modified, or were essentially created, by human activities. This change was particularly evident with respect to the ecology of urban areas, with numerous research studies, commentaries and reviews being published (e.g. Gilbert 1989; McDonnell & Pickett 1993; Grimm et al. 2000, 2008a, 2008b; Marzluff et al. 2001, 2008; Pickett et al. 2001; Kelcey & Rheinwald 2005; Forman 2008). Indeed, not only did ecologists begin to study urban systems in earnest, but urban ecology also began to influence the study of ecology more widely. In short, urban ecology came of age.

The rise of urban ecology

Ecology has variously been described as the scientific study of the processes determining the abundance and distribution of organisms, of the interactions between organisms, of the interactions between organisms and the environment, and of the flows of energy and materials through ecosystems. Urban ecology is quite simply therefore the study of these issues within urban systems.

The rise of urban ecology has arguably been fuelled by seven things. First, there has been mounting recognition that much of the world is now covered by human-dominated ecosystems, and that humans are integral to any general models and understanding of ecosystems. Urban systems are simply at one extreme of the spectrum from areas which are entirely uninfluenced by human activities (now very scarce, if existing at all) through to those which are predominantly shaped by interactions with people.

Second, a growing proportion of the Earth and its human population has become urbanised, and in more and more regions almost everybody has come to live in towns and cities. Thus, an understanding of the ecology of urban systems has become much more obviously relevant to people's lives and to solving the environmental problems that they face. This has resulted in particular in attention being given to the ecosystem services (the benefits humans obtain from ecosystems) associated with urban areas, the influence of different forms of urban development and management on them, and how these services can best be maintained and improved (e.g. Bolund & Hunhammar 1999; Alberti 2005).

Third, urban areas have been studied more closely in order to elucidate the connections between urban and rural landscapes. Indeed, urban areas are disproportionately responsible for many of the pressures that more natural ecosystems elsewhere experience. In large part as a consequence of the high proportion of the human population that they account for, urban systems make major local, regional and global demands for resources and for waste assimilation (e.g. Rees 1992; Folke et al. 1997; Millennium Ecosystem Assessment 2005; Bagliani et al. 2008). They are thus the primary drivers of habitat loss and fragmentation (particularly through demands for food and timber) and the multitude of consequences that follow, are the principal source of greenhouse gas emissions (and hence have significant implications for climate change) and many other atmospheric and aquatic pollutants, and are major sources for biological invasions (Grimm et al. 2008a; Trusilova & Churkina 2008).

Fourth, the potential significance for people's health and wellbeing of their having direct interactions with the natural world has become progressively apparent (e.g. Ulrich et al. 1991; Kaplan 1995; Health Council of the Netherlands 2004; Maas et al. 2006; Fuller et al. 2007). Because such interactions are often at their most sparse and least frequent in urban ecosystems, these areas have also become the places in which this significance is most easily demonstrated, in which observational studies and experiments can most readily be conducted, in which a need to understand how these can be improved becomes most apparent, and in which actions to do so are most pressing.

Fifth, the ecology of urban areas is rather different from that of other systems, owing to the pervasive influence that human activities have on ecosystem structure and function. This makes them particularly interesting subjects of ecological study because it means on the one hand that models cannot necessarily simply be borrowed from more traditional ecology and applied successfully to urban systems, and on the other hand that the development of models appropriate for urban ecology may enrich ecological research at large, especially as the human influence on ecological systems rises more generally.

Sixth, and more methodologically, urban ecology has been fuelled by the recognition that an understanding of urban ecosystems requires an approach that draws on multiple disciplines, and by a generally more positive and constructive attitude to interdisciplinary research than has typified much of the recent history of academic discourse. Studies of urban ecosystems require inputs from social, behavioural and economic sciences, as they are profoundly shaped by human activities, attitudes and choices (McIntyre et al. 2000; Alberti et al. 2003; Hope et al. 2003; Kinzig et al. 2005; Grove et al. 2006).

One final stimulus to the study of urban ecology has been provided by the discovery, which at the time was surprising, that in many regions of the world areas that have experienced urban development often coincide with those that support high native species richness and endemism (Kerr & Currie 1995; Balmford et al. 2001; Chown et al. 2003; Gaston & Evans 2004; Gaston 2005). Conservation biology has thus had to engage not simply with generic issues of human population growth, but more particularly with understanding just where that growth is taking place, what its consequences are, and how to maintain habitats and species of conservation concern in areas with high human densities.

Ecology in and of urban areas

Urban ecology can be viewed from two, closely related, perspectives (Grimm et al. 2000; Pickett et al. 2001). The first addresses ecology *in* urban areas. It is closely allied to traditional ecology, in that it typically investigates ecological patterns and processes in urban areas in much the same way as they might be examined in any other environment. That said, increasingly there is a much greater socioeconomic component, acknowledging the dominance of human activities in shaping urban ecosystems.

Studies in urban areas tend to be of four broad types: (i) comparison of different land-use types within an urban setting; (ii) comparison of an urban area with a nearby 'natural' area; (iii) monitoring of an urban area through time; and (iv) gradient analysis (McIntyre et al. 2000). A particularly dominant theme has been the study of rural–urban gradients, in which typically the ecological characteristics of sites are compared along a, usually spatial, transect of increasing urbanisation (e.g. McDonnell & Pickett 1990; Blair 1996;

Guntenspergen & Levenson 1997; McDonnell *et al.* 1997; Pouyat *et al.* 1997; Carreiro *et al.* 1999; Niemelä *et al.* 2002). However, having resulted in many valuable insights, this dominance is fading with a greater diversification of approaches. Indeed, even studies treading this same route now seem to be considering multi-dimensional environmental gradients more frequently. Temporal patterns have received far less attention in urban environments than have spatial patterns, largely because of a lack of appropriate data, although this is also beginning to change (e.g. Zapparoli 1997; Morneau *et al.* 1999; Chocholoušková & Pyšek 2003; Puth & Burns 2009).

The second perspective on urban ecology is a more system-oriented approach, the ecology *of* urban areas. The distinction from ecology *in* urban areas is not so much one of methodology per se, but rather something of a conceptual shift. The ecology of urban areas concerns how those areas function as aggregated wholes. There has thus been a growing number of studies particularly of aspects of the ecology of entire cities (e.g. Hadidian *et al.* 1997; Kent *et al.* 1999; Grimm *et al.* 2000; Turner 2003; Pickett & Cadenasso 2006; Davies *et al.* 2008; Walton 2008). A logical extension of such work is then the comparison of the ecologies of different cities, and this is indeed developing rapidly, predominantly in relation to patterns of land cover and assemblage composition (e.g. Pyšek 1998; Turner *et al.* 2004; La Sorte *et al.* 2007).

Neither the ecology *in* urban areas nor the ecology *of* urban areas perspective has particular precedence, and knowledge will advance most swiftly when both are employed. An analogy might perhaps usefully be drawn between local ecology and macroecology (Gaston & Blackburn 2000). Local studies have dominated ecology for much of its existence and have contributed much to understanding of the dynamics of populations, communities and ecosystems. However, a great deal remains unexplained without also taking a much broader view, and placing local study sites in the context of landscapes and the processes that shape the spatial flows of individuals and materials. It is when the two views are combined that the greatest insights and predictive power can be achieved. Likewise, the combination of studies of ecology *in* and ecology *of* urban areas is likely to bring about the most rapid advances in understanding, particularly because of the inherent complexity of urban systems.

To date, urban ecology has predominantly been an observational science, given the practical constraints on conducting field experiments posed by issues of land ownership and access, security from interference and the spatial variation of urban landscapes. However, some small-scale experiments have been conducted (e.g. Denys & Schmidt 1998; Gaston *et al.* 2005), and there may be considerable potential to do so at much larger scales through collaboration with urban designers (Felson & Pickett 2005).

'Wicked' problems

More so than in research into many other ecosystems, studies in urban ecology are almost invariably conducted against the background of a concern to understand the consequences of particular human-wrought changes to the environment. Whether the focus be a particular species, habitat or ecosystem process, and whatever motivation drove the original investigator, such studies are apt to be interpreted in terms of the successes or failures of past urban planning, how the structures that resulted can best be exploited or modified, and how future planning should be conducted. A measure of the maturity of the field of urban ecology will be the extent to which these considerations are realistic, not only about the complexity of urban ecosystems, but also about the formidable array of pressures, constraints and compromises, and the interactions thereof, that shape the planning process and contribute to (without being able entirely to dictate) the resultant emergent urban systems. As the Royal Commission on Environmental Pollution (2007, p. 5) states, urban systems 'have evolved in particular ways as agglomerations of people; accretions of buildings and roads; infrastructures for water and energy supply and the removal of sewage and waste; public and private spaces; places of business and residence; locations for the production and consumption of goods and services; facilities for entertainment, education and health; and so forth'.

It has been argued that urban environmental management presents a classic case of a 'wicked problem' (Royal Commission on Environmental Pollution 2007); 'wicked' in the sense of nasty or vicious, rather than passing any ethical judgement. Rittel and Webber (1973) observed that the kinds of societal problems that planners deal with are intrinsically different from archetypal problems in science in that they are ill-defined and cannot be definitively solved. Their characterisation of the wickedness of planning problems provides valuable context for studies in urban ecology, but arguably also extends to some of the issues which those studies themselves set out to address: (i) there is no definitive formulation of a wicked problem – the process of formulating the problem and of conceiving the solution are essentially the same; (ii) wicked problems have no stopping rule – because there is no definitive formulation of the problem there is no point at which *the* solution has been found; (iii) solutions to wicked problems are not true-or-false, but good-or-bad – the particular solutions are likely to depend on who provides them; (iv) there is no immediate and no ultimate test of a solution to a wicked problem – any solution is likely to have many consequences, some of which will be unintended and unexpected, which play out over potentially long periods of time; (v) every solution to a wicked problem is a 'one-shot operation'; because there is no opportunity to learn by trial and error, every attempt counts significantly – this is because planning actions are seldom reversible but have so many consequences; (vi) wicked problems do not have an enumerable (or an exhaustively describable) set of potential solutions, nor is there a well-described set of permissible operations that may be incorporated into the plan – the set of potential solutions and the extent to which they are permissible will depend on who provides them; (vii) every wicked problem is essentially unique – there are always particularities to a problem which may override the commonalities with other problems; (viii) each wicked problem can be considered a symptom of another problem; (ix) the existence of a discrepancy representing a wicked problem can be explained in numerous ways – the choice of explanation determines the nature of the problem's resolution; and (x) the planner has no right to be wrong – the objective is to improve a situation.

Prospects

Despite its relatively late emergence as an important component of the ecological research agenda, understanding of urban ecology has developed rapidly. The numbers of papers, books, symposia and conferences dedicated to the topic continue on steeply rising trajectories. Contributions appear increasingly frequently in high-profile general science journals as well as more subject-specific ones and, perhaps more tellingly, key findings are more regularly being cross-referenced in other fields of ecological and environmental science. All of this reflects a vibrant and fast moving research agenda.

Of course, much work remains to be done in urban ecology. Almost all of the subsequent, more detailed, chapters in this volume highlight questions for which answers are urgently required, and important topics that have remained poorly explored or are as yet unexplored. More generically, obvious issues for future work include (i) the ecology of towns and cities in developing countries; (ii) the relationship between the ecology of towns and cities and how urban areas grow; (iii) improved understanding of the role of history and culture in shaping the ecology of different urban areas; (iv) how the ecology of urban areas will alter as a consequence of climate change and other global change pressures; (v) more comparative work between different towns and cities; and (vi) more experimental studies.

Acknowledgements
I am grateful to Z. Davies for helpful comments.

References

Alberti, M. (2005). The effects of urban patterns on ecosystem function. *International Regional Science Review*, **28**, 168–92.

Alberti, M., Marzluff, J. M., Shulenberger, E. *et al.* (2003). Integrating humans into ecology: opportunities and challenges for studying urban ecosystems. *BioScience*, **53**, 1169–79.

Bagliani, M., Galli, A., Niccolucci, V. and Marchettini, N. (2008). Ecological footprint analysis applied to a sub-national area: the case of the Province of Siena (Italy). *Journal of Environmental Management*, **86**, 354–64.

Balmford, A., Moore, J. L., Brooks, T. *et al.* (2001). Conservation conflicts across Africa. *Science*, **291**, 2616–19.

Blair, R. B. (1996). Land use and avian species diversity along an urban gradient. *Ecological Applications*, **6**, 506–19.

Bolund, P. and Hunhammar, S. (1999). Ecosystem services in urban areas. *Ecological Economics*, **29**, 293–301.

Carreiro, M. M., Howe, K., Parkhurst, D. F. and Pouyat, R. V. (1999). Variation in quality and decomposability of red oak leaf litter along an urban–rural gradient. *Biology and Fertility of Soils*, **30**, 258–68.

Chocholoušková, Z. and Pyšek, P. (2003). Changes in composition and structure of urban flora over 120 years: a case study of the city of Plzeň. *Flora*, **198**, 366–76.

Chown, S. L., van Rensburg, B. J., Gaston, K. J., Rodrigues, A. S. L. and van Jaarsveld, A. S. (2003). Species richness, human population size and energy: conservation implications at a national scale. *Ecological Applications*, **13**, 1223–41.

Davies, R. G., Barbosa, O., Fuller, R. A. *et al.* (2008). City-wide relationships between green spaces, urban form and topography. *Urban Ecosystems*, **11**, 269–87.

Denys, C. and Schmidt, H. (1998). Insect communities on experimental mugwort (*Artemisia vulgaris* L.) plots along an urban gradient. *Oecologia*, **113**, 269–77.

Felson, A. J. and Pickett, S. T. A. (2005). Designed experiments: new approaches to studying urban ecosystems. *Frontiers in Ecology and the Environment*, **3**, 549–56.

Folke, A., Jansson, J., Larsson, J. and Costanza, R. (1997). Ecosystem appropriation by cities. *Ambio*, **26**, 167–72.

Forman, R. T. T. (2008). *Urban Regions: Ecology and Planning Beyond the City*. Cambridge: Cambridge University Press.

Fuller, R. A., Irvine, K. N., Devine-Wright, P., Warren, P. H. and Gaston, K. J. (2007). Psychological benefits of greenspace increase with biodiversity. *Biology Letters*, **3**, 390–4.

Gaston, K. J. (2005). Biodiversity and extinction: species and people. *Progress in Physical Geography*, **29**, 239–47.

Gaston, K. J. and Blackburn, T. M. (2000). *Pattern and Process in Macroecology*. Oxford: Blackwell Science.

Gaston, K. J. and Evans, K. L. (2004). Birds and people in Europe. *Proceedings of the Royal Society London B*, **271**, 1649–55.

Gaston, K. J., Smith, R. M., Thompson, K. and Warren, P. H. (2005). Urban domestic gardens (II): experimental tests of methods for increasing biodiversity. *Biodiversity and Conservation*, **14**, 395–413.

Gilbert, O. L. (1989). *The Ecology of Urban Habitats*. London: Chapman & Hall.

Grimm, N. B., Faeth, S. H., Golubiewski, N. E. *et al.* (2008a). Global change and the ecology of cities. *Science*, **319**, 756–60.

Grimm, N. B., Foster, D., Groffman, P. *et al.* (2008b). The changing landscape: ecosystem responses to urbanization and pollution across climatic and societal gradients. *Frontiers in Ecology and the Environment*, **6**, 264–72.

Grimm, N. B., Grove, J. M., Pickett, S. T. A. and Redman, C. L. (2000). Integrated approaches to long-term studies of urban ecological systems. *BioScience*, **50**, 571–84.

Grove, J. M., Troy, A. R., O'Neil-Dunne, J. P. M. *et al.* (2006). Characterization of households and its implications for the vegetation of urban ecosystems. *Ecosystems*, **9**, 578–97.

Guntenspergen, G. R. and Levenson, J. B. (1997). Understory plant species composition in remnant stands along an urban-to-rural land-use gradient. *Urban Ecosystems*, **1**, 155–69.

Hadidian, J., Sauer, J., Swarth, C. *et al.* (1997). A citywide breeding bird survey for Washington, D.C. *Urban Ecosystems*, **1**, 87–102.

Health Council of the Netherlands (2004). *Nature and Health. The Influence of Nature on Social, Psychological and Physical Well-being.* The Hague, Netherlands: Health Council of the Netherlands and Dutch Advisory Council for Research on Spatial Planning, Nature and the Environment.

Hope, D., Gries, C., Zhu, W. *et al.* (2003). Socioeconomics drive urban plant diversity. *Proceedings of the National Academy of Sciences of the USA*, **100**, 8788–92.

Kaplan, S. (1995). The restorative benefits of nature: toward an integrative framework. *Journal of Environmental Psychology*, **15**, 169–82.

Kelcey, J. G. and Rheinwald, G. (eds.) (2005). *Birds in European Cities*. St Katharinen: Ginster Verlag.

Kent, M., Stevens, R. A. and Zhang, L. (1999). Urban plant ecology patterns and processes: a case study of the flora of the City of Plymouth, Devon, UK. *Journal of Biogeography*, **26**, 1281–98.

Kerr, J. T. and Currie, D. J. (1995). Effects of human activity on global extinction risk. *Conservation Biology*, **9**, 1528–38.

Kinzig, A. P., Warren, P., Martin, C., Hope, D. and Katti, M. (2005). The effects of human socioeconomic status and cultural characteristics on urban patterns of biodiversity. *Ecology and Society*, **10**, 23.

La Sorte, F., McKinney, M. L. and Pyšek, P. (2007). Compositional similarity among urban floras within and across continents: biogeographical consequences of human-mediated biotic interchange. *Global Change Biology*, **13**, 913–21.

Luck, G. W., Ricketts, T. H., Daily, G. C. and Imhoff, M. (2004). Alleviating spatial conflict between people and biodiversity. *Proceedings of the National Academy of Sciences of the USA*, **101**, 182–6.

Maas, J., Verheij, R. A., Groenewegen, P. P., de Vries, S. and Spreeuwenberg, P. (2006). Green space, urbanity, and health: how strong is the relation? *Journal of Epidemiology and Community Health*, **60**, 587–92.

Marzluff, J. M., Bowman, R. and Donnelly, R. (eds.) (2001). *Avian Ecology and Conservation in an Urbanizing World*. Boston: Kluwer Academic.

Marzluff, J. M., Shulenberger, E., Endlicher, W. *et al.* (eds.) (2008). *Urban Ecology: An International Perspective on the Interaction between Humans and Nature*. New York: Springer.

McDonnell, M. J. and Pickett, S. T. A. (1990). Ecosystem structure and function along urban–rural gradients: an unexploited opportunity for ecology. *Ecology*, **71**, 1232–7.

McDonnell, M. J. and Pickett, S. T. A. (eds.) (1993). *Humans as Components of Ecosystems: The Ecology of Subtle Human Effects and Populated Areas*. New York: Springer-Verlag.

McDonnell, M. J., Pickett, S. T. A., Groffman, P. *et al.* (1997). Ecosystem processes along an urban-to-rural gradient. *Urban Ecosystems*, **1**, 21–36.

McIntyre, N. E., Knowles-Yanez, K. and Hope, D. (2000). Urban ecology as an interdisciplinary field: differences in the use of 'urban' between the social and natural sciences. *Urban Ecosystems*, **4**, 5–24.

Millennium Ecosystem Assessment (2005). *Ecosystems and Human Well-being: Current State and Trends*, Vol. I. Washington, DC: Island Press.

Morneau, F., Décarie, R., Pelletier, R. *et al.* (1999). Changes in breeding bird richness and abundance in Montreal parks over a period of 15 years. *Landscape and Urban Planning*, **44**, 111–21.

Niemelä, J., Kotze, D. J., Venn, S. *et al.* (2002). Carabid beetle assemblages (Coleoptera, Carabidae) across urban–rural gradients: an international comparison. *Landscape Ecology*, **17**, 387–401.

Pickett, S. T. A. and Cadenasso, M. L. (2006). Advancing urban ecological studies: frameworks, concepts, and results from the Baltimore Ecosystem Study. *Austral Ecology*, **31**, 114–25.

Pickett, S. T. A., Cadenasso, M. L., Grove, J. M. *et al.* (2001). Urban ecological systems: linking terrestrial ecological, physical, and socioeconomic components of metropolitan areas. *Annual Review of Ecology and Systematics*, **32**, 127–57.

Pouyat, R. V., McDonnell, M. J. and Pickett, S. T. A. (1997). Litter decomposition and nitrogen mineralization in oak stands along an urban-rural land use gradient. *Urban Ecosystems*, **1**, 117–31.

Puth, L. M. and Burns, C. E. (2009). New York's nature: a review of the status and trends in species richness across the metropolitan area. *Diversity and Distributions*, **15**, 12–21.

Pyšek, P. (1998). Alien and native species in Central European urban floras: a quantitative comparison. *Journal of Biogeography*, **25**, 155–63.

Rees, W. E. (1992). Ecological footprints and appropriated carrying capacity: what urban economics leaves out. *Environment and Urbanization*, **4**, 121–30.

Rittel, H. W. J. and Webber, M. M. (1973). Dilemmas in a general theory of planning. *Policy Sciences*, **4**, 155–69.

Royal Commission on Environmental Pollution (2007). *The Urban Environment*. London: The Stationery Office.

Trusilova, K. and Churkina, G. (2008). The response of the terrestrial biosphere to urbanization: land cover conversion, climate, and urban pollution. *Biogeosciences*, **5**, 1505–15.

Turner, W. R. (2003). Citywide biological monitoring as a tool for ecology and conservation in urban landscapes: the case of the Tucson Bird Count. *Landscape and Urban Planning*, **65**, 149–66.

Turner, W. R., Nakamura, T. and Dinetti, M. (2004). Global urbanization and the separation of humans from nature. *BioScience*, **54**, 585–90.

Ulrich, R. S., Simons, R. F., Losito, B. D. *et al.* (1991). Stress recovery during exposure to natural and urban environments. *Journal of Environmental Psychology*, **11**, 201–30.

Walton, J. T. (2008). Difficulties with estimating city-wide urban forest cover change from national, remotely-sensed tree canopy maps. *Urban Ecosystems*, **11**, 81–90.

Zapparoli, M. (1997). Urban development and insect biodiversity of the Rome area, Italy. *Landscape and Urban Planning*, **38**, 77–86.

CHAPTER TWO

Urbanisation

KEVIN J. GASTON

Unsurprisingly, knowledge of the pattern and process of urbanisation is key background to an understanding of urban ecology. However, of course, the necessary material lies largely in other fields of study. Moreover, it is widely scattered and rather poorly synthesised. This chapter therefore provides a broad but selective overview of some of the most important issues of especial relevance to urban ecology. In particular, it addresses what constitutes an urban area, how urbanisation occurs, the history of urbanisation, the scale of urbanisation (in terms of population and household numbers, land cover and ecological footprints), the structure of urban areas (in terms of location, size, form and composition) and, finally, the sustainability of cities. Whilst an attempt has been made to draw on an extensive range of examples, there is an inevitable and unfortunate bias towards the developed regions on which the vast majority of research, and the associated literature, has been focused.

What is urban?

Urbanisation is the process by which a rural area becomes an urban one, or the degree to which an area is urbanised (although some differentiate the former process as urbanisation and the latter as urbanicity). But what is an urban area? There is no simple answer to this question, although most of us would have an intuitive sense of what constitutes such an area and what does not, probably based on the numbers of people and the level of cover by buildings, transport networks and other such infrastructure. More formally, urban areas have, in practice, been distinguished in a wide variety of ways. These include using administrative or geopolitical boundaries (e.g. towns, cities, metropolitan districts), functional boundaries (based on flows of people or resources), human population size or density (of varying accuracy, depending on census techniques), housing numbers or density, land cover (commonly derived from satellite imagery) and/or other indicators of human activity (e.g. artificial

Urban Ecology, ed. Kevin J. Gaston. Published by Cambridge University Press.
© British Ecological Society 2010.

night time lights; Theobald 2002; Millennium Ecosystem Assessment 2005; Small *et al.* 2005; Kasanko *et al.* 2006; UN-HABITAT 2006; Decker *et al.* 2007). Almost invariably, the cut-offs between what is and is not regarded as urban are essentially arbitrary, and arise from applied rather than theoretical considerations. They variously depend, for example, on precisely which administrative boundary scheme is used (and how through time it is amended), what levels of flows are employed to identify functional boundaries, what levels of population or housing are regarded as urban, and how urban land cover is distinguished from others (including how small any associated green spaces and water bodies need to be if they are to be treated as urban cover). These thresholds may, in turn, be influenced by variations in the perceptions of users as to what constitutes an urban area. Those living in regions in which human population densities and building coverage are generally lower often regard lower thresholds of such variables as being more appropriate cut-offs for defining urban areas.

Obviously, the definitions employed to distinguish urban areas can make a substantial difference to their measured extent. Different authorities use different definitions, and these may also alter through time (Cohen 2004; Montgomery 2008). For instance, the administrative definition of an urban area in New Zealand is a settlement of more than 1000 people, with no reference to the population density (Statistics New Zealand 2006). In contrast, in Canada it is defined as land with more than 400 people km^{-2} (Statistics Canada 2001), and in Japan as areas with a population density greater than 4000 people km^{-2} that contain at least 5000 people (Japanese Statistics Bureau 2005). As a consequence of such differences, estimates of the level of urbanisation of a particular region or country may differ markedly between different sources, and often may not be strictly comparable. There seems to be little apparent concerted effort to resolve this formidable problem, although this may perhaps change in response to research analyses requiring more standardised measures of urbanisation. This variation in approaches to measuring urbanisation needs to be borne in mind when considering much of what follows in this and other chapters in this volume. Indeed, in this spirit, I shall subsequently treat urban areas rather broadly to include towns, cities and associated infrastructure. Arguably, even quite substantial variation in estimates of levels of urbanisation would not markedly alter many of the generalisations reported.

How does urbanisation occur?

The urbanisation of a region can take place through three distinctly different routes, and combinations thereof. First, it may result from the migration of people from rural areas (or urban areas elsewhere) to existing urban areas, leading to their intensification and/or expansion. This currently accounts for a substantial proportion of urban population growth. Second, differences in birth and death rates may result in greater natural population growth in existing urban than rural areas, again leading to the intensification and/or expansion of the former. At the present time, although fertility rates tend to be lower, health is generally better and longevity greater in urban areas than rural ones (Dye 2008; Mace 2008). Third, natural or planned (e.g. through the location of specific new developments or initiatives) population growth in rural areas may result in their becoming urban, either through the expansion of existing urban areas into their periphery or the development of independent urban areas. These different routes to urbanisation can individually, or in combination, lead to three distinct urban growth types: (i) infilling, in which unurbanised land that has been surrounded by previously urban land itself becomes urbanised; (ii) edge-expansion, in which existing urban areas spatially expand; and (iii) spontaneous growth, in which new urban areas are formed with no direct spatial connection to existing urban areas (Xu *et al.* 2007).

Models of urbanisation, both informal and formal, have a long history (Berling-Wolff & Wu 2004a). Traditionally, urban areas have been pictured as comprising a series of concentric circles of different land-use, radiating from the central business district, that over time progressively expand outwards (e.g. Burgess 1925). More recent models acknowledge that this is unduly simplistic, and particularly highlight (i) leap-frog development, in which fresh housing in particular is established beyond the pre-existing urban fringe; (ii) coalescence, in which large cities emerge as a consequence of development between pre-existing towns; and (iii) polycentric developments, in which urban regions emerge as collections of historically distinct and independent cities (e.g. Manrubia & Zanette 1998; Meijers 2005; Decker *et al.* 2007; Batty 2008). In addition, more complex approaches to the processes involved in urbanisation have been incorporated, including, for example, reaction-diffusion, based on simple assumptions about human reproduction and migration, and/or fractal structures in which cities show self-similarity or hierarchical clustering in patterns of land coverage at multiple spatial scales.

One of the most frequently used approaches to modelling urban systems spatially employs cellular automata, typically comprising a lattice of cells, each of which takes a particular state of land-use or land cover which may change through time following a set of transition rules that reflect the influence of the states of neighbouring cells. Complex spatial patterns can result even when the transition rules are rather simple. Such approaches have been used both to study the general interaction between particular sets of processes and resultant urban forms, and to simulate the growth of specific cities (e.g. Li & Yeh 2000; Barredo *et al.* 2003; Berling-Wolff & Wu 2004b; Divigalpitiya *et al.* 2007; He *et al.* 2008). They have also been used to predict the future growth of particular urban areas (e.g. He *et al.* 2008).

A brief history of urbanisation

Relative to the time since the human species first appeared, urbanisation is a rather recent phenomenon. Urban societies are thought initially to have developed about 5000 to 6000 years ago, emerging as a consequence of humans organising themselves into larger groups, establishing sedentary settlements in which tools and facilities could be increased and agriculture pursued, and developing cultural mechanisms for assembling and redistributing knowledge (Redman 1999). Mesopotomia is argued to have been the first region of the world to become urbanised (Leick 2001). The environmental impacts of some early urban societies were severe. Typically resulting from regional degradation as a consequence of the overexploitation of resources associated with rising population levels, these led to the abandonment of entire cities and the collapse of some civilisations (Redman 1999; Ehrlich & Ehrlich 2004; Diamond 2006).

Even by 1700, only a small proportion (<10%) of the world's population inhabited urban areas, with these individuals distributed amongst a relatively limited number of small urban centres (Figure 2.1; Berry 1990). Parts of western Europe in the eighteenth century were the first to achieve more sizeable urban populations, with capital cities in particular accounting for much of this growth. Although throughout the history of urbanisation cities have risen and fallen in size, both in absolute and relative terms (Lanaspa *et al.* 2003; Batty 2006), only one city (Beijing) had a population of more than one million in 1800 (there were 16 in 1900, and more than 400 by 2000; Chandler 1987; United Nations Centre for Human Settlements 2002).

Up until the early twentieth century, at least in Europe, urban dwellers experienced higher rates of infant mortality than those living in rural environments, with cities apparently being maintained in large part through immigration (Woods 2003; Birchenall 2007). In the first half of the century, urbanisation was predominantly confined to countries with the highest levels of per capita income (Cohen 2004). Indeed, rapid urbanisation began in what are today the more developed regions of the world, with around 30% of their population being urban in 1920, and more than a half by 1950 (United Nations 2008). In 2007, more than 80% of the populations of Australia, New Zealand and North America were urbanised, and 72% of the population of Europe (which, perhaps surprisingly, is the least urbanised major region of the developed world). The level of urbanisation of a region, as reflected in the proportion of the population living in urban areas, has thus historically tended to increase with levels of wealth, although there may be no simple causal link (Bloom *et al.* 2008). The entire abandonment or collapse of cities has been relatively infrequent over the past two centuries, even as a consequence of conflict and natural disasters.

The urban transition that is currently taking place across much of the world is rather different from those that have gone before, being characterised as (i) unprecedented in spatial scale; (ii) rapid (albeit not unprecedented) in pace; (iii) occurring more rapidly in countries that have relatively middle to low levels of per capita income (indeed, in some countries the growth of cities has not had concomitant growth in economic activity); (iv) heavily dependent on the global economy; (v) accompanied by a convergence in rural and urban

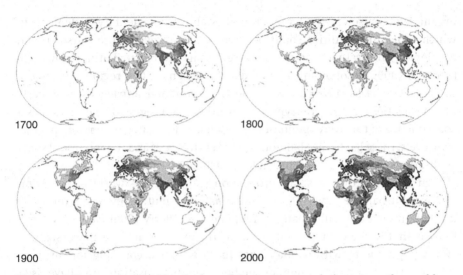

Figure 2.1 Estimates of the distribution of the human population across the world between 1700 and 2000. Reproduced with permission from Klein Goldewijk (2005).

lifestyles, particularly as a consequence of ease of transport and communication; and (vi) occurring under a different set of demographic regimes, particularly abrupt declines in fertility and mortality (Cohen 2004).

The scale of urbanisation

The present and predicted scales of urbanisation are, by multiple metrics, extraordinary. Indeed, to many minds, this is one key reason why rapid continued development of the research field of urban ecology is essential.

Population and household numbers

The percentage of the population that is urban varies markedly from one country to another. Estimates for 2007 give a mean of 58%, ranging between 10 and 100% (Figure 2.2; United Nations 2008). However, in 2008, for the first time in history, the global urban population was estimated to have equalled the rural population, and has subsequently exceeded it (Figure 2.3; United Nations 2008). That makes this the urban century, and a key point of transition in human history, at least in terms of simple descriptive statistics. Whilst this may conjure visions of modern cities and suburbs, it is important to recognise that 2007 saw the number of slum dwellers – those living in urban areas in inadequate housing with few or no basic services – exceed one billion, equating to one in every three urban dwellers (UN-HABITAT 2006). This number is also predicted to continue to grow.

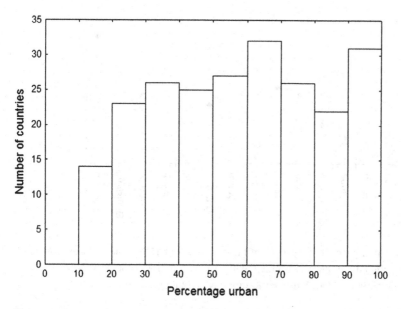

Figure 2.2 The percentage of the population of each of the countries of the world that resides in urban areas. From data in United Nations (2008).

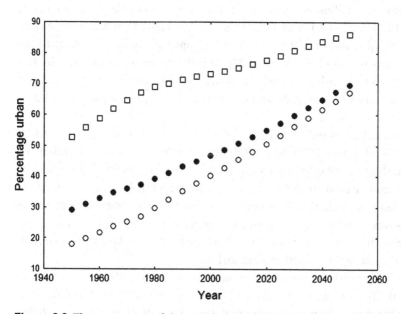

Figure 2.3 The percentage of the population of the world (filled circles), more developed regions (open squares) and less developed regions (open circles) estimated (1950–2005) and projected (2010–2050) to reside in urban areas. From data in United Nations (2008).

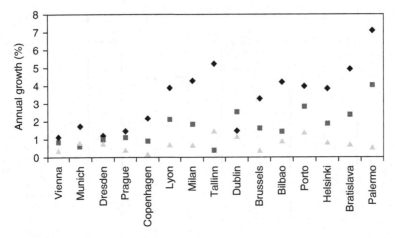

Figure 2.4 Annual percentage growth of built-up areas of 15 European cities from the mid 1950s to the late 1990s. Diamonds, 1950s–1960s; squares, 1960s–1980s; and triangles, 1980s–1990s. Modified from Kasanko *et al.* (2006).

The growth rates of even well-established cities can be high (Figures 2.4–2.6). Cairo (Egypt) grew from 6.7 million people in 1976 to more than 10 million in 2002 (49% increase; Stewart *et al.* 2004), Las Vegas (USA) from 557 000 people in 1985 to nearing 1.7 million in 2004 (205% increase; UNEP 2007), Shanghai (China) from 10.8 million people in 1975 to 13.4 million in 2003 (24% increase) and from 159.1 km^2 in 1975 to 1179.3 km^2 in 2005 (641% increase; Zhao *et al.* 2006), and Nanjing metropolitan area (China) from 128 km^2 in 1979 to 461 km^2 in 2003 (260% increase; Xu *et al.* 2007). Significant differences exist in the growth patterns of urban areas, variously as a consequence of available resources, juxtaposition to trade and transportation routes, investment, innovation and cultural factors.

The global urban population is predicted to increase from 3.29 billion in 2007 to 4.58 billion in 2025, and 6.40 billion in 2050. This is equivalent to almost doubling by 2050 to approximately the same size as the entire population (rural and urban) in 2004 (United Nations 2008). Indeed, urban areas in less developed regions will account for most of the world's population growth. However, whilst the majority of people have lived in urban areas in more developed regions since at least 1950, they will not be in the majority in less developed regions until around 2019.

The rate of growth in numbers of households or dwellings is currently greater than that of the populations in much of the developed world, even increasing in some regions in which population size is declining (Liu *et al.* 2003; Lepczyk *et al.* 2007). Foremost, this has arisen as a consequence of social changes, including increases in longevity, single occupancy and divorce rates, as well as in some regions a growing number of second homes.

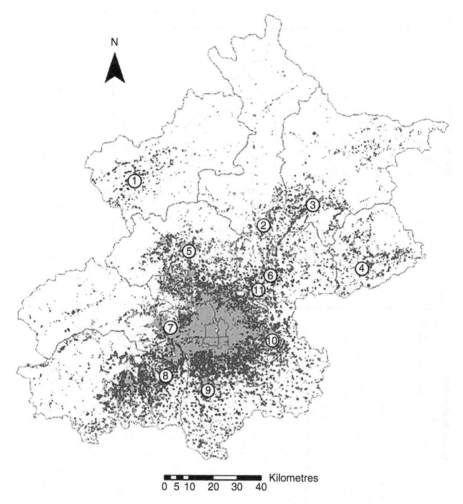

N

Kilometres
0 5 10   20   30   40

Figure 2.5 Urban expansion of Beijing. Light grey indicates urban area in 1991; dark grey, urban area in 1991–2004. 1, Yanqing; 2, Huairou; 3, Miyun; 4, Pinggu; 5, Changping; 6, Shunyi; 7, Mengtougou; 8, Fangshan; 9, Daxing; 10, Tongzhou; 11, Capital airport. Modified from He *et al.* (2008).

Land cover

Various attempts have been made to determine the global coverage of urban or built-up areas. Estimates include 260 092 km^2 (0.2%; Hansen *et al.* 2000), 260 117 km^2 (0.2%; Loveland *et al.* 2000), 3 million km^2 (2.3%; Wackernagel *et al.* 2002), 3 673 155 km^2 (2.8%; Millennium Ecosystem Assessment 2005), 2 213 496 km^2 (1.7%; Salvatore *et al.* 2005), 3 485 596 km^2 (2.7%; Salvatore *et al.* 2005) and 500 000 km^2 (0.38%; Sterling & Ducharne 2008; all percentages calculated using a global land extent, excluding permanent ice cover, of *c*.130 million km^2). Much of the marked variation in these figures results from differences in the spatial resolution and approach of analyses, particularly in

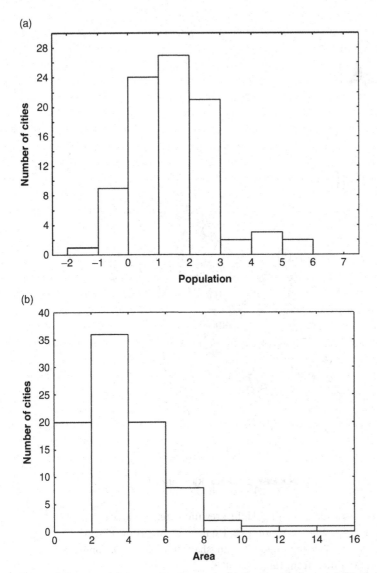

Figure 2.6 Annual percentage change in (a) the population and (b) the built-up area (km^2) of 90 cities of over 100 000 individuals, calculated for various combinations of years between 1984 and 2002. From data in Angel *et al.* (2005).

the extent to which smaller settlements and infrastructure (e.g. roads) are included. Nonetheless, whichever figures are used, the percentage coverages are low in comparison with croplands covering *c.*18 million km^2 and grazing lands comprising 36 million km^2 (Sterling & Ducharne 2008). However, the consequences of this urban coverage, particularly in terms of resource demands, are far reaching (see below).

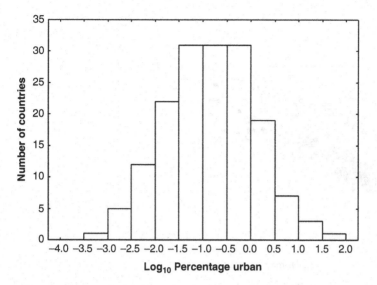

Figure 2.7 Percentage of land area in 2000 that is urban and built-up for 163 countries for which this percentage is greater than zero. From data in World Resources Institute (2007).

Angel *et al.* (2005) estimated that in 2000, cities with populations in excess of 100 000 covered *c.* 200 000 km^2 both in developing countries and in industrialised countries. These numbers are expected to increase to 600 000 km^2 and 500 000 km^2, respectively, by 2030.

The percentage of the areal extent of individual countries that is covered by urban areas is hugely variable, ranging from close to zero to about 32% (Figure 2.7). Unsurprisingly, the coverage of urban areas generally increases with the size of the associated human population. However, the intercept and slope of this relationship vary between regions. For example, the intercept is greater and the slope is shallower for large cities in Europe than for large cities in the USA (Figure 2.8). Thus, whilst the former house more people for a given urban area, they accrue people at a slower rate for a given increase in that area. The first of these outcomes is likely to be a consequence of the suburban sprawl that is particularly marked around many cities in the USA (Theobald 2005).

Not only are population size and urban area positively correlated for particular points in time, they also tend to scale positively through time for any given urban area. For much of the world this relationship is commonly such that the area occupied increases faster than the population size, with new residents effectively occupying more land per capita than do existing residents (Marshall 2007). This follows from the previously mentioned tendency for the rate of growth in numbers of households to outstrip that in population

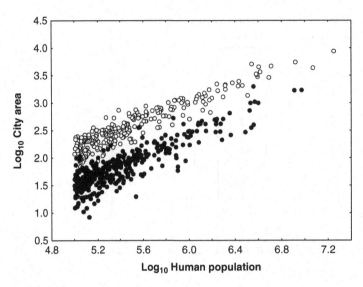

Figure 2.8 The relationship between city area (km^2) and the number of human inhabitants, for cities exceeding 100 000 inhabitants in Europe (filled circles) and the USA (open circles). From unpublished data kindly provided by R. A. Fuller.

size, and is exacerbated by rising living standards (with demands for more space per person) and changing land-use policies (e.g. attitudes to urban sprawl; Kasanko *et al.* 2006).

Ecological footprint

Industrialised cities have energy fluxes of between 100 000 and 3 000 000 kcal m^{-2} yr^{-1}, in contrast to natural ecosystems which flux between 1000 and 40 000 kcal m^{-2} yr^{-1} (Odum 1997). This is one reflection of the influence of urban areas reaching far beyond their geographic limits, often drawing on resources and ecosystem services from distant lands. The ecological footprint of a city is the total area of biologically productive land and water required to generate the renewable resources that it consumes and assimilate the waste it produces, using prevailing technology (Rees 1992, 2001; Wackernagel *et al.* 2006). Although the details of the accounting vary between studies, and the approach has its limitations (Chapter 12), cities commonly have ecological footprints one or two orders of magnitude larger than the urban areas themselves (Rees 1992, 1999; Wackernagel *et al.* 2006). For example, one estimation of the ecological footprint of Greater London in 2000 was 293 times its geographical area, which is twice the size of the UK, and roughly the same size as Spain (Best Foot Forward 2002). Likewise, in 1996, metropolitan Toronto had an ecological footprint that was 290 times its geographical area, with each citizen having a footprint of 7.7 ha (Rees 2001). It is important to remember, however, that in many cases per capita footprints may be smaller for urban than rural areas, because of the efficiencies of urban living.

At a global scale the ecological footprint of humanity is estimated already to exceed the available capacity, which if sustained will lead to the loss of ecological capital (Wackernagel *et al.* 2006; Kitzes *et al.* 2008). This demand arises disproportionately from the urban segment of the population, particularly that in developed nations.

The structure of urban areas
Location
The human population is not randomly distributed across the Earth's terrestrial surface (Figure 2.1). Although the precise figures are dependent on the spatial resolution of analysis, on a grid of 2.5 arc-minute quadrilaterals, half of the potentially habitable area contains less than 2% of the population, whilst 50% of the population occurs in less than 3% of the potentially habitable area (Small & Cohen 2004). The density of people is greatest at sea level and *c.* 2300 m elevation (the densely populated volcanic highlands of Mexico), increases at high levels of variability in precipitation, peaks at tropical and temperate mean temperatures and towards arid (but not the driest) climates, and declines with distance from coastlines and major rivers (Small & Cohen 2004). The physiographic variables appear to have a much stronger influence than do the climatic ones, although one consequence of the combination of influences is that, at reasonably coarse spatial resolutions, in many regions of the world human density tends to increase with levels of net primary productivity (or at least correlates thereof; Chown *et al.* 2003; Evans & Gaston 2005; Gaston 2005). Many of these variables tend also to be correlated with native and endemic species richness, resulting in important patterns of covariation with human density at these low resolutions (Chapter 5).

Looking at different broad ecosystem types, the highest proportion of the human population is urbanised within coastal and inland water zones, with these also having the highest urban population densities, and the highest proportions of the areas of coastal and cultivated lands being urbanised (Table 2.1). Conversely, the lowest proportion of the human population is urbanised within mountain and forest lands, which also have the lowest proportions of their areas urbanised (Table 2.1). The proportion of the population that is urbanised is quite similar across Latin America, North America, Europe and Oceania (*c.* 70–80%, depending on source), and much lower in Africa and Asia (*c.* 35–40%; Millennium Ecosystem Assessment 2005).

The probability that an area becomes urbanised and the intensity of urbanisation have been shown to be a function of a large number of variables, including topography, soil type, water availability, population density,

Table 2.1 *Estimated population size, population density and land area of different broad ecosystem types, divided into urban and rural. Note that population numbers for ecosystems do not sum to totals, as systems are not mutually exclusive.*

	Population				Population density			Land area			
	Total	Urban	Rural	Urban share	Overall	Urban	Rural	Total	Urban	Rural	Urban share
	(million)			(%)	(person per square km)			(square km)			(%)
Coastal zone	1147	744	403	64.9	175	1119	69	6538097	664816	5873281	10.2
Cultivated	4233	1914	2309	45.3	119	793	70	35475983	2412618	33063350	6.8
Dryland	2149	963	1185	44.8	36	749	20	59990129	1286421	58703698	2.1
Forest	1126	401	725	35.6	27	478	18	42092529	839094	41253435	2.0
Inland water	1505	780	726	51.8	51	826	25	29439286	943518	28495767	3.2
Mountain	1154	349	805	30.3	36	636	26	32083873	548559	31535242	1.7
World	6052	2828	3224	46.7	46	770	25	130669507	3673155	126996316	2.8

Source: Modified from Millennium Ecosystem Assessment (2005)

proximity to population centres, markets and transportation networks, and land-use policies (e.g. Kline *et al.* 2001; Angel *et al.* 2005; Aguayo *et al.* 2007; Conway & Hackworth 2007; Divigalpitiya *et al.* 2007; Xu *et al.* 2007; Davies *et al.* 2008). The particular weighting given to different factors varies between regions and through time (particularly as the dependence of urban populations on local food production declines, and the role of transportation of goods and services, and thus the importance of good transportation pathways, increases). However, within reasonably homogeneous regions empirical models can be built to predict which areas have the greatest probability of being developed in the near future.

Size

The size of a city or urban area is dependent on a complex set of factors, including its history, regional topography and environment, and socioeconomics (including production patterns, income distribution and economic growth). There are many more small urban areas than large ones. Indeed, the frequency distribution of the sizes of cities or urban areas more generally, as reflected in the numbers of people, is strongly right-skewed.

As has been the case with species-abundance distributions (the frequency of species of different abundances) in community ecology, there has been much vigorous debate as to the precise form taken by the frequency distribution of settlement or city sizes. It has principally been argued to follow either a power function or a lognormal distribution (Zipf 1949; Manrubia & Zanette 1998; Eeckhout 2004; Decker *et al.* 2007; Batty 2008). Zipf's law describes the distribution of city sizes in terms of a power function in which, on logarithmic scales, the population size of a city is linearly and inversely related to the rank of that population size, with an exponent that approximately equals unity (Zipf 1949). However, this seems to apply best to larger cities, with the full distribution being better described as lognormal (Eeckhout 2004; Decker *et al.* 2007). An approximate lognormal would accord with Gibrat's law (Gibrat 1931), that such distributions emerge when the growth rates of entities, in this case cities, are random (or at least independent of entity size) and so is their variance. The extent to which this is actually true for cities is, however, contentious (Batty 2008; Garmestani *et al.* 2008). Although there is some empirical support, at least for some periods and regions of the world (Eeckhout 2004; Resende 2004), there is also contradictory evidence for declines in growth rates of cities at larger sizes (i.e. density dependence; e.g. Garmestani *et al.* 2007).

Whatever its detailed form, the frequency distribution of settlement sizes is such that most people live in smaller towns and cities (Figure 2.9). In 2007, more than half of the world's urban population lived in cities or towns with fewer than half a million inhabitants (United Nations 2008).

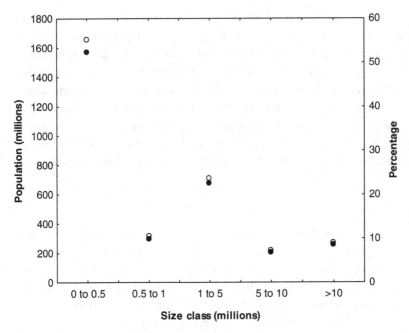

Figure 2.9 The population of the world residing in each of five size classes of urban settlement in 2005 (filled circles; left axis), and the percentage of total urban population in each class (open circles; right axis). Size classes of settlement are as follows: fewer than 500 000 individuals; 500 000 to 1 million; 1 to 5 million; 5 to 10 million; 10 million or more. From data in United Nations (2008).

These are likely to absorb much of the urban growth that is projected to occur in the next few decades, in part because they are destinations for many rural migrants.

The largest cities (or urban agglomerations) are extremely large. Tokyo (35.3 million) is at present the only so-called hypercity or metacity of more than 20 million inhabitants, although a number of others are projected to reach this level in the near future (e.g. Mumbai, São Paulo, Ciudad de México (Mexico City), Delhi, New York–Newark; United Nations 2008). Megacities of more than 10 million inhabitants currently comprise Ciudad de México, New York–Newark, São Paulo, Mumbai, Delhi, Shanghai, Kolkata (Calcutta), Dhaka, Buenos Aires, Los Angeles–Long Beach–Santa Ana, Karachi, Cairo, Rio de Janeiro, Osaka–Kobe, Manila, Beijing and Moskva (Moscow) (United Nations 2008). However, these largest cities account for only a relatively small proportion of the total urban population (<9%; United Nations 2008). Many important demographic, socioeconomic and behavioural characteristics, such as housing, employment, energy and water consumption, gross domestic product, wages, bank deposits, numbers of patents and numbers of crimes, are scaling functions of city size (Bettencourt et al. 2007).

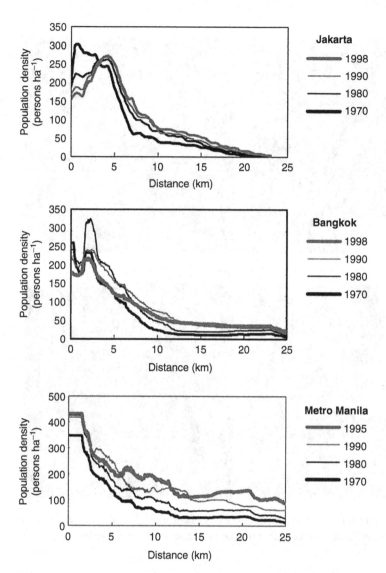

Figure 2.10 Population density as a function of distance from the centre of the cities of Jakarta (Indonesia), Bangkok (Thailand) and Metro Manila (Philippines). Reproduced with permisson from Murakami *et al.* (2005).

Form

Urban areas vary greatly in structure or form (the patterning of land cover and transport networks). Particular focus has fallen on the sprawl that epitomises the fringes of many metropolitan areas in the USA, and on the notion of compact cities, with a high density of buildings and limited interstitial space, which is more typical of parts of Europe (Breheny 1997; Radeloff *et al.* 2005; Tsai 2005; Irwin & Bockstael 2007). However, these reflect only some of the major differences.

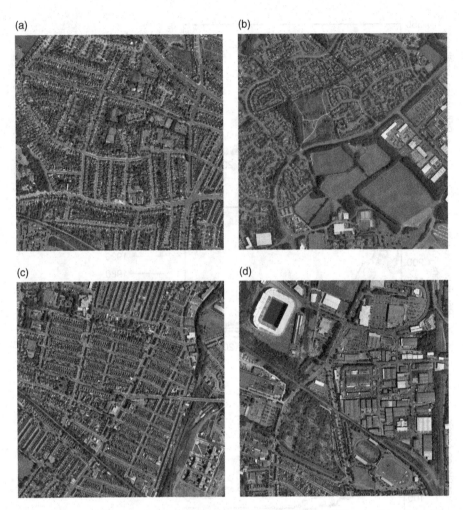

Figure 2.11 Variation in the urban form of the city of Leicester, UK, as reflected in eight 1 km × 1 km areas. Images © Bluesky International and © Infoterra 2006.

Differences in urban form result from the complex variation in, and interplays between, early settlement, industrialisation, land ownership, planning, regulation and infrastructure development (Iverson & Cook 2000; Huang *et al.* 2007). Nonetheless, some patterns do exist. The structure of many cities has traditionally been characterised in terms of a curvilinear decline in human density with distance from the central business district (a distance-decay function, commonly represented as a negative exponential), albeit with the precise form of this gradient varying greatly from city to city (Clark 1951; Edmonston *et al.* 1985). However, the explanatory power of this pattern has clearly been declining with the suburbanisation (decentralisation) of cities particularly in western countries, including both the loss of population in the more built-up

(e)

(f)

(g)

(h)

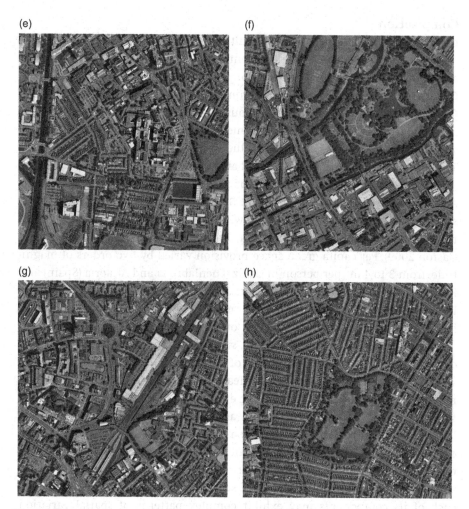

Figure 2.11 (*cont.*)

areas and growth in low-density peripheral developments (Zheng 1991; Bunting *et al.* 2002; Stewart *et al.* 2004). In terms of these density gradients, available data suggest that the Asian megacities, for example, are at different stages of urbanisation, with Jakarta undergoing suburbanisation, Metro Manila at an earlier stage and Bangkok at an intermediate stage (Figure 2.10).

Cities of developed regions tend to be more complex (i.e. irregular in shape), less compact and more porous (greater ratio of open space to urban area), and have less dense human populations than those in developing regions (Huang *et al.* 2007). Much of this difference is associated with variation in levels of mobility, and particularly car transport, which facilitates low-density living.

Composition

Urban areas are heterogeneous at multiple spatial resolutions (Figure 2.11). A wide variety of schemes have been employed to characterise the land cover and land-use of urban areas (e.g. Iverson & Cook 2000; Pauleit & Duhme 2000; Cadenasso et al. 2007; Gill et al. 2007). From the perspective of urban ecology, the first key distinction is between impervious surfaces (buildings, roads etc.), 'green spaces' (parks, gardens/backyards, allotments, road verges, school playing fields, sports fields, green landscaping) and 'blue spaces' (streams, rivers, ponds, lakes). The coverage of these different kinds of areas varies enormously between cities.

Published estimates of the extent of green space in different urban areas are rather scarce. However, a survey of 386 cities across Europe using remote-sensed land cover data revealed that whilst averaging 18.6%, green coverage ranged from 1.9% (Reggio di Calabria, Italy) to 46% (Ferrol, Spain; Fuller & Gaston 2009). Per capita green space provision varied by two orders of magnitude, from 3 to 4 m^2 per person in Cádiz, Fuenlabrada and Almería (Spain) and Reggio di Calabria (Italy) to more than 300 m^2 in Liège (Belgium), Oulu (Finland) and Valenciennes (France). Coverage was lowest in the south and east, increasing to the north and northwest of Europe.

Owing to their much larger number, small parcels of green space contribute disproportionately to overall coverage of urban areas (Fuller et al. 2010), so the spatial grain at which green space is measured will have a large impact on the resulting figures. This is an asymmetric effect, such that if one starts with areas that at a coarse resolution are scored as urban/built-up, increasing the resolution of the mapping will result in more green spaces being found. The converse would occur if mapping a rural area, but such areas are of course excluded from urban studies by definition.

Each of the different categories of surface (impervious, green, blue) and each of its components may exhibit complex patterns of spatial variation (Figure 2.12). Surprisingly, however, these have been characterised in rather few studies.

Of course, patterns of land cover are not static. For example, in the UK available evidence suggests that the extent of green space within urban areas has declined substantially in recent years (Pauleit et al. 2005). This has resulted chiefly from a combination of the loss of brownfield sites (including abandoned industrial areas) and municipal green spaces (e.g. school playing fields) to development (brownfield sites are commonly treated as green spaces in this region), and the increasing density of housing within new developments and within existing housing areas through infilling and backland development (Royal Commission on Environmental Pollution 2007). This has been further exacerbated by the extension of existing buildings, the paving over of front gardens to provide off-road parking, road widening schemes and the general progressive erosion of smaller green spaces.

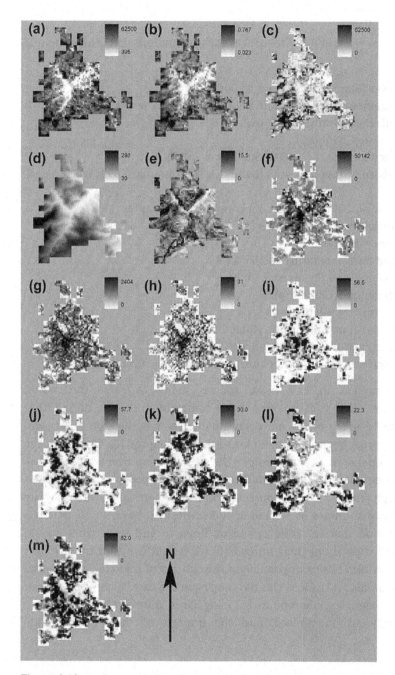

Figure 2.12 Environmental surfaces for Sheffield at 250 m × 250 m cell resolution: (a) extent of green space (m²); (b) vegetation cover Normalised Difference Vegetation Index (NDVI); (c) tree cover (m²); (d) elevation (m); (e) degree angle of slope; (f) buildings area (m²); (g) total road length (m); (h) road nodes (junctions); (i) density of households in blocks of flats; (j) terraced housing density; (k) semi-detached housing density; (l) detached housing density; and (m) total housing density. All housing densities were measured as households per hectare. Reproduced with permission from Davies *et al.* (2008).

Sustainable cities

Urbanisation has profound consequences for the environment. These include changes in:

- habitat cover (Blair 1996; Zhao *et al.* 2006);
- habitat fragmentation (Radeloff *et al.* 2005; Irwin & Bockstael 2007; Di Giulio *et al.* 2009; Evans *et al.* 2009);
- nutrient flows (Chapter 3);
- light levels (Cinzano *et al.* 2001; Small *et al.* 2005);
- noise levels and other acoustics (Bolund & Hunhammar 1999; Warren *et al.* 2006);
- atmospheric and aquatic chemical composition (Chapter 3);
- carbon stocks and flows (Chapter 3);
- water cycles (Chapter 3);
- ground and air temperatures (Chapter 3);
- disturbance regimes (Evans *et al.* 2009);
- species phenology (Chapter 4);
- species life history (Chapter 4);
- species dispersal and migration (Chapter 4);
- species composition and abundance (Chapter 5); and
- biotic interactions (including parasites and diseases; Chapter 4).

Subsequent chapters in this book address many of these differences. Concerns about such consequences have contributed to the notion that cities need to be made much more sustainable than is the case at present. A variety of approaches are available for so doing (Chapter 12). Critical to the notion of sustainable cities is the issue of the spatial extent over which sustainability is considered. With their high concentrations of people, cities necessarily draw large amounts of resources from elsewhere and export waste materials. This will always be so, and as such cities will never be sustainable. Whilst the notion that the inflows and outflows to and from cities should not exceed the capacity of their immediate hinterlands is an attractive one, this is obviously difficult or impossible in the context of regional, continental and global markets, and indeed this has long been the case. What is important is that the efficiencies inherent in bringing large numbers of people into close proximity are maximised, and the impacts of their resource use minimised.

Acknowledgements
Z. G. Davies, F. Eigenbrod, K. L. Evans, R. A. Fuller, S. Gaston and P. Johnson are kindly thanked variously for their advice, assistance and comments.

References

Aguayo, M. I., Wiegand, T., Azócar, G. D., Wiegand, K. and Vega, C. E. (2007). Revealing the driving forces of mid-cities urban growth patterns using spatial modeling: a case study of Los Ángeles, Chile. *Ecology and Society*, **12**, 13.

Angel, S., Sheppard, S. C. and Civco, D. L. (2005). *The Dynamics of Global Urban Expansion*. Washington, DC: World Bank.

Barredo, J. I., Kasanko, M., McCormick, N. and Lavalle, C. (2003). Modelling dynamic spatial processes: simulation of urban future scenarios through cellular automata. *Landscape and Urban Planning*, **64**, 145–60.

Batty, M. (2006). Rank clocks. *Nature*, **444**, 592–6.

Batty, M. (2008). The size, scale, and shape of cities. *Science*, **319**, 769–71.

Berling-Wolff, S. and Wu, J. (2004a). Modeling urban landscape dynamics: a review. *Ecological Research*, **19**, 119–29.

Berling-Wolff, S. and Wu, J. (2004b). Modeling urban landscape dynamics: a case study in Phoenix, USA. *Urban Ecosystems*, **7**, 215–40.

Berry, B. J. L. (1990). Urbanization. In B. L. Turner, II, W. C. Clark, R. W. Kates *et al*, eds., *The Earth as Transformed by Human Action: Global and Regional Changes in the Biosphere over the Past 300 Years*. Cambridge: Cambridge University Press, pp. 103–19.

Best Foot Forward (2002). *City Limits: A Resource Flow and Ecological Footprint Analysis of Greater London*. Available at http://www.citylimitslondon.com.

Bettencourt, L. M. A., Lobo, J., Helbing, D., Kühnert, C. and West, G. B. (2007). Growth, innovation, scaling, and the pace of life in cities. *Proceedings of the National Academy of Sciences of the USA*, **104**, 7301–6.

Birchenall, J. A. (2007). Economic development and the escape from high mortality. *World Development*, **35**, 543–68.

Blair, R. B. (1996). Land use and avian species diversity along an urban gradient. *Ecological Applications*, **6**, 506–19.

Bloom, D. E., Canning, D. and Fink, G. (2008). Urbanization and the wealth of nations. *Science*, **319**, 772–5.

Bolund, P. and Hunhammar, S. (1999). Ecosystem services in urban areas. *Ecological Economics*, **29**, 293–301.

Breheny, M. (1997). Urban compaction: feasible and acceptable? *Cities*, **14**, 209–17.

Bunting, T., Filion, P. and Priston, H. (2002). Density gradients in Canadian metropolitan regions, 1971–96: differential patterns of central area and suburban growth and change. *Urban Studies*, **39**, 2531–52.

Burgess, E. W. (1925). The growth of the city: an introduction to a research project. In R. E. Park, E. W. Burgess and R. D. McKenzie, eds., *The City*. Chicago: University of Chicago Press, pp. 47–62.

Cadenasso, M. L., Pickett, S. T. A. and Schwarz, K. (2007). Spatial heterogeneity in urban ecosystems: reconceptualizing land cover and a framework for classification. *Frontiers in Ecology and the Environment*, **5**, 80–8.

Chandler, T. (1987). *Four Thousand Years of Urban Growth. An Historical Census*, 2nd edn. Lewiston/Queenston: Edwin Mellen Press.

Chown, S. L., van Rensburg, B. J., Gaston, K. J., Rodrigues, A. S. L. and van Jaarsveld, A. S. (2003). Species richness, human population size and energy: conservation implications at a national scale. *Ecological Applications*, **13**, 1233–41.

Cinzano, P., Falchi, F. and Elvidge, C. D. (2001). The first world atlas of the artificial night sky brightness. *Monthly Notices of the Royal Astronomical Society*, **328**, 689–707.

Clark, C. (1951). Urban population densities. *Journal of the Royal Statistical Society Series A*, **114**, 490–6.

Cohen, B. (2004). Urban growth in developing countries: a review of current trends and a caution regarding existing forecasts. *World Development*, **32**, 23–51.

Conway, T. and Hackworth, J. (2007). Urban pattern and land cover variation in the greater Toronto area. *The Canadian Geographer*, **51**, 43–57.

Davies, R. G., Barbosa, O., Fuller, R. A. *et al*. (2008). City-wide relationships between green spaces, urban land use and topography. *Urban Ecosystems*, **11**, 269–87.

Decker, E. H., Kerkhoff, A. J. and Moses, M. E. (2007). Global patterns of city size distributions and their fundamental drivers. *PLoS One*, **9**, e934.

Diamond, J. (2006). *Collapse: How Societies Choose to Fail or Succeed*. London: Penguin.

Di Giulio, M., Holderegger, R. and Tobias, S. (2009). Effects of habitat and landscape fragmentation on humans and biodiversity in densely populated landscapes. *Journal of Environmental Management*, **90**, 2959–68.

Divigalpitiya, P., Ohgai, A., Tani, T., Watanabe, K. and Gohnai, Y. (2007). Modeling land conversion in the Colombo metropolitan area using cellular automata. *Journal of Asian Architecture and Building Engineering*, **6**, 291–8.

Dye, C. (2008). Health and urban living. *Science*, **319**, 766–9.

Edmonston, B., Goldberg, M. A. and Mercer, J. (1985). Urban form in Canada and the United States: an examination of urban density gradients. *Urban Studies*, **22**, 209–17.

Eeckhout, E. (2004). Gibrat's law for (all) cities. *American Economic Review*, **94**, 1429–51.

Ehrlich, P. R. and Ehrlich, A. H. (2004). *One with Nineveh: Politics, Consumption, and the Human Future*. Washington, DC: Island Press.

Evans, K. L. and Gaston, K. J. (2005). People, energy and avian species richness. *Global Ecology and Biogeography*, **14**, 187–96.

Evans, K. L., Newson, S. E. and Gaston, K. J. (2009). Habitat influences on urban avian assemblages. *Ibis*, **151**, 19–39.

Fuller, R. A. and Gaston, K. J. (2009). The scaling of green space coverage in European cities. *Biology Letters*, **5**, 352–5.

Fuller, R. A., Tratalos, J., Warren, P. H. *et al.* (2010). Environment and biodiversity. In M. Jenks and C. Jones, eds., *Dimensions of the Sustainable City*. Dordrecht, Netherlands: Springer, pp. 75–103.

Garmestani, A. S., Allen, C. R. and Gallagher, C. M. (2008). Power laws, discontinuities and regional city size distributions. *Journal of Economic Behavior and Organization*, **68**, 209–16.

Garmestani, A. S., Allen, C. R., Gallagher, C. M. and Mittelstaedt, J. D. (2007). Departures from Gibrat's law, discontinuities and city size distributions. *Urban Studies*, **44**, 1997–2007.

Gaston, K. J. (2005). Biodiversity and extinction: species and people. *Progress in Physical Geography*, **29**, 239–47.

Gibrat, R. (1931). *Les Inégalités Économiques: applications: aux inégalités des richesses, à la concentration des enterprises, aux populations des villes, aux statistiques des familles, etc., d'une loi nouvelle, la loi de l'effet proportionnel*. Paris: Sirey.

Gill, S. E., Handley, J. F., Ennos, A. R. and Pauleit, S. (2007). Adapting cities for climate change: the role of the green infrastructure. *Built Environment*, **33**, 115–33.

Hansen, M. C., Defries, R. S., Townshend, J. R. G. and Sohlberg, R. (2000). Global land cover classification tree approach. *International Journal of Remote Sensing*, **21**, 1331–64.

He, C., Okada, N., Zhang, Q., Shi, P. and Li, J. (2008). Modelling dynamic urban expansion processes incorporating a potential model with cellular automata. *Landscape and Urban Planning*, **86**, 79–91.

Huang, J., Lu, X. X. and Sellers, J. M. (2007). A global comparative analysis of urban form: applying spatial metrics and remote sensing. *Landscape and Urban Planning*, **82**, 184–97.

Irwin, E. G. and Bockstael, N. E. (2007). The evolution of urban sprawl: evidence of spatial heterogeneity and increasing land fragmentation. *Proceedings of the National Academy of Sciences of the USA*, **104**, 20672–7.

Iverson, L. R. and Cook, E. A. (2000). Urban forest cover of the Chicago region and its relation to household density and income. *Urban Ecosystems*, **4**, 105–24.

Japanese Statistics Bureau (2005). *Information on the 2005 Population Census of Japan*. Available at http://www.stat.go.jp/english/data/kokusei/index.htm (accessed 5 Jan 2009).

Kasanko, M., Barredo, J. I., Lavalle, C. *et al.* (2006). Are European cities becoming dispersed? A comparative analysis of 15 European urban areas. *Landscape and Urban Planning*, **77**, 111–30.

Kitzes, J., Wackernagel, M., Loh, J. *et al.* (2008). Shrink and share: humanity's present and future Ecological Footprint. *Philosophical Transactions of the Royal Society B*, **363**, 467–75.

Klein Goldewijk, K. (2005). Three centuries of global population growth: a spatial referenced population (density) database for 1700–2000. *Population and Environment*, **26**, 343–67.

Kline, J. D., Moses, A. and Alig, R. J. (2001). Integrating urbanization into landscape-level ecological assessments. *Ecosystems*, **4**, 3–18.

Lanapsa, L., Pueyo, F. and Sanz, F. (2003). The evolution of Spanish urban structure during the twentieth century. *Urban Studies*, **40**, 567–80.

Leick, G. (2001). *Mesopotamia: The Invention of the City*. London: Penguin.

Lepczyk, C. A., Hammer, R. B., Stewart, S. I. and Radeloff, V. C. (2007). Spatiotemporal dynamics of housing growth hotspots in the North Central U.S. from 1940 to 2000. *Landscape Ecology*, **22**, 939–52.

Li, X. and Yeh, A. G. (2000). Modelling sustainable urban development by the integration of constrained cellular automata and GIS. *International Journal of Geographical Information Science*, **14**, 131–52.

Liu, J., Daily, G. C., Ehrlich, P. R. and Luck, G. W. (2003). Effects of household dynamics on resource consumption and biodiversity. *Nature*, **421**, 530–3.

Loveland, T. R., Reed, B. C., Brown, J. F. *et al.* (2000). Development of a global land cover characteristics database and IGBP DISCover from 1 km AVHRR data. *International Journal of Remote Sensing*, **21**, 1303–30.

Mace, R. (2008). Reproducing in cities. *Science*, **319**, 764–6.

Manrubia, S. C. and Zanette, D. H. (1998). Intermittency model for urban development. *Physical Review E*, **58**, 295–413.

Marshall, J. D. (2007). Urban land area and population growth: a new scaling relationship for metropolitan expansion. *Urban Studies*, **44**, 1889–904.

Meijers, E. (2005). Polycentric urban regions and the quest for synergy: is a network of cities more than the sum of its parts? *Urban Studies*, **42**, 765–81.

Millennium Ecosystem Assessment (2005). *Ecosystems and Human Well-being: Current State and Trends*, Vol. I. Washington, DC: Island Press.

Montgomery, M. R. (2008). The urban transformation of the developing world. *Science*, **319**, 761–4.

Murakami, A., Zain, A. M., Takeuchi, K., Tsunekawa, A. and Yokota, S. (2005). Trends in urbanization and patterns of land use in the Asian mega cities Jakarta, Bangkok, and Metro Manila. *Landscape and Urban Planning*, **70**, 251–9.

Odum, E. P. (1997). *Ecology: A Bridge Between Science and Society*. Massachusetts: Sinauer Associates.

Pataki, D. E., Xu, T., Luo, Y. Q. and Ehleringer, J. R. (2007). Inferring biogenic and anthropogenic carbon dioxide sources across an urban to rural gradient. *Oecologia*, **152**, 307–22.

Pauleit, S. and Duhme, F. (2000). Assessing the environmental performance of land cover types for urban planning. *Landscape and Urban Planning*, **52**, 1–20.

Pauleit, S., Ennos, R. and Golding, Y. (2005). Modeling the environmental impacts of urban land use and land cover change – a study in Merseyside, UK. *Landscape and Urban Planning*, **71**, 295–310.

Radeloff, V. C., Hammer, R. B. and Stewart, S. I. (2005). Rural and suburban sprawl in the U.S. midwest from 1940 to 2000 and its relation to forest fragmentation. *Conservation Biology*, **19**, 793–805.

Redman, C. L. (1999). *Human Impact on Ancient Environments*. Tuscon: University of Arizona Press.

Rees, W. E. (1992). Ecological footprints and appropriated carrying capacity: what urban economics leaves out. *Environment and Urbanization*, **4**, 121–30.

Rees, W. E. (1999). The built environment and the ecosphere: a global perspective. *Building Research and Information*, **27**, 206–20.

Rees, W. E. (2001). Ecological footprint, concept of. In S. A. Levin, ed., *Encyclopedia of Biodiversity*, Vol. **2**. San Diego: Academic Press.

Resende, M. (2004). Gibrat's law and the growth of cities in Brazil: a panel data investigation. *Urban Studies*, **41**, 1537–49.

Royal Commission on Environmental Pollution (2007). *The Urban Environment*. London: The Stationery Office.

Salvatore, M., Pozzi, F., Ataman, E., Huddleston, B. and Bloise, M. (2005). *Mapping Global Urban and Rural Population Distributions*. Rome: FAO.

Small, C. and Cohen, J. E. (2004). Continental physiography, climate and the global distribution of human population. *Current Anthropology*, **45**, 269–77.

Small, C., Pozzi, F. and Elvidge, C. D. (2005). Spatial analysis of global urban extent from DMSP-OLS night lights. *Remote Sensing of Environment*, **96**, 277–91.

Statistics Canada (2001). *2001 Census*. Available at http://www12.statcan.ca/english/census01/Products/ (accessed 20 Sept 2008).

Statistics New Zealand (2006). *2006 Census data*. Available at http://www.stats.govt.nz/census/ (accessed 20 Sept 2008).

Sterling, S. and Ducharne, A. (2008). Comprehensive data set of global land cover change for land surface model applications. *Global Biochemical Cycles*, **22**, GB3017.

Stewart, D. J., Yin, Z.-Y., Bullard, S. M. and MacLachlan, J. T. (2004). Assessing the spatial structure of urban and population growth in the Greater Cairo area, Egypt: a GIS and imagery analysis approach. *Urban Studies*, **41**, 95–116.

Theobald, D. M. (2002). Land-use dynamics beyond the American urban fringe. *The Geographical Review*, **91**, 544–64.

Theobald, D. M. (2005). Landscape patterns of exurban growth in the USA from 1980 to 2020. *Ecology and Society*, **10**, 32.

Tsai, Y.-H. (2005). Quantifying urban form: compactness versus 'sprawl'. *Urban Studies*, **42**, 141–61.

United Nations (2008). *World Urbanization Prospects. The 2007 Revision*. New York: United Nations,

United Nations Centre for Human Settlements (2002). *The State of the World's Cities Report 2001*. New York: United Nations.

United Nations Environment Programme (2007). *Global Environment Outlook GEO4*. Valletta, Malta: Progress Press.

UN-HABITAT (2006). *The State of the World's Cities Report 2006/2007*. London: Earthscan.

Wackernagel, M., Kitzes, J., Moran, D., Goldfinger, S. and Thomas, M. (2006). The Ecological Footprint of cities and regions: comparing resource availability with resource demand. *Environment and Urbanization*, **18**, 103–12.

Wackernagel, M., Schulz, N. B., Deumling, D. et al. (2002). Tracking the ecological overshoot of the human economy. *Proceedings of the National Academy of Sciences of the USA*, **99**, 9266–71.

Warren, P. S., Katti, M., Ermann, M. and Brazel, A. (2006). Urban bioacoustics: it's not just noise. *Animal Behaviour*, **71**, 491–502.

Woods, R. (2003). Urban–rural mortality differentials: an unresolved debate. *Population and Development Review*, **29**, 29–46.

World Resources Institute (2007). http://earthtrends.wri.org/index.php

Xu, C., Liu, M., Zhang, C. et al. (2007). The spatiotemporal dynamics of rapid urban growth in the Nanjing metropolitan region of China. *Landscape Ecology*, **22**, 925–37.

Zhao, S., Da, L., Tang, Z. et al. (2006). Ecological consequences of rapid urban expansion: Shanghai, China. *Frontiers in Ecology and the Environment*, **4**, 341–6.

Zheng, X.-P. (1991). Metropolitan spatial structure and its determinants: a case-study of Tokyo. *Urban Studies*, **28**, 87–104.

Zipf, G. K. (1949). *Human Behavior and the Principle of Least Effort*. Cambridge: Addison-Wesley.

CHAPTER THREE

Urban environments and ecosystem functions

KEVIN J. GASTON, ZOE G. DAVIES
AND JILL L. EDMONDSON

Urbanisation profoundly changes both the abiotic and biotic properties of ecosystems. It does so not just within urban areas, but also in surrounding landscapes and often much further afield. Traditionally, the foremost focus for research has been on the negative impacts of these changes, particularly for human health and wellbeing, and how these can most effectively be mitigated. This is readily understandable given that, for much of their history, urban environments have often been associated with high rates of infant mortality, disease outbreaks and a generally poor quality of life, and that this still remains true for many of those living in cities today (Woods 2003; UN-HABITAT 2006; Birchenall 2007). From a broader ecological perspective, urban areas have also long been considered depauperate in comparison to their rural counterparts in terms of flora and fauna, with the exception of a few notable species that were widely categorised as pests.

More recently, research into urban environments has increasingly shifted towards examining the positive contributions that such areas can make both to the human population and to other species. Of course, viewing urban environments in terms of the benefits they can provide or the costs they can exact are essentially two sides of the same coin. However, closer consideration of the potential advantages has served to broaden the range of environmental issues that have received emphasis; rather than focusing almost exclusively on the undoubtedly vital human health concerns resulting from poor air quality, unclean water and inadequate sanitation, there is now a growing appreciation of the benefits that greener and more ecologically diverse urban areas have on the mental and physical status of residents, on economic markets and for biological conservation (Chapters 7–11; Fuller *et al.* 2010; Gaston & Evans 2010; Irvine *et al.* 2010).

This focus on the positive contributions that urban areas can make to human health and wellbeing is closely associated with the concept of ecosystem service

Urban Ecology, ed. Kevin J. Gaston. Published by Cambridge University Press.

provision, which can be broadly defined as the benefits people obtain from ecosystems (e.g. Bolund & Hunhammar 1999; Pataki *et al.* 2006; Tratalos *et al.* 2007). Such services can be divided into four main categories (Millennium Ecosystem Assessment 2005): (i) supporting services (e.g. soil formation, photosynthesis, primary production, nutrient cycling, water cycling); (ii) regulating services (e.g. air quality regulation, climate regulation, water regulation, erosion regulation, water purification and waste treatment, disease regulation, pest regulation, pollination, natural hazard regulation); (iii) provisioning services (e.g. food, fibre, genetic resources, biochemicals, natural medicines, pharmaceuticals, fresh water); and (iv) cultural services (e.g. cultural diversity, spiritual and religious values, knowledge systems, educational values, inspiration, aesthetic values, social relations, sense of place, cultural heritage values, recreation and ecotourism). Plainly, urban ecosystems supply services from each of these different categories, although some are obviously more pertinent than others. In certain cases this is foremost a matter of provision to local human populations (e.g. pollination). Alternatively, for services such as carbon sequestration, although the immediate benefits to the local population may be less significant, urban centres can contribute substantially to provision at a regional scale.

The transition towards appreciating the benefits that can be obtained from urban areas, rather than simply mitigating their negative influences, may also increasingly be driven by the recognition that the long-held belief that rural landscapes are better for native biota is breaking down in some parts of the world. This is particularly true in regions where agriculture has become very intensive and has restricted natural or semi-natural vegetation to small and highly fragmented areas, resulting in some components of urban environments (especially suburban locations with extensive coverage by green space and high habitat heterogeneity) acting as havens for wildlife (Beebee 1997; Gregory & Baillie 1998; Gehrt & Chelsvig 2004; Peach *et al.* 2004; Gaston & Evans 2010).

This chapter provides a broad and selective overview of the changes that urbanisation can have on ecosystem functions and services. Issues specifically relating to biodiversity and cultural services are addressed in other chapters (Chapters 5 and 8).

Temperature

Urban environments across a wide range of latitudes and climatic regimes commonly experience 'heat island' effects, in which air temperatures are elevated compared with surrounding landscapes (e.g. Karl *et al.* 1988; Taha *et al.* 1999; Pickett *et al.* 2001; Baker *et al.* 2002; Arnfield 2003). For larger cities the maximum difference between urban and rural ambient temperatures is 6–12 °C (Kaye *et al.* 2006), and air temperatures are often 2–10 °C higher in urban areas than neighbouring ones (Shepherd 2005). Indeed, in general, mean

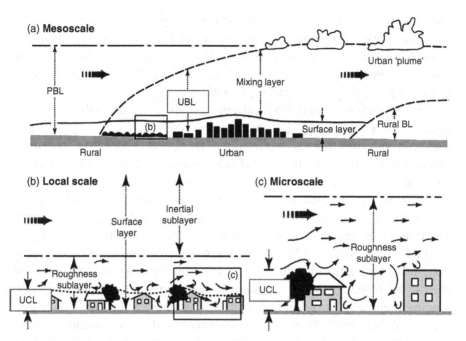

Figure 3.1 Urban environments at different spatial scales, with an emphasis on atmospheric components: (a) mesoscale; (b) local scale; (c) microscale. PBL, planetary boundary layer; UBL, urban boundary layer; UCL, urban canopy layer; BL, boundary layer. Reproduced with permission from Shepherd (2005), modified after Oke (1987).

and minimum temperatures are often increased, and daily periods of cool temperatures, cool seasons and frost days are all reduced. These effects are exhibited at ground level, in the urban canopy layer (which extends from ground to roof level, and within which are urban street-canyon flows) and in the urban boundary layer (above roof level, and developing downwind of the leading edge of a city; Figure 3.1), although the patterns may not be entirely congruent (Arnfield 2003) and may be spatially complex (Gaffin et al. 2008). Ordinarily, the heat island phenomenon is greatest within large cities, during the summer and at night, and lessens with rising wind speed and more extensive cloud cover.

Urban heat island effects are exacerbated by a number of different factors, but principally tend to increase with proportional coverage by impervious surfaces (e.g. buildings and roads, which often have albedos and heat capacities that result in the storage of incoming solar radiation as sensible heat during the day and its release as long wave radiation at night). Conversely, they will decrease relative to the proportion of unsealed surface, and more particularly green space (Henry & Dicks 1987; Lo et al. 1997; Chen & Wong 2006; Chen et al. 2006; Jenerette et al. 2007; Weng et al. 2007; but see Gaffin et al. 2008). The benefits accrued from planting urban vegetation tend to be non-linear, such that the addition of even small amounts of green material where previously there was none makes a much greater difference to ambient temperatures than small increases in areas where the coverage was already extensive. This vegetation can be effective in a wide diversity of forms, including green roofs and green walls (Alexandri & Jones 2008).

In regions where urban temperatures become uncomfortably warm, the demand for air-conditioning in buildings is likely to rise. The planting of vegetation can therefore also contribute to reductions in fossil fuel consumption and resultant carbon dioxide (CO_2) emissions (Akbari et al. 1997; Taha et al. 1999). In particular, levels of tree cover can have a marked influence on energy use in buildings, as well as outside air temperatures, by (i) reducing the heat gain by lowering ambient temperatures through evapotranspiration (the transfer of water to the atmosphere by plants); (ii) increasing the latent air-conditioning load by adding moisture to the air through evapotranspiration; (iii) reducing the outside air infiltration rate by lowering ambient wind speeds; (iv) reducing solar heat gain through windows, walls and roofs by shading them; and (v) reducing radiant heat gain from the surroundings by shading (Akbari 2002). Trees can also improve the thermal comfort of urban dwellers more generally (Stone & Rodgers 2001; Georgi & Zafiriadis 2006), with the positive impacts tending substantially to outweigh any negative influences on temperatures during the winter (e.g. through shading buildings). Given that the determinants of tree cover can be both biophysical (e.g. soil type, drainage, herbivores) and socioeconomic (e.g. wealth, cultural norms), this provides an example of how in urban areas most ecosystem functions routinely change in response to these two sets of factors, and the limited applicability of models derived from more natural environments.

The influence of urbanisation on air temperatures appears to be much smaller than the effects of global climate change, although not insignificant when combined with other land cover changes (Jones et al. 1990; Kalnay & Cai 2003). Nonetheless, the temperature increases associated with urbanisation can have a variety of ecological consequences, some of which extend beyond the urban area itself, including adjustments relative to rural areas in the timing of germination, leaf flush, leaf drop, flowering and length of growing season of plants, and in the breeding of animals (Chapter 4; White et al. 2002; Partecke et al. 2004; Zhang et al. 2004a, 2004b; Neil & Wu 2006). The potential knock-on impacts that these phenological differences may have on species survival, assemblage structures and community dynamics are clearly diverse.

Water quality

Urbanisation is commonly, and perhaps almost universally, associated with declines in the quality of the water carried in streams and rivers (e.g. Paul & Meyer 2001; Walsh et al. 2005; Zhao et al. 2006). Indeed, it has been argued that urbanisation is second only to agriculture in its negative impacts on water quality (Paul & Meyer 2001). This quality is typically measured by some combination of (i) temperature, which may be altered in urban areas as a consequence of the loss of riparian vegetation, decreased groundwater discharge, industrial outputs and the heat island effect, and affects oxygen solubility and the rate of biochemical processes; (ii) pH, which influences the dissolving of ions; (iii) dissolved oxygen, which is important for aquatic fauna and the taste of water; (iv) biochemical oxygen demand; (v) faecal coliform bacteria, indicative of inadequate wastewater treatment or sewage plant overflow; (vi) phosphates and nitrates, which may be at heightened concentrations over hundreds of kilometres from the point of origin and are agents of eutrophication; and (vii) turbidity and total suspended solids, both of which have an effect on aquatic fauna and the treatment cost of drinking water (Paul & Meyer 2001; Duh et al. 2008). Water pollutants arise from a variety of sources, including wastewater, fertilizers and sewage discharges. Stormwater runoff washes pollutants from impervious surfaces, providing a significant non-point source of pollution in urban areas (Characklis & Wiesner 1997; Bibby & Webster-Brown 2005). Nonetheless, there is no evidence to suggest that, in general, the level of water pollutants increases in relation to the size of an individual city (Duh et al. 2008). There is potential within urban areas to improve water quality by construction of, or improved management of existing, wetland ecosystems.

Water flows

Cities almost invariably draw the bulk of their water supplies from outside their boundaries, as they typically have only a small area of standing water, directing large quantities in and out through often heavily engineered infrastructure. In most cases, these quantities are considerably greater than would otherwise have passed through such environments, although occasionally they may be reduced. Urbanisation thus changes the hydrology of an area by altering the water supply, drainage and outflow. Indeed, in simple terms of mass, water is by far the largest component of the metabolism of an urban centre, vastly exceeding the amounts of food and fuel used by the human population (Kennedy et al. 2007). It is, therefore, perhaps unsurprising that although the natural density of water channels tends to be much reduced in urban ecosystems, with many streams being filled in, paved over or culverted, the overall density may actually be higher owing to the number of artificial ones that are created. In addition, the development and structure of such channels tends to differ between natural and urban settings (Paul & Meyer 2001).

To date, much of the research into urban water flows has focused on the relationship between impervious surface coverage and water regulation (Arnold & Gibbons 1996). The construction of such sealed areas, which cover

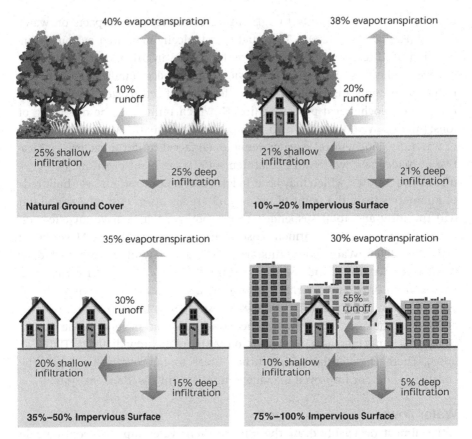

Figure 3.2 Relationships between impervious surface cover and surface runoff, infiltration and evapotranspiration. Reproduced from Federal Interagency Stream Restoration Working Group (1998).

~9% of Europe alone (Scalenghe & Marsan 2009), results in marked increases in surface runoff and decreases in infiltration (Figure 3.2). The outcome of this, especially when associated with infrastructure to drain paved surfaces as quickly and efficiently as possible (e.g. piped stormwater drainage systems), is a substantially faster flow of water through systems and higher peak discharges during storm events. In turn, this can lead to a greater frequency and severity of flooding. At the same time, base flow rates may be influenced in a variety of ways, sometimes becoming much diminished because of the lower infiltration rates of water (reducing groundwater recharge). However, this may be offset by irrigation, leakage from water supply or sewerage infrastructure, and waste-water treatment plant outputs (Paul & Meyer 2001; Walsh et al. 2005).

Urbanisation can also influence precipitation both within and downwind of cities, although usually to a smaller and less consistent extent than is the case with temperature (Shepherd 2005; Collier 2006; Hand & Shepherd 2009).

Deep, moist convection within urban environments is primarily enhanced by the heat island effect (which destabilises the urban boundary layer; Figure 3.1), the surface roughness of the area caused by buildings, and potentially by vegetation. Conversely, precipitation may be suppressed by atmospheric pollution, particularly aerosols which change the microstructure of clouds by providing sources of condensation nuclei. The magnitude of some of these effects may be related to the size of a city, and it has been suggested that the expansion of urban areas may be having non-trivial influences on global patterns of precipitation (Shepherd 2005).

The distribution of green space is crucial in explaining variation in water regulation across the urban landscape. Increasing the extent of vegetated areas enhances water regulation and improves water loss from the ground through evapotranspiration. Generalised models of the impacts of various land covers on runoff tend to rely on simple coefficients for rainwater infiltration of different surfaces, usually obtained from the literature and not tailored to particular urban areas (e.g. Pauleit & Duhme 2000; Whitford *et al.* 2001; Pauleit *et al.* 2005; Tratalos *et al.* 2007). Nonetheless they frequently reveal non-linear relationships with the extent of green space; initially, runoff rapidly increases as green space declines, but at low levels of vegetative cover, changes in the remaining green space have little effect (Tratalos *et al.* 2007). This is a consequence of the relatively low infiltration rates of even many of the non-sealed surfaces, due to factors such as extensive compaction of soil surfaces (Scalenghe & Marsan 2009), when building densities become high.

Carbon cycles

Urbanisation influences local carbon cycles in two main ways. First, it changes land cover into a mosaic of green spaces and impervious surfaces. Second, it generates CO_2 emissions as a consequence of the importation and subsequent combustion of fossil fuels (especially oil, gas and coal) for residential and industrial energy production, as well as for transportation.

The two major biological carbon pools in urban green spaces are vegetation and soils. Most attention given to vegetation tends to fall on trees, which have been shown to provide significant levels of carbon storage and sequestration in urban environments, albeit substantially less than equivalent areas of forest stands (Nowak & Crane 2002). Estimates of carbon sequestration by urban trees tend to be generated by using simple functions of tree cover, which is logical since this is both likely to have a dominant influence and is extremely variable amongst different urban areas (Rowntree & Nowak 1991; Whitford *et al.* 2001; Tratalos *et al.* 2007). However, in practice, the relationship will depend on a variety of factors, including growth rate, size at maturity, demographic structure and species composition (Nowak *et al.* 2002). These in turn will be influenced by land-use, planning policies, management and more immediate environmental conditions, such as soil quality, water potential and coverage by impervious surfaces (Quigley 2004). In particular, on public lands, trees may be removed or subject to surgery in response to concerns over issues of subsidence or human safety. For instance, although over a 5-year period there was no overall change in the number of trees in a study conducted in London, UK, there was nonetheless a rapid turnover of individual trees and a disproportionate loss of mature native trees (London Assembly Environment Committee 2007). Despite the potential to use these removed trees and waste wood products as an alternative energy source to fossil fuels, such bioenergy generation still produces CO_2 emissions (MacFarlane 2009). The degree to which urban trees provide net carbon sequestration will also depend on the intensity of management which they receive using carbon-emitting maintenance equipment (e.g. chainsaws, chippers, motor vehicles), and the extent to which decomposition is limited by long-term timber storage (e.g. wood products, landfills; Nowak et al. 2002; MacFarlane 2009). Similar concerns have been previously expressed about the environmental costs of management of other urban land covers (e.g. grasslands; Falk 1976; Golubiewski 2006).

Soil carbon pools naturally comprise both labile organic carbon (with relatively quick turnover rates) and black carbon, formed from incomplete combustion of organic matter or fossil fuel (which is highly recalcitrant with a long soil residence time). The black carbon constituent within urban soils is often enhanced by vehicle and industry emissions (Lorenz et al. 2006; Rawlins et al. 2008). Below-ground carbon storage in urban environments has variously been projected to be equal to, or considerably bigger than, that associated with aboveground vegetation, depending partially on the nature of this vegetation (Figure 3.3; Jo & McPherson 1995; Pataki et al. 2006; Pouyat et al. 2006). On a per unit area basis, there is also evidence that the carbon pool in urban soils is often high compared with other types of ecosystem, and may be greater than that in equivalent rural soils because of the profound differences in management (Pouyat et al. 2002, 2006, 2009; Golubiewski 2006; Pataki et al. 2006; Lorenz & Lal 2009). Indeed, it seems probable that urban soils will often be markedly richer in organic carbon than those in rural landscapes with a significant history of intensive agriculture (this is less likely to be true in forest-dominated regions). Unfortunately, understanding of urban soil carbon pools and fluxes is limited, primarily because the soils are often spatially very heterogeneous, are poorly mapped and characterised, in substantial part lie beneath impervious surfaces, and are often within mosaics of different aged urban green spaces (Effland & Pouyat 1997; Lorenz & Lal 2009). Nonetheless, it is clear that the patterns and rates of decomposition and nutrient cycling may be profoundly altered in urban areas compared with rural ones, owing to differences in leaf litter quality, because of altered biotic, chemical and physical environments (often including high levels of heavy

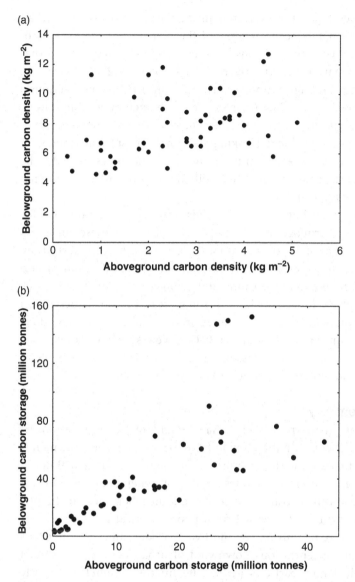

Figure 3.3 Relationships between estimated belowground and aboveground (a) urban carbon density and (b) urban carbon storage estimated for different states in the USA. From data in Pouyat *et al.* (2006).

metals), and owing to management practices and levels of disturbance (McDonnell *et al.* 1997; Pouyat *et al.* 1997, 2009; Carreiro *et al.* 1999; Pickett *et al.* 2001; Shen *et al.* 2008).

Given the relatively limited coverage by urban areas worldwide (see Chapter 2), the immediate influence of urbanisation on vegetation and soils has arguably also been reasonably small globally. Hence this is also true at a regional scale (e.g. Erb 2004), although of course less so for those regions that have become more significantly developed. However, in many regions the expansion of urban areas has disproportionately been on some of the best soils at the expense of agricultural and forested land, and thus the ecosystem goods and services that they provided (Nizeyimana et al. 2001; Alig et al. 2004; Zhao et al. 2006; Zhang et al. 2008). There are, therefore, significant concerns that this has become a seriously problematic addition to the other pressures on such high-value land covers, especially where agricultural land can be very productive. For example, in the USA urbanisation is disproportionately taking place on the most fertile lands, with a substantial overall negative effect on net primary production estimated to be equivalent to the caloric requirement of 16.5 million people or c. 6% of the human population (Imhoff et al. 2004).

Cities are major contributors to CO_2 emissions, being responsible for c. 75–90% of global anthropogenic CO_2. Clearly, the levels of these emissions are such that they cannot be offset by local natural carbon sequestration. Nonetheless, the per unit area and the gross sizes of biological urban carbon pools can be substantial (Nowak & Crane 2002; Pouyat et al. 2002; Kaye et al. 2005; Lorenz et al. 2006). Indeed, at least in some areas, carbon sequestration within urban environments may not be an entirely trivial consideration (Jo & McPherson 1995; Johnson & Gerhold 2003; Golubiewski 2006; Pataki et al. 2006), and there has even been discussion of the potential for these pools to be cost effective in carbon credit markets (McHale et al. 2007).

Atmospheric chemistry

The atmospheres of urban areas tend to have heightened concentrations of CO_2, nitrous oxides (NO_x), sulphur dioxide (SO_2), ozone (O_3), aerosols, metals and suspended particulates, with profound implications for nutrient flows in particular. Nitrous and sulphurous oxides are mainly generated through fossil fuel combustion, and can be transformed in the atmosphere into acidic precipitation. Ozone is principally derived from photochemical reactions in the atmosphere that involve nitrous oxides and hydrocarbons, and tends to occur in particularly high concentrations downwind of urban centres. Suspended particulates include larger particles (c. 2.5–100 microns diameter), usually consisting of smoke and dust caused by industrial processes, agriculture, construction and road traffic, and smaller particles (<2.5 microns) mainly from the burning of fossil fuels (World Resources Institute, the United Nations Environment Programme, the United Nations Development Programme and The World Bank 1998).

The temporal dynamics of these pollutants are often complex, depending on patterns in the history, growth and composition of an individual city, the balance between increasing demand and output in power generation, industry and transport, and the effectiveness of any active attempts to reduce the levels of pollutants (including through new technologies). Thus, for example, in Shanghai, China, concentrations of SO_2, total suspended particulates and acid precipitation all fell from 1983 to 2004, mainly as a consequence of reductions in the usage of coal, but levels of NO_x have risen, primarily with growing numbers of motor vehicles (Zhao et al. 2006). However, SO_2 and acid rain have tended to increase in the suburbs of Shanghai, in response to a policy to move factories from urban centres to suburban areas, and NO_x has decreased in more urbanised areas as a consequence of policies towards cleaner vehicle emissions.

Generally, atmospheric pollution in the urban areas of industrialised nations has declined over the past few decades whereas, conversely, there have been increases in pollutants within the developing world (Mayer 1999). However, even where the levels of atmospheric pollutants have fallen in urban environments in recent years, they commonly still remain above desirable (albeit variable) regional standards. The changes made to atmospheric chemistry by urban areas can have influences on multiple scales. Positively, they can affect nutrient cycling and plant production, particularly through fertilisation effects, both locally and regionally (Trusilova & Churkina 2008). More negatively, air pollution from urban centres is a major contributor to global climate change and acid deposition, and many studies have shown that such pollution and its biogeochemical consequences influence both terrestrial and aquatic organisms differentially, leading to shifts in assemblage structure (Lovett et al. 2009). Minimisation of atmospheric pollutants within urban areas is also of utmost importance for human health (Brunekreef & Holgate 2002). For example, in Beijing, China, increases in atmospheric air pollutants were associated with a rise in emergency room admissions of patients with cardiovascular disease (Guo et al. 2009). Indeed, the World Health Organization has identified particulate pollution as one of the most important contributors to poor human health, with high concentrations being responsible for triggering a wide range of respiratory diseases, exacerbating heart disease and aggravating a variety of other conditions. In many cities across the world, the concentration of suspended particulate matter greatly exceeds the recommended air quality standard of less than 90 micrograms m^{-3} (Figure 3.4).

Urban vegetation may help to improve air quality through the uptake of gaseous, aerosol and particulate pollutants (Freer-Smith et al. 1997), depending on the type and the location of the planting. In urban ecosystems, the presence of trees increases the surface area and roughness onto which pollutants can be deposited (Beckett et al. 1998), meaning that heavily forested cities reap considerable benefits (Bolund & Hunhammar 1999). Likewise, in Guangzhuo, China, areas of the city with the highest tree cover were responsible for the largest proportion of pollutant removal (Jim & Chen 2008).

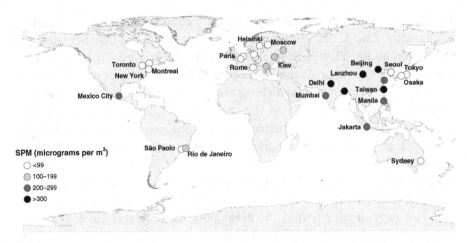

Figure 3.4 Levels of particulate air pollution in 1995 within the world's largest cities. SPM, Suspended Particulate Matter. From data in Soubbotina (2004).

Species interactions

The effective functioning of any ecosystem depends on the interactions that occur between species. These take many forms, including competition, preda-tion, parasitism and mutualism. It is clear that urbanisation can have profound effects on such relationships, although there is a lack of information regarding how alterations in species abundances across urban landscapes will affect the interactions between them (Faeth *et al.* 2005). For example, the urbanisation of Phoenix, Arizona, has led to a dramatic increase in available water, a limiting resource in the surrounding landscape, which has resulted in a shift in trophic dynamics including a greater top-down influence of predators on trophic interactions (Faeth *et al.* 2005).

In much of the world, an important element in maintaining an urban green infrastructure (e.g. parks, gardens) is an adequate level of insect pollination, without which many plant species cannot produce fertile seeds and the yields of others are substantially reduced. Unfortunately, several studies have documented declines in the species richness and abundance of key groups of insect pollinators in response to urbanisation (e.g. McIntyre & Hostetler 2001; Zanette *et al.* 2005; Matteson *et al.* 2008). However, others have actually found elevated richness and abundance in urban areas, which has been attributed to heat island effects, reduced exposure to agricultural chemicals and the wide variety of microhabitats present in urban landscapes (Eremeeva & Sushchev 2005; Fetridge *et al.* 2008).

In conclusion

Key to an understanding of urban ecology is the recognition that ecosystem function in urban areas has been profoundly shaped by human activities, such that there are often only weak parallels with natural ecosystems. This has two major consequences. First, it requires the continued development of a specifically urban approach to studying ecosystem function, recognising that the opportunities to generalise from what is known about other ecosystems may be very limited. Second, it provides an opportunity, as yet largely unexploited, to test broad ideas about ecosystem function in an arena in which the forces are very different.

Acknowledgements
K. L. Evans, S. Gaston and P. Johnson are kindly thanked variously for their advice, assistance and comments. Z.G.D. and J.L.E. were supported by EPSRC grant EP/F007604/1 to the 4M consortium: Measurement, Modelling, Mapping and Management: an Evidence Based Methodology for Understanding and Shrinking the Urban Carbon Footprint. The consortium has five UK partners: Loughborough University, De Montfort University, Newcastle University, the University of Sheffield and the University of Leeds.

References
Akbari, H. (2002). Shade trees reduce building energy use and CO_2 emissions from power plants. *Environmental Pollution*, **116**, S119–26.

Akbari, H., Kurn, D. M., Bretz, S. E. and Hanford, J. W. (1997). Peak power and cooling energy savings of shade trees. *Energy and Buildings*, **25**, 139–48.

Alexandri, E. and Jones, P. (2008). Temperature decrease in an urban canyon due to green walls and green roofs on diverse climates. *Building and Environments*, **43**, 480–93.

Alig, R. J., Kline, J. D. and Lichtenstein, M. (2004). Urbanization on the US landscape: looking ahead in the 21st century. *Landscape and Urban Planning*, **69**, 219–34.

Arnfield, A. J. (2003). Two decades of urban climate research: a review of turbulence, exchanges of energy and water, and the urban heat island. *International Journal of Climatology*, **23**, 1–26.

Arnold, C. L. Jr and Gibbons, C. J. (1996). Impervious surface coverage: the emergence of a key environmental indicator. *Journal of the American Planning Association*, **62**, 243–58.

Baker, L. A., Brazel, A. J., Selover, N. *et al.* (2002). Urbanization and warming of Phoenix (Arizona, USA): impacts, feedbacks and mitigation. *Urban Ecosystems*, **6**, 183–203.

Beckett, K. P., Freer-Smith, P. H. and Taylor, G. (1998). Urban woodlands: their role in reducing the effects of particulate pollution. *Environmental Pollution*, **99**, 347–60.

Beebee, T. J. C. (1997). Changes in dewpond numbers and amphibian diversity over 20 years on chalk downland in Sussex, England. *Biological Conservation*, **81**, 215–19.

Bibby, R. L. and Webster-Brown, J. G. (2005). Characterisation of urban catchment suspended particulate matter (Auckland region, New Zealand); a comparison with non-urban SPM. *Science of the Total Environment*, **343**, 177–97.

Birchenall, J. A. (2007). Economic development and the escape from high mortality. *World Development*, **35**, 543–68.

Bolund, P. and Hunhammar, S. (1999). Ecosystem services in urban areas. *Ecological Economics*, **29**, 293–301.

Brunekreef, B. and Holgate, S. T. (2002). Air pollution and health. *The Lancet*, **360**, 1233–42.

Carreiro, M. M., Howe, K., Parkhurst, D. F. and Pouyat, R. V. (1999). Variation in quality and decomposability of red oak leaf litter along an urban–rural gradient. *Biology and Fertility of Soils*, **30**, 258–68.

Characklis, G. W. and Wiesner, M. R. (1997). Particles, metals and water quality in runoff from a large urban watershed. *Journal of Environmental Engineering*, **123**, 753–9.

Chen, X.-L., Zhao, H.-M., Li, P.-X. and Yin, Z.-Y. (2006). Remote sensing image-based analysis of the relationship between urban heat island and land use/cover changes. *Remote Sensing of the Environment*, **104**, 133–46.

Chen, Y. and Wong, N. H. (2006). Thermal benefits of city parks. *Energy and Buildings*, **38**, 105–20.

Collier, C. G. (2006). The impact of urban areas on weather. *Quarterly Journal of the Royal Meteorological Society*, **132**, 1–25.

Duh, J.-D., Shandas, V., Chang, H. and George, L. A. (2008). Rates of urbanisation and the resiliency of air and water quality. *Science of the Total Environment*, **400**, 238–56.

Effland, W. R. and Pouyat, R. V. (1997). The genesis, classification, and mapping of soils in urban areas. *Urban Ecosystems*, **1**, 217–28.

Erb, K.-H. (2004). Land-use related changes in aboveground carbon stocks of Austria's terrestrial ecosystems. *Ecosystems*, **7**, 563–72.

Eremeeva, N. I. and Sushchev, D. V. (2005). Structural changes in the fauna of pollinating insects in urban landscapes. *Russian Journal of Ecology*, **36**, 259–65.

Faeth, S. H., Warren, P. S., Stochat, E. and Marussich, W. A. (2005). Trophic dynamics in urban communities. *BioScience*, **55**, 399–407.

Falk, J. H. (1976). Energetics of a suburban lawn ecosystem. *Ecology*, **57**, 141–50.

Federal Interagency Stream Restoration Working Group (1998). Disturbance affecting streams. In *Stream Corridor Restoration: Principles, Processes, and Practices*. Federal Interagency Stream Restoration Working Group (FISRWG). Available at http://www.nrcs.usda.gov/technical/stream_restoration/newgra.html

Fetridge, E. D., Ascher, J. S. and Langellotto, G. A. (2008). The bee fauna of residential gardens in a suburb of New York City (Hymenoptera: Apoidea). *Annals of the Entomological Society of America*, **101**, 1067–77.

Freer-Smith, P. H., Holloway, S. and Goodman, A. (1997). The uptake of particulates by an urban woodland: site description and particulate composition. *Environmental Pollution*, **95**, 27–35.

Fuller, R. A., Tratalos, J., Warren, P. H. *et al.* (2010). Environment and biodiversity. In M. Jenks and C. Jones, eds., *Dimensions of the Sustainable City*. Dordrecht, the Netherlands: Springer, pp. 75–103.

Gaffin, S. R., Rosenzweig, C., Khanbilvardi, R. *et al.* (2008). Variations in New York City's urban heat island strength over time and space. *Theoretical and Applied Climatology*, **94**, 1–11.

Gaston, K. J. and Evans, K. L. (2010). Urbanisation and development. In N. Maclean, ed., *Silent Summer: The State of Wildlife in Britain and Ireland*. Cambridge: Cambridge University Press, in press.

Gehrt, S. D. and Chelsvig, J. E. (2004). Species-specific patterns of bat activity in an urban landscape. *Ecological Applications*, **14**, 625–35.

Georgi, N. J. and Zafiriadis, K. (2006). The impact of park trees on microclimate in urban areas. *Urban Ecosystems*, **9**, 195–209.

Golubiewski, N. E. (2006). Urbanization increases grassland carbon pools: effects of landscaping in Colorado's front range. *Ecological Applications*, **16**, 555–71.

Gregory, R. D. and Baillie, S. R. (1998). Large-scale habitat use of some declining British birds. *Journal of Applied Ecology*, **35**, 785–99.

Guo, Y., Jia, Y., Pan, X., Liu, L. and Wichmann, H.-E. (2009). The association between fine particulate air pollution and hospital emergency room visits for cardiovascular disease in Beijing, China. *Science of the Total Environment*, **407**, 4826–30.

Hand, L. M. and Shepherd, J. M. (2009). An investigation of warm-season spatial rainfall variability in Oklahoma City: possible linkages to urbanization and prevailing wind. *Journal of Applied Meteorology and Climatology*, **48**, 251–69.

Henry, J. A. and Dicks, S. E. (1987). Association of urban temperatures with land use and surface materials. *Landscape and Urban Planning*, **14**, 21–9.

Imhoff, M. L., Bounoua, L., DeFries, R. *et al.* (2004). The consequences of urban land transformation on net primary productivity in the United States. *Remote Sensing of the Environment*, **89**, 434–43.

Irvine, K. N., Fuller, R. A., Devine-Wright, P. *et al.* (2010). Ecological and psychological value of urban green space. In M. Jenks and C. Jones, eds., *Dimensions of the Sustainable City*. Dordrecht, the Netherlands: Springer, pp. 215–37.

Jenerette, G. D., Harlan, S. L., Brazel, A. *et al.* (2007). Regional relationships between surface temperature, vegetation, and human settlement in a rapidly urbanizing ecosystem. *Landscape Ecology*, **22**, 353–65.

Jim, C. Y. and Chen, W. Y. (2008). Assessing the ecosystem service of air pollutant removal by urban trees in Guangzhou (China). *Journal of Environmental Management*, **88**, 665–76.

Jo, H.-K. and McPherson, E. G. (1995). Carbon storage and flux in urban residential greenspace. *Journal of Environmental Management*, **45**, 109–33.

Johnson, A. D. and Gerhold, H. D. (2003). Carbon storage by urban tree cultivars, in roots and above-ground. *Urban Forestry and Urban Greening*, **2**, 65–72.

Jones, P. D., Groisman, P. Y., Coughlan, M. *et al.* (1990). Assessment of urbanization effects in time series of surface air temperature over land. *Nature*, **347**, 169–72.

Kalnay, E. and Cai, M. (2003). Impact of urbanization and land-use change on climate. *Nature*, **423**, 528–31.

Karl, T. R., Diaz, H. F. and Kukla, G. (1988). Urbanization: its detection and effect in the United States climate record. *Journal of Climate*, **1**, 1099–123.

Kaye, J. P., Groffman, P. M., Grimm, N. B., Baker, L. A. and Pouyat, R. V. (2006). A distinct urban biogeochemistry? *Trends in Ecology and Evolution*, **21**, 192–9.

Kaye, J. P., McCulley, R. L. and Burke, I. C. (2005). Carbon fluxes, nitrogen cycling, and soil microbial communities in adjacent urban, native and agricultural ecosystems. *Global Change Biology*, **11**, 575–87.

Kennedy, C., Cuddihy, J. and Engel-Yan, J. (2007). The changing metabolism of cities. *Journal of Industrial Ecology*, **11**, 43–59.

Lo, C. P., Quattrochi, D. A. and Luvall, J. C. (1997). Application of high-resolution thermal infrared remote sensing and GIS to assess the urban heat island effect. *International Journal of Remote Sensing*, **18**, 287–304.

London Assembly Environment Committee (2007). *Chainsaw Massacre: A Review of London's Street Trees*. London: Greater London Authority.

Lorenz, K. and Lal, R. (2009). Biogeochemical C and N cycles in urban soils. *Environment International*, **35**, 1–8.

Lorenz, K., Preston, C. M. and Kandeler, E. (2006). Soil organic matter in urban soils: estimation of elemental carbon by thermal oxidation and characterization of organic matter by solid-state ^{13}C nuclear magnetic resonance (NMR) spectroscopy. *Geoderma*, **130**, 312–23.

Lovett, G. M., Tear, T. H., Evers, D. C. *et al.* (2009). Effects of air pollution on ecosystems and biological diversity in the eastern United States. *Annals of the New York Academy of Sciences*, **1162**, 99–135.

MacFarlane, D. W. (2009). Potential availability of urban wood biomass in Michigan: implications for energy production, carbon sequestration and sustainable forest management in the U.S.A. *Biomass and Bioenergy*, **33**, 628–34.

Matteson, K. C., Ascher, J. S. and Langellotto, G. A. (2008). Bee richness and abundance in New York City urban gardens. *Annals of the Entomological Society of America*, **101**, 140–50.

Mayer, H. (1999). Air pollution in cities. *Atmospheric Environment*, **33**, 4029–37.

McDonnell, M. J., Pickett, S. T. A., Groffman, P. et al. (1997). Ecosystem processes along an urban-to-rural gradient. *Urban Ecosystems*, **1**, 21–36.

McHale, M. R., McPherson, E. G. and Burke, I. C. (2007). The potential of urban tree plantings to be cost effective in carbon credit markets. *Urban Forestry and Urban Greening*, **6**, 49–60.

McIntyre, N. E. and Hostetler, M. E. (2001). Effects of urban land use on pollinator (Hymenoptera: Apoidea) communities in a desert metropolis. *Basic and Applied Ecology*, **2**, 209–18.

Millennium Ecosystem Assessment (2005). *Ecosystems and Human Well-being: Synthesis*. Washington, DC: Island Press.

Neil, K. and Wu, J. (2006). Effects of urbanization on plant flowering phenology: a review. *Urban Ecosystems*, **9**, 243–57.

Nizeyimana, E. L., Petersen, G. W., Imhoff, M. L. et al. (2001). Assessing the impact of land conversion to urban land use on soils with different productivity levels in the USA. *Soil Science Society of America Journal*, **65**, 391–402.

Nowak, D. J. and Crane, D. E. (2002). Carbon storage and sequestration by urban trees in the USA. *Environmental Pollution*, **116**, 381–9.

Nowak, D. J., Stevens, J. C., Sisinni, S. M. and Luley, C. J. (2002). Effects of urban tree management and species selection on atmospheric carbon dioxide. *Journal of Arboriculture*, **28**, 113–22.

Oke, T. R. (1987). *Boundary Layer Climates*, 2nd edn. Methuen Co.

Partecke, J., Van't Hof, T. and Gwinner, E. (2004). Differences in the timing of reproduction between urban and forest European

Blackbirds (*Turdus merula*): result of phenotypic flexibility or genetic differences? *Proceedings of the Royal Society of London Series B*, **271**, 1995–2001.

Pataki, D. E., Alig, R. J., Fung, A. S. et al. (2006). Urban ecosystems and the North American carbon cycle. *Global Change Biology*, **12**, 2092–102.

Paul, M. J. and Meyer, J. L. (2001). Streams in the urban landscape. *Annual Review of Ecology and Systematics*, **32**, 333–65.

Pauleit, S. and Duhme, F. (2000). Assessing the environmental performance of land cover types for urban planning. *Landscape and Urban Planning*, **52**, 1–20.

Pauleit, S., Ennos, R. and Golding, Y. (2005). Modeling the environmental impacts of urban land use and land cover change – a study in Merseyside, UK. *Landscape and Urban Planning*, **71**, 295–310.

Peach, W. J., Denny, M., Cotton, P. A. et al. (2004). Habitat selection by song thrushes in stable and declining farmland populations. *Journal of Applied Ecology*, **41**, 275–93.

Pickett, S. T. A., Cadenasso, M. L., Grove, J. M. et al. (2001). Urban ecological systems: linking terrestrial ecological, physical, and socioeconomic components of metropolitan areas. *Annual Review of Ecology and Systematics*, **32**, 127–57.

Pouyat, R., Groffman, P., Yesilonis, I. and Hernandez, L. (2002). Soil carbon pools and fluxes in urban ecosystems. *Environmental Pollution*, **116**, S107–18.

Pouyat, R. V., McDonnell, M. J. and Pickett, S. T. A. (1997). Litter decomposition and nitrogen mineralization in oak stands along an urban-rural land use gradient. *Urban Ecosystems*, **1**, 117–31.

Pouyat, R. V., Yesilonis, I. D. and Golubiewski, N. E. (2009). A comparison of soil organic carbon stocks between residential turf grass and native soil. *Urban Ecosystems*, **12**, 45–62.

Pouyat, R. V., Yesilonis, I. D. and Nowak, D. J. (2006). Carbon storage by urban soils in the United States. *Journal of Environmental Quality*, **35**, 1566–75.

Quigley, M. F. (2004). Street trees and rural conspecifics: will long-lived trees reach full size in urban conditions? *Urban Ecosystems*, **7**, 29–39.

Rawlins, B. G., Vane, C. H., Kim, A. W. *et al.* (2008). Methods for estimating types of organic soil carbon and their application to surveys of UK urban areas. *Soil Use and Management*, **24**, 47–59.

Rowntree, R. A. and Nowak, D. (1991). Quantifying the role of urban forests in removing atmospheric carbon dioxide. *Journal of Arboriculture*, **17**, 269–75.

Scalenghe, R. and Marsan, F. A. (2009). The anthropogenic sealing of soils in urban areas. *Landscape and Urban Planning*, **90**, 1–10.

Shen, W., Wu, J., Grimm, N. B. and Hope, D. (2008). Effects of urbanization-induced environmental changes on ecosystem functioning in the Phoenix Metropolitan Region, USA. *Ecosystems*, **11**, 138–55.

Shepherd, J. M. (2005). A review of current investigations of urban-induced rainfall and recommendations for the future. *Earth Interactions*, **9**, 1–27.

Soubbotina, T. P. (2004). *Beyond Economic Growth: An Introduction to Sustainable Development*, 2nd edn. Washington, DC: The World Bank.

Stone, B. Jr and Rodgers, M. O. (2001). Urban form and thermal efficiency. *Journal of the American Planning Association*, **67**, 186–98.

Taha, H., Konopacki, S. and Gabersek, S. (1999). Impacts of large-scale surface modifications on meteorological conditions and energy use: a 10-region modelling study. *Theoretical and Applied Climatology*, **62**, 175–85.

Tratalos, J., Fuller, R. A., Warren, P. H., Davies, R. G. and Gaston, K. J. (2007). Urban form, biodiversity potential and ecosystem services. *Landscape and Urban Planning*, **83**, 308–17.

Trusilova, K. and Churkina, G. (2008). The response of the terrestrial biosphere to urbanization: land cover conversion, climate, and urban pollution. *Biogeosciences*, **5**, 1505–15.

UN-HABITAT (2006). *The State of the World's Cities Report 2006/2007*. London: Earthscan.

Walsh, C. J., Roy, A. H., Feminella, J. W. *et al.* (2005). The urban stream syndrome: current knowledge and the search for a cure. *Journal of the North American Benthological Society*, **24**, 706–23.

Weng, Q., Liu, H. and Lu, D. (2007). Assessing the effects of land use and land cover patterns on thermal conditions using landscape metrics in city of Indianapolis, United States. *Urban Ecosystems*, **10**, 203–19.

White, M. A., Nemani, R. R., Thornton, P. E. and Running, S. W. (2002). Satellite evidence of phenological differences between urbanized and rural areas of the eastern United States deciduous broadleaf forest. *Ecosystems*, **5**, 260–77.

Whitford, V., Ennos, A. R. and Handley, J. F. (2001). 'City form and natural process' – indicators for the ecological performance of urban areas and their application to Merseyside, UK. *Landscape and Urban Planning*, **57**, 91–103.

Woods, R. (2003). Urban–rural mortality differentials: an unresolved debate. *Population and Development Review*, **29**, 29–46.

World Resources Institute, the United Nations Environment Programme, the United Nations Development Programme and the World Bank (1998). *World Resources 1998–99: Environmental Change and Human Health*. New York: Oxford University Press.

Zanette, L. R. S., Martins, R. P. and Ribeiro, S. P. (2005). Effects of urbanization on Neotropical wasp and bee assemblages in a Brazilian metropolis. *Landscape and Urban Planning*, **71**, 105–21.

Zhang, C., Tian, H., Pan, S. *et al.* (2008). Effects of forest regrowth and urbanization on ecosystem carbon storage in a rural-urban gradient in the southeastern United States. *Ecosystems*, **11**, 1211–22.

Zhang, X. Y., Friedl, M. A., Schaaf, C. B. and Strahler, A. H. (2004a). Climate controls on vegetation phenological patterns in northern mid- and high latitudes inferred from MODIS data. *Global Change Biology*, **10**, 1133–45.

Zhang, X. Y., Friedl, M. A., Strahler, A. H. and Schneider, A. (2004b). The footprint of urban climates on vegetation phenology. *Geophysical Research Letters*, **31**, L12209.

Zhao, S., Da, L., Tang, Z. *et al.* (2006). Ecological consequences of rapid urban expansion: Shanghai, China. *Frontiers in Ecology and the Environment*, **4**, 341–6.

CHAPTER FOUR

Individual species and urbanisation

KARL L. EVANS

Intraspecific variation in ecological and life history traits has interested ecologists since at least the nineteenth century (Bergmann 1847; Allen 1878; Gloger 1883; Jordan 1891). Such variation is still the focus of intensive research, but somewhat surprisingly the environmental factors driving large-scale patterns have rarely been firmly established and are intensely debated (Gaston *et al.* 2008). Turning to finer spatial scales, local variation in environmental conditions can promote marked trait divergence between adjacent populations. This raises important evolutionary questions regarding the role of phenotypic plasticity in generating such divergence, and the interactions between genetic adaptation and gene flow (Postma & van Noordwijk 2005; Senar *et al.* 2006; Ghalambor *et al.* 2007; McCormack & Smith 2008). The patterns and processes associated with trait divergence have, with the exception of a few classic studies concerning the effects of pollution, traditionally been assessed in relatively natural rural systems. Recently, however, there has been an explosion of research focusing on how the marked ecological differences between rural and urban areas influence the traits of conspecific populations. So far this work has largely emerged as a series of isolated case studies of trait differences, with the few attempts to synthesise the available information being confined to a subset of taxa or traits (Marzluff 2001; Bradley & Altizer 2007; Chamberlain *et al.* 2009). Assessments of intraspecific variation among rural and urban populations have also rarely been placed in the broader context of trait variation in more natural systems.

Against the background of the profound influence of urbanisation on ecosystem form and function (Chapters 2 and 3), this chapter provides a detailed synthetic overview of trait differentiation between conspecific rural and urban populations, and its consequences. Studies assessing which traits are associated with interspecific variation in species responses to urbanisation (e.g. Croci *et al.* 2008; Thompson & McCarthy 2008; Evans *et al.* in press, a) are not covered here

Urban Ecology, ed. Kevin J. Gaston. Published by Cambridge University Press.

(a)

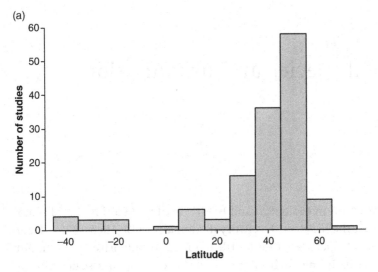

(b)

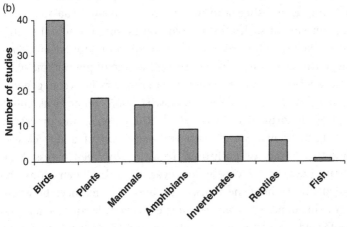

Figure 4.1 Assessment of the ecological and life history traits in urban populations is biased towards (a) northern temperate regions and (b) certain taxonomic groups, particularly birds. Data concern those studies cited in this chapter; multiple papers reporting divergence in more than one trait for the same population are considered as a single study. Positive and negative latitudes indicate northern and southern hemispheres, respectively.

but are discussed in Chapter 5. I consider a number of broad groupings of ecological traits and, taking each trait in turn, summarise intraspecific patterns of divergence between rural and urban populations. Where possible such variation is assessed in relation to the mechanisms proposed to drive spatial patterns in intraspecific trait variation in more natural areas. The aim is to illustrate the predominant patterns that have been documented to date and their causal mechanisms. Examples are drawn from as large a taxonomic and geographical scope as possible, although the dominance of research conducted in northern temperate regions and on birds creates inevitable biases (Figure 4.1). Finally, I assess the implications of trait divergence between urban and rural populations, and highlight a number of issues that have received insufficient attention.

Species traits
Abundance

Population density is an important trait in its own right, and its response to urbanisation may influence the nature of other patterns of trait divergence between urban and rural populations if these traits are density dependent effects. In addition, the effect of urbanisation on demographic and other traits must be viewed in the context of the suitability of urban environments for the focal species, which can be assessed by the density of urban populations, especially in comparison to rural densities (Evans *et al.* in press, a). Urbanisation typically influences population density in one of three ways. Abundance may be lowest in urban areas, peak in areas of intermediate development or be greatest in highly urbanised sites (Blair 1996; Tratalos *et al.* 2007; Grimm *et al.* 2008; Figure 4.2). Species exhibiting these patterns have been termed respectively urban avoiders, exploiters and adapters (Blair 1996).

The relative proportion of species exhibiting each of the three main ways in which population density responds to urbanisation varies taxonomically. Amongst plants, urban adapters appear particularly frequent, but this pattern is largely driven by the positive responses to urbanisation of many introduced species (Grimm *et al.* 2008). Amongst birds, it appears that a larger proportion of species, compared with most other taxonomic groups, have peak densities at intermediate levels of urbanisation; this is perhaps in response to increased habitat diversity in such areas (Marzluff 2005; Tratalos *et al.* 2007). Insufficient studies have been conducted in other taxonomic groups to formulate generalisations, but as relatively few species occur in highly urbanised areas (Chapter 5) it seems highly likely that in most taxonomic groups the majority of species will occur at higher densities in rural environments than highly urbanised ones.

That urbanisation influences species densities is unsurprising. What is more interesting is that the nature of a species' response to urbanisation can vary spatially, the most dramatic examples of which concern those species which are absent from urban areas in parts of their range despite occurring at high densities in urban areas elsewhere. One such example is the dunnock *Prunella modularis*, which has high population densities in British cities but is largely confined to rural areas elsewhere in its European range (Vogel & Tuomenpuro 1997). Similarly, wild boar *Sus scrofa* are urbanised in parts of Europe, such as Germany, but are absent from some urban

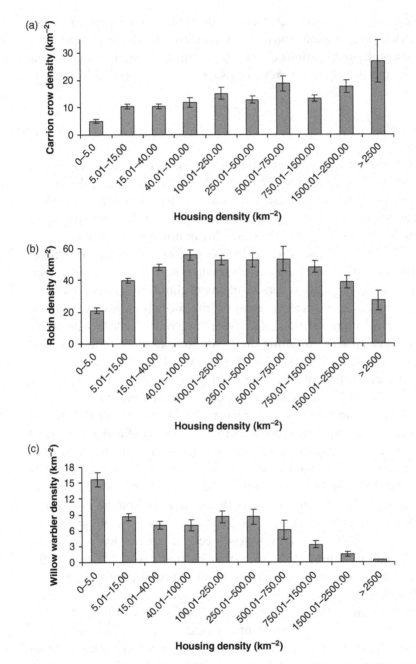

Figure 4.2 Urbanisation influences abundance in one of three ways, (a) density may peak in highly urbanised areas, (b) density may peak in moderately urbanised areas, and (c) density may peak in rural areas. These patterns are illustrated respectively by the changes in carrion crow *Corvus corone*, European robin *Erithacus rubecula* and willow warbler *Phylloscopus trochilus* densities in 1-kilometre squares in the UK across a gradient in housing density. Error bars represent standard errors. Modified from Tratalos *et al.* (2007).

areas in the Baltic states despite occurring in the nearby countryside (Zorenko & Leontyeva 2003). Even species that are frequently highly urbanised may be absent from towns and cities in part of their range. The feral pigeon *Columba livia* is amongst one of the most urbanised birds in the world, but in central Asia the species is lacking from some towns despite the occurrence of nearby rural populations (Obukhova 2001).

The mechanisms driving spatial variation in intraspecific responses to urban environments have seldom been identified, but four main mechanisms are likely to generate such patterns (Evans *et al.* in press, b). Differences in the history of urban development will affect the opportunity for urban populations to arise by influencing both the time available for colonisation and the number of sites that can be colonised. The occurrence of urban populations will also be linked to biotic factors, such as the availability of potential colonists, which may vary spatially according to rural population size and the pressure on such individuals to disperse. The quality of urban habitats will also change with location owing to differences in factors such as human attitudes to urban wildlife, soil type, pollution levels, vegetation structure and composition, and the occurrence of competitors, pathogens, predators and pollinators. Finally, rural populations may differ in their ability to acquire specific traits that increase fitness in urban environments owing to variation in either their genetic composition or their ability to exhibit sufficient phenotypic plasticity. There is certainly increasing evidence that the contrast in the selection pressures of rural and urban environments is large enough to generate marked intraspecific trait variation between urban and rural populations. The rest of this chapter assesses the nature and consequences of such divergence.

Body size

It is relatively straightforward, in comparison with many other types of traits, to assess the effect of urbanisation on body size, and a number of studies have thus done so. Reduced body size has been reported in urban populations from a wide range of taxonomic groups including invertebrates (Weller & Ganzhorn 2004), snakes (Luiselli *et al.* 2002) and birds (Richner 1989; Ruiz *et al.* 2002; Rasner *et al.* 2004; Liker *et al.* 2008; Rodewald & Shustack 2008). In contrast, a smaller number of studies have reported larger body size in urban populations (amphibians: Ovaska 1991; snakes: Savidge 1991; mammals: Liro 1985). Moreover, the impacts of urbanisation on body size may be population-specific (Evans *et al.* 2009a) or gender-specific (Auman *et al.* 2008), and urbanisation certainly does not have a universal impact on body size.

Most studies of body size divergence in rural and urban populations have focused on describing patterns, rather than assessing the driving mechanisms. A large literature has, however, arisen concerning factors that may drive spatial patterns in body size across larger spatial scales. Three broad groups of mechanisms have been proposed, and it is useful to assess their potential role in driving the impacts of urbanisation on body size. First, thermodynamic hypotheses that relate body size to heat conservation and dissipation predict reduced body size in warmer environments (Bergmann 1847; Brown & Lee 1969), and thus that urbanisation will generate smaller bodied individuals because of the urban heat island effect. Second, amongst ectotherms, the effects of temperature on cell division and replication, or on growth and differentiation, may generate larger cells and therefore larger individuals in cold environments (Partridge *et al.* 1994; Walters & Hassall 2006). This mechanism thus also predicts reduced body size in urban populations. Finally, a suite of alternative hypotheses states that the optimum body size is determined by trade-offs between rates of reproduction and mortality that are mediated by factors related to food availability, such as primary productivity, starvation resistance and seasonality (Roff 1980; Lindstedt & Boyce 1985; Zeveloff & Boyce 1988; Kozłowski *et al.* 2004). These trade-off mechanisms thus suggest that rural and urban populations will diverge in body size in a manner dependent on numerous factors, but with relative resource availability in rural and urban areas playing a major role.

The thermodynamic and physiological hypotheses may drive changes in body size in some urban populations, but they cannot provide a general mechanism as they fail to explain the increased body size of some urban populations. The trade-off mechanisms predict that temporal increases in food supply in urban areas as a consequence of increased human activity is likely to generate increased body size, at least in commensal species, and such patterns have been documented (Yom-Tov 2003). The trade-off mechanisms also predict that body size will exhibit contrasting responses to urbanisation in closely related species that experience different levels of food availability in urban environments. Such a pattern is exhibited by two sympatric cobra species. Urbanisation reduces food availability for *Naja melanoleuca*, and urban populations of this species are smaller bodied than rural ones, whilst *N. nigricollis* experiences equally high food availability in urban and rural areas, and its body size does not vary with habitat type (Luiselli *et al.* 2002). Whilst the effects of urbanisation on body size will often arise through changes in food availability, this is probably not a universal mechanism. Indeed, the smaller body size of urban house sparrows *Passer domesticus*, relative to rural ones, is maintained when individuals are raised under identical conditions with an unconstrained food supply (Liker *et al.* 2008).

Additional mechanisms, other than those proposed to explain large-scale patterns in body size, probably also contribute to the impacts of urbanisation on body size. First, increased predation pressure may favour lighter individuals, at least in birds, as this facilitates predator avoidance (Witter & Cuthill 1993). Urban populations frequently coexist with higher densities of some

predators (Sims *et al.* 2008); predator density may not always determine predation risk, but it strongly influences the perceived risk of predation, which is
often the most important factor in determining species' responses to predators
(Cresswell 2008). Second, body size often has high heritability (Glazier 2002),
and therefore in isolated urban populations changes in body size, or any other
genetically inherited trait, could simply be a stochastic consequence of the
characteristics of founding individuals. Finally, body size is also often a correlate of individual quality, so intraspecific competition may result in smaller
individuals occupying less preferred habitats, which may include urban ones.
The Acadian flycatcher *Empidonax virescens* is a possible example of this mechanism; urban populations occur at lower densities and have less breeding
success than rural ones, suggesting that urban habitats are of poorer quality,
and urban birds are also smaller than rural ones (Rodewald & Shustack 2008).

Urban and rural populations frequently differ in their body size, but insufficient studies have been conducted for general trends to emerge. Additional
studies are clearly required, and these should focus on testing driving mechanisms, rather than just describing patterns, and the adaptive value of any
documented changes. Ideally, such studies should assess patterns of body size
divergence in multiple paired rural and urban populations, and do so in
species for which urban environments provide both high and low quality
habitats.

Communication

A number of theoretical and empirical studies have documented how populations in rural areas have adapted acoustic signals to the divergent sound
transmission properties of contrasting habitat types in order to reduce attenu-
ation (loss of volume), degradation (change in acoustic structure) and masking
by background noise (Morton 1975; Nottebohm 1975; Hunter & Krebs 1979;
Ryan *et al.* 1990). Anthropogenic noise has been demonstrated to reduce the
effectiveness of acoustic communication in many taxa, including fish
(McCauley *et al.* 2003; Popper *et al.* 2003), frogs (Bee & Swanson 2007), birds
(Leader *et al.* 2005; Habib *et al.* 2007) and mammals (Nowacek *et al.* 2007;
Weilgart 2007). As urban environments are typically noisier than rural ones
(Butler 2004), individuals occupying such habitats are likely to experience
selection pressures that will promote divergence of their vocalisations from
those of rural conspecifics.

The Lombard effect predicts that in urban areas the masking effects of
background noise can be reduced by increasing the amplitude (i.e. volume) of
vocalisations (Lombard 1911). It is thus predicted that urban birds will sing
louder on weekdays, when traffic noise is greater, than at weekends; this
pattern has been documented in urban nightingales *Luscinia megarhynchos*
(Brumm & Todt 2002; Brumm 2004), but not urban great tits *Parus major*

(B. Hawkins & A. G. Gosler, unpublished data). This contrast may arise because the focal populations may differ in their ability to pay the energetic cost of louder vocalisations.

Californian ground squirrels *Spermophilus beecheyi* in areas subject to noise from wind-turbines had higher-pitched vocalisations than those in quieter areas (Rabin *et al.* 2003); such changes are typically considered to be adaptive because of the low pitch of most anthropogenic noise which may otherwise mask signals. Increased traffic noise is also associated with increased pitch of the mating calls of some frogs (e.g. *Litoria ewingii*, Parris *et al.* 2009), but other anurans do not exhibit such responses (e.g. *Hyla arborea*, Lengagne 2008). The increased pitch of vocalisations of urban populations has been documented most frequently in birds (great tit: Slabbekoorn & den Boer-Visser 2006; Slabbekoorn & Ripmeester 2008; Mockford & Marshall 2009; house finch *Carpodacus mexicanus*: Fernández-Juricic *et al.* 2005; Bermúdez-Cuamatzin *et al.* 2009; song sparrow *Melospiza melodia*: Wood & Yezerinac 2006; blackbird *Turdus merula*: Nemeth & Brumm 2009). Male great tits appear to exhibit reduced territorial responses when hearing song from a territory where background noise differed from their own, which at the very least suggests that these changes in vocalisations influence their effectiveness as a signal (Mockford & Marshall 2009).

Despite the frequent interpretation of the increased pitch of urban populations' vocalisations as being adaptive, there are alternative perspectives. Changes in pitch in response to urbanisation may simply be a consequence of release from the constraints of sound transmission in densely vegetated habitats, which readily absorb high-pitched sounds (Morton 1975), or may be an unintended consequence of other differences between urban and rural populations (Nemeth & Brumm 2009). The higher population densities of urban populations (see above) may result in more intraspecific competition and thus greater arousal status, which can result in higher-frequency avian vocalisations (Dabelsteen & Pedersen 1985). Moreover, some urban bird populations are characterised by lower testosterone levels than their rural counterparts (e.g. blackbird; Partecke 2005), and these hormonal changes can increase the pitch of vocalisations (Cynx *et al.* 2005). Similarly, the pitch of anuran vocalisations increases with temperature (Gerhardt & Mudry 1980), and thus changes in vocalisations of urban populations may be a consequence of the urban heat island effect. Switches to higher-frequency vocalisations in urban environments may also be maladaptive because such signals degrade more rapidly (Leader *et al.* 2005). Experimental studies are thus required to confirm that changes in the pitch of urban populations' vocalisations are adaptive.

Increasing the effectiveness of acoustic communication in noisy environments may also be achieved by switching the timing of vocalisation to quieter periods. A number of frog species reduce calling rates following exposure to anthropogenic noise (Sun & Narins 2005; Lengagne 2008), and many bird species commence singing earlier in the day in urban areas than in forest (Bergen & Abs 1997; Nemeth & Brumm 2009); the extent to which such changes are adaptive mechanisms for avoiding peaks in ambient noise is, however, debatable. More convincingly, European robins *Erithacus rubecula* breeding in noisier urban sites increase the amount of singing during quiet periods (i.e. at night), and this effect is not attributable to increased ambient light pollution in such territories (Fuller *et al.* 2007). Moreover, urban great tits and house finches tend to produce shorter songs than conspecifics in more rural areas (Fernández-Juricic *et al.* 2005; Slabbekoorn & den Boer-Visser 2006). This pattern may arise because shorter songs are more likely to transmit fully during brief and unpredictable lulls in anthropogenic noise, but the driving mechanisms are uncertain and not all urban bird populations have reduced their song length (e.g. blackbirds: Nemeth & Brumm 2009).

It would be useful to extend studies of urbanisation impacts on avian song to other vocalisation types, such as begging calls, the structure of which can be markedly influenced by the acoustic environment (Leonard & Horn 2008). The influence of urbanisation on non-acoustic communication systems has also received very little attention. As an example, the amount of white in the tail of dark-eyed juncos *Junco hyemalis* is a very important signalling trait in rural environments, but although urban and rural populations differ in the amount of white in the tail, the causes and consequences of this variation are unknown (Yeh 2004; Price *et al.* 2008). More generally, high rates of background movement can alter the efficiency of visual displays (Peters 2008). Therefore, in areas where wind frequently moves vegetation, lizards with rapid displays signal territory boundaries more effectively than conspecifics with slower displays, leading to habitat-specific divergence of display characteristics (Ord *et al.* 2007). Relatively rapidly moving objects, such as traffic and people, may be more frequent in urban areas than rural ones. This could generate equivalent divergent selective pressures on the rapidity of visual displays in the two habitats, but this has not yet been assessed. Similarly, pollution of aquatic systems through eutrophication and increased turbidity reduces the effectiveness of visual signals, thus reducing the intensity of sexual selection (Seehausen *et al.* 1997; Järvenpää & Lindström 2004; Wong *et al.* 2007) and promoting the use of non-visual signals (Heuschele & Candolin 2007). Water pollution in urban areas may generate equivalent effects, but again this remains untested.

Research conducted to date has tended to assume that changes in the vocalisations of urban individuals are adaptive, but this may not always be the case. Future research should thus determine the consequences of alterations in species' vocalisations associated with urbanisation, and this should include further assessment of how rural individuals respond to the altered vocalisations of urban conspecifics. As case studies accumulate, it is also becoming apparent that all species do not alter their vocalisations in the same manner when occupying urban localities, and the species' traits and environmental factors driving this variation need to be determined. Much more work on species other than birds is also needed. Despite these requirements for further research, it is apparent that the communication signals of many urbanised populations have changed in the direction predicted by theory (Patricelli & Blickley 2006; Warren et al. 2006), providing a clear example of how urban ecology interlinks with that in more natural areas.

Physiology

Physiological traits can determine spatial patterns in biodiversity and its response to environmental degradation (Chown & Gaston 2008). Adjustment of physiological traits may thus make it easier to exploit urban environments, and whilst such divergence between rural and urban populations has seldom been explored, a number of very different suites of physiological traits have received initial investigation.

The markedly altered climatic regimes of urban areas may generate divergence in the physiological tolerances of rural and urban conspecific populations. As an example, urban populations of *Aedes* mosquitoes have greater tolerance to desiccation than rural conspecifics, presumably because the former tend to breed in more temporary waterbodies (Mogi et al. 1996). Similarly, urban leafcutter ants *Atta sexdens* have much greater tolerance to high temperatures than their rural conspecifics (Angilletta et al. 2007). Despite the general nature of the urban heat island, it is unclear whether most urban populations will exhibit increased thermal tolerances, and such divergence is perhaps particularly unlikely to occur in cooler temperate regions.

Stress physiology can differ in rural and urban populations, at least in birds. Urban blackbirds have reduced concentrations of the stress hormone corticosterone than rural individuals; this appears to be genetically controlled, and may be an adaptive response that enables urban individuals to tolerate closer approach by humans (Partecke et al. 2006). Reduced fear of humans is certainly a common trait of urban populations (reptiles: Burger 2001; birds: Cooke 1980; Møller 2008; mammals: Shargo 1988; Gliwicz et al. 1994; Smith & Engeman 2002), suggesting that reduced levels of stress hormones in urban populations may be a general pattern (Partecke et al. 2006). Other factors will, however, also influence corticosterone concentrations in urban populations; they may be lowered by increased food availability in urban areas arising from anthropogenic supplementary feeding (Schoech et al. 2007), or increased in urban adapters because the higher population densities elevate sexual competition (Bonier et al. 2007). Furthermore, the rural and urban populations of some bird species have very similar baseline corticosterone levels (Fokidis et al. 2009). General trends in the stress physiology of urban populations and their driving factors are thus highly uncertain, although they illustrate the potential for rural and urban populations to exhibit divergent physiologies.

The frequently divergent diets of rural and urban populations (e.g. Harris 1984; Kiat *et al.* 2008) may have physiological consequences. In particular, urban populations tend to ingest more food from anthropogenic sources than do rural ones. Such shifts have been documented in a wide range of birds (e.g. rufous-collared sparrow *Zonotrichia capensis*: Ruiz *et al.* 2002; Eurasian starling *Sturnus vulgaris*: Mennechez & Clergeau 2006; Eurasian kestrel *Falco tinnunculus*: Kubler *et al.* 2005) and mammals (e.g. long-nosed bandicoot *Perameles nasuta*: Scott *et al.* 1999; raccoon *Procyon lotor*: Zeveloff 2002; red fox *Vulpes vulpes*: Contesse *et al.* 2004; coyote *Canis latrans*: Morey *et al.* 2007). The biochemical implications of these changes in diet have rarely been explored, but increased blood cholesterol and glucose levels have been reported, and are likely to have adverse physiological effects that compromise fitness (Ruiz *et al.* 2002; Ishigame *et al.* 2006).

Urbanisation, mainly via its high levels of anthropogenic pollution, increases oxidative stress in both avian and mammalian urban populations, and thus the demand for antioxidants (C. Isaksson, unpublished meta-analysis). Despite these challenges to the immune system, it appears that levels of oxidative damage in urban populations are similar to those in conspecific rural populations (Isaksson *et al.* 2009; Isaksson, unpublished meta-analysis). This is perhaps because urban individuals tend to invest more in immune defence than rural ones. Such patterns have certainly been documented in urban birds (Isaksson *et al.* 2007), and there is some evidence for this in urban *Rana* frog spp. (Romanova & Egorikhina 2006). In birds, this increased investment is mainly due to internal up-regulation, since many urban populations experience reduced food quality such as lower availability of dietary carotenoids (Hõrak *et al.* 2001, but see Isaksson & Andersson 2007), which are extremely important immuno-stimulants and may play additional roles in avian immune systems (Costantini & Møller 2008, 2009). The population-level consequences of these patterns are not well understood. Indeed, little research has addressed the impacts of urbanisation on the immune physiology of (non-human) animals. Many more studies are required that investigate how the environmental characteristics of urban environments, such as alterations in nutrition and disease risk (see below), interact with each other to determine how investment in the immune system diverges between urban and rural populations.

Finally, and turning to plants, traits related to photosynthesis can be markedly altered by urbanisation. In urban areas, the greater concentrations of photochemical pollutants, such as ozone, nitrous oxides and sulphur dioxide, and of heavy metals frequently reduce photosynthetic rates (NEGTAP 2001; Baycu *et al.* 2006). In less polluted urban sites experiments suggest that photosynthetic rates may be greater than in rural areas, perhaps partly in response to increased carbon dioxide concentrations (Gregg *et al.* 2003). Moreover, urban plant populations can exhibit adaptations to pollution that minimise its impacts on photosynthesis; examples of such trait differentiation include genetic adaptations to heavy metals (Wilson & Bell 1985; Shaw 1999), increased chlorophyll concentrations (Gratani *et al.* 2000; but see Joshi & Swami 2009) and higher stomatal densities (Alves *et al.* 2008). The increase in stomatal densities has occurred despite higher carbon dioxide concentrations in urban areas which, over longer timescales, are associated with a reduction in stomatal densities (Kouwenberg *et al.* 2003). However, the extent to which the impact of urbanisation on stomatal densities depends upon water availability rather than carbon dioxide concentrations remains untested. Indeed, much further work is required to ascertain how the numerous changes in photosynthetic traits induced by urbanisation interact to determine relative photosynthetic rates in urban and rural plant populations.

The above studies clearly demonstrate that conspecific rural and urban populations can differ in a diverse range of physiological traits. The variation documented so far is, however, likely to be a small proportion of the physiological traits that are affected by urbanisation, and insufficient studies have been conducted for general trends to emerge.

Disease risk

There is increasing evidence that urbanisation can dramatically alter parasite loads and disease risk (Bradley & Altizer 2007). One example concerns urban blackbirds, which have lower abundances of *Ixodes* ticks, ectoparasites that impose direct fitness costs and transmit pathogens, and also often have reduced prevalence of avian malaria than rural birds (Figure 4.3; Grégoire *et al.* 2002; Evans *et al.* 2009b). Such reductions in disease risk seem likely to arise largely because the abundance of parasites and pathogens is adversely affected by urban environments; this may happen through three mechanisms. First, pollution may kill pathogens and parasites. This has frequently been documented in freshwater systems (Marcogliese & Cone 1997; King *et al.* 2007), and plant pathogens also often respond negatively to air pollution (Saunders 1966; Jarrauld 2000). Second, alterations in urban habitats or climatic conditions may directly reduce parasite abundance by increasing mortality (Lafferty 1997). Finally, parasite and pathogen abundance may be lowered indirectly through disruption of their life cycle, because urban environments contain fewer intermediate hosts or vectors (Reperant *et al.* 2007; Page *et al.* 2008).

Conversely, a number of studies have reported increased disease risk in urban populations (Bradley & Altizer 2007). Four mechanisms seem likely to

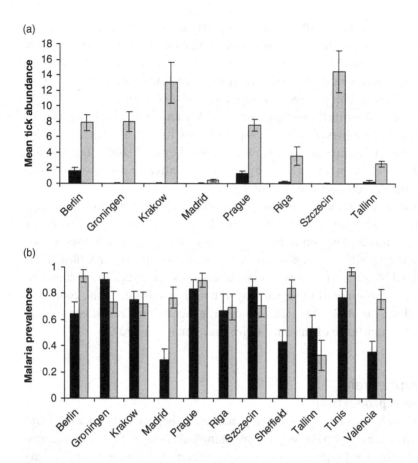

Figure 4.3 Urbanisation can markedly influence exposure to parasites and pathogens, but the nature of this influence may vary across cities and with the identity of the parasite/pathogen. For example, (a) urbanisation has consistent impacts on exposure to *Ixodes* ticks (no ticks were detected in urban or rural blackbird populations in Sheffield, Tunis or Valencia), but (b) the impact of urbanisation on avian malaria prevalence in paired European urban and rural blackbird *Turdus merula* populations is very variable. In both panels, black represents urban. Error bars represent standard errors. Modified from Evans *et al.* (2009b).

generate such patterns. First, and counter to the effects described above, susceptibility to pathogens and their vectors can be increased by the immunosuppressive effects of pollutants (Dohmen *et al.* 1984; deSwart *et al.* 1996; Galloway & Depledge 2001; Kiesecker 2002), which tend to be commoner in urban areas. This may be one factor driving observations of increased virulence and density of ranaviruses and trematode parasites in urban amphibian populations (Johnson *et al.* 1999; Carey *et al.* 2003). Second, pathogen encounter rates may increase in urban areas owing to increased introductions of exotic species through the horticultural trade and other human activities (Brasier 2008; Hamer & McDonnell 2008). Third, urbanisation may alter assemblage diversity and composition in a manner which increases the transmission of some vector-borne diseases owing to reduced abundance of less competent reservoir hosts, which would otherwise dilute pathogen transmission between vectors and more competent hosts. This is the converse of the dilution effect (Ostfeld & Keesing 2000), and may contribute to the greater prevalence of West Nile virus in North American urban bird populations compared with rural ones (Bradley et al. 2008). Finally, the higher densities of many urban populations, relative to rural ones, seem likely to increase disease transmission rates.

Urbanisation can clearly alter biotic interactions through a number of pathways that can generate both increases and decreases in disease prevalence in urban populations. No trends have yet emerged regarding whether urban populations typically experience reduced or increased disease risk (Bradley & Altizer 2007). The occurrence of spatial variation in such patterns (Evans et al. 2009b) and variation with the taxonomy of both hosts (Fokidis et al. 2008) and parasites (Fenoglio et al. 2009) does, however, generate opportunities to assess the relative importance of the alternative causal mechanisms.

Demographic traits
Timing of reproduction

The timing of reproduction can vary markedly between urban and rural populations. Many aspects of vegetation phenology, such as flower emergence and bud burst, are frequently advanced in urban environments as a consequence of the urban heat island effect (Roetzer et al. 2000; Zhang et al. 2004; Lu et al. 2006; Neil & Wu 2006). However, this pattern is not universal; it is particularly unusual in the tropics, where vegetation phenology is often dependent upon humidity, which is less uniformly influenced by urbanisation than is temperature (Gazal et al. 2008).

Turning to animals, the relative timing of reproduction in urban populations has most frequently been studied in birds, which typically breed earlier in towns and cities (Chamberlain et al. 2009). The mechanisms driving this pattern warrant further investigation, but improved body condition of urban individuals due to anthropogenic provision of additional food is considered to be a major factor (Schoech & Bowman 2003; Chamberlain et al. 2009). Although empirical data are lacking, it also seems likely that the urban heat island effect will result in earlier availability of invertebrate food sources in towns and cities, at least in temperate regions where humidity is typically unimportant, further contributing to earlier reproduction. Advancement of avian reproduction in urban environments is not, however, a universal pattern (Rodewald & Shustack 2008). Indeed, for many species urban environments are less suitable than rural ones, which may delay the acquisition of sufficient resources for breeding. Furthermore, in those species for which urban environments provide poorer-quality habitat than rural ones, intraspecific competition may result in urban populations comprising a greater proportion of low-quality individuals which may thus breed later in the year.

Reproductive success

Urbanisation can also exert strong influences on reproductive output. Pollination, and thus seed production, may be disrupted in towns and cities because of reduced abundances of animal pollinators (Cheptou & Avendaño 2006; R. A. Fuller et al., unpublished data), or the fragmented nature of urban habitats (Andrieu et al. 2009). An increase in alternative food sources can also disrupt pollinator behaviour, leading to reduced pollination even when pollinator abundance is unaltered. The ornamental and exotic plants that typically dominate urban floras can have this effect (Kandori et al. 2009; Morales & Traveset 2009), and in the New World, the provision of hummingbird feeders can have similar consequences (Arizmendi et al. 2007). Alternative pollinators can, however, sometimes compensate for reduced availability of core pollinators, resulting in altered biotic interactions but no reduction in reproductive success (Arizmendi et al. 2007; Roberts et al. 2007). Effects of urbanisation on wind pollination have not yet been studied. It seems possible though that, at least in highly developed areas, buildings and other infrastructure will intercept pollen, and may reduce pollination rates and thus the connectivity of fragmented populations. Lower rates of successful cross-pollination in urban plant populations can also be associated with increased self-compatibility (Cheptou & Avendaño 2006).

The impacts of urbanisation on reproductive success have been most rigorously studied in birds. Clutch size, nestling weight and productivity per nesting attempt are frequently, but not universally, reduced in urban populations (Chamberlain et al. 2009). These patterns appear to be driven mainly by reduced abundance of natural food, although the role of density dependence in urban adapters has not been adequately explored. Nest predation risk is often considered to be influenced by urbanisation, but there is little evidence for a general trend (Chamberlain et al. 2009). Some studies provide support for the 'safe nesting zone' hypothesis (i.e. reduced nest predation rates in urban areas; Gering & Blair 1999; Kosinski 2001; Newhouse et al. 2008); the mechanisms generating such reductions in predation are poorly understood, although urban noise can mask the acoustic signals used by predators to locate nests (Francis et al. 2009). In contrast, other studies suggest that in towns and cities the increased abundance of key nest predators, such as cats and many corvid species (Sims et al. 2008) increases nest predation rates (Thorington & Bowman 2003;

Jokimäki *et al.* 2005; Beck & Heinshon 2006; Marzluff *et al.* 2007). Many other factors, including habitat structure, the responses of predators to human disturbance, and human attitudes, are also likely to influence the relative rates of nest predation in urban and rural landscapes. Data from other taxa are much rarer and provide conflicting results, with urbanisation being associated with increased (amphibians: Severtsova *et al.* 2002), decreased (reptiles: Rubin *et al.* 2004) or little significant variation (reptiles: Endriss *et al.* 2007; mammals: Scott *et al.* 1999; McCleery 2009) in reproductive success and recruitment relative to rural populations. It is thus not yet possible to ascertain if the trends documented for urban bird populations generalise to other groups.

Survival

It is remarkable how few studies have assessed the effect of urbanisation on survival. It is clear that increased predation or pollution can reduce survival markedly in some urban populations, and such reductions can even result in local extinction (e.g. Scott *et al.* 1999; Hamer & McDonnell 2008). Other studies have, however, found that survival rates differ little in rural and urban populations (McCleery *et al.* 2008; Rodewald & Shustack 2008), or are higher in urban populations (Liro 1985; Endriss *et al.* 2007). Indeed, supplementary feeding seems likely to increase survival rates of many urban bird populations, although conclusive data are not yet available (Brittingham & Temple 1998; Robb *et al.* 2008). The nature of differences in rural and urban survival rates may also be dependent on spatial and temporal variation in mortality agents, such as predation rates and disease outbreaks (Gosselink *et al.* 2007).

Many more case studies are required across a wide range of taxonomic groups, and across species that exhibit both positive and negative responses to urbanisation, before broad generalisations can be made concerning the effects of urbanisation on demographic traits and the mechanisms driving these patterns. Survival rates in urban populations, relative to rural conspecifics, are especially poorly known. Most species that occur in urban environments do not reach markedly higher densities in towns and cities than in rural areas, which increases the probability that most urban populations will not have higher survival rates or reproductive output. This prediction receives further support from evidence that, even in species that are abundant in towns and cities, urbanisation can adversely affect demographic traits (Mennechez & Clergeau 2006), suggesting that some urban populations may be sinks maintained by immigration from surrounding rural areas.

Population genetic structure

Urban populations may arise through development of a site that converts an existing rural population into an urban one. In such cases the genetic structure of the urban population will partly reflect underlying patterns that were present before urbanisation commenced (Swei *et al.* 2003). In addition, however, habitat alteration is likely to alter dispersal rates between the recently created urban population and surrounding rural ones, resulting in changes to genetic structure that are similar to those created by habitat fragmentation (i.e. reduced genetic diversity and increased genetic drift and divergence; Keyghobadi 2007).

Alternatively, urban populations may arise through colonisation (Wandeler *et al.* 2003; Marzluff 2005; Abs & Bergen 2008; Evans *et al.* in press, b). These cases provide rarely documented examples of internal range expansion, the colonisation of unoccupied habitat types within a species' original geographic range, rather than the more frequently studied cases of colonisation of novel areas external to a range (Evans *et al.* in press, b). During colonisation, founder effects are likely to promote reduced genetic diversity in novel populations, and genetic drift and divergence from ancestral populations. The magnitude of such changes will be influenced by whether colonisation occurs from a surrounding rural population, or in a leap-frog manner in which urban adapted, or imprinted, individuals colonise other towns and cities (Evans *et al.* in press, b). In the latter situation the source individuals will be derived from an urban population that has already experienced at least one founder event, thus increasing the loss of genetic diversity and divergence from rural populations (Evans *et al.* 2009c).

The effects of urbanisation on population genetic structure have been assessed in a wide range of species, but these still represent a very small proportion of those that are urbanised. The plethora of metrics used to measure genetic structure hinders direct comparisons between studies, but it is clear that although urbanisation typically reduces genetic diversity there is marked variation in the magnitude of such reductions (Table 4.1). Three factors are likely to have a strong influence on the genetic diversity of urban populations. First, and perhaps most obviously, the magnitude of bottlenecks arising through founder effects will exert a large effect on genetic diversity, at least in recently established urban populations. Second, loss of genetic diversity is likely to be greater in native species with limited dispersal capacity (Desender *et al.* 2005; Aruda & Morielle-Versuta 2008). Gene flow between horticultural specimens and native plants can, however, result in increased genetic diversity of urban populations (Roberts *et al.* 2007), and similar effects may arise if urban populations are founded from multiple rural ones (Aruda & Morielle-Versuta 2008). Finally, reductions in genetic diversity are likely to be marked in species that require genetic adaptation to urban environments, as not all genotypes will contain the necessary genetic adaptations (Keane *et al.* 2005).

Most urban populations exhibit genetic divergence from rural ones, albeit again with considerable variation in the magnitude of differentiation (Table 4.1). In part, this is simply a consequence of reduced genetic diversity,

Table 4.1 Studies comparing the population genetic structure of rural and urban populations. Differences that are not statistically significant are indicated by ns. Genetic diversity is measured by a – number of alleles, a_R – allelic richness, GD – Nei's gene diversity, GI – mean number of genotypes per individual, h' – Shannon's diversity index, Ha – number of haplotypes, Ha_D – haplotype diversity, h_o – observed heterozygosity, PB – percentage of polymorphic bands, and PL – percentage of polymorphic loci. Genetic divergence is measured by d_S – Nei's genetic distance, d_{NL} – Nei and Li's genetic distance, DI_S – Dice's similarity value, f_{ST} and its derivative Φ_{PT} (see the cited sources for how these metrics were calculated).

Species	Region	Change in genetic diversity of urban populations	Divergence of urban and rural populations	Divergence within urban populations	Source
Plants					
[moss] *Leptodon smithii*	Italy	Ha −36% PL −20%	F_{ST} 0.370	Pop A F_{ST} 0.242 Pop B F_{ST} ns	Spagnuolo et al. (2007)
McGillivray *Grevillea macleayana*	Australia	A +95% H_o +27%			Roberts et al. (2007)
Common dandelion *Taraxacum officinale*	USA	GI −41%			Keane et al. (2005)
Broad-leaved helleborine *Epipactis helleborine*	Scotland	A −10% H_o −29%		F_{ST} 0.328	Hollingsworth & Dickson (1997)
Rue-leaved saxifrage *Saxifraga tridactlites*	Germany	GD H' and PB ns	Φ_{PT} 0.30		Reisch (2007)
Downy yellow violet *Viola pubescens*	USA	H' and PL ns	D_{NL} 0.453		Culley et al. (2007)

Species	Country	Genetic measures	Value	Reference
Oxlip *Primula elatior*	Belgium		F_{ST} 0.054	van Rossum (2008)
Insects				
Small white butterfly *Pieris rapae*	Japan	PL ns		Takami et al. (2004)
[butterfly] *P. melete*	Japan	PL ns		Takami et al. (2004)
[ground beetle] *Pterostichus madidus*	UK	A_R and H_o ns		Desender et al. (2005)
[ground beetle] *P. madidus*	Belgium	A_R and H_o ns		Desender et al. (2005)
[ground beetle] *Abax ater*	UK	A_R −15% H_o −88%		Desender et al. (2005)
[ground beetle] *Abax ater*	Belgium	A_R and H_o ns		Desender et al. (2005)
Amphibians				
[frog] *Leptodactylus ocellatus*	Brazil	GD −46% PB −32%	D_S 0.21	Aruda & Morielle-Versuta (2008)
[frog] *L. fuscus*	Brazil	GD −10% PB −8%	D_S 0.04	Aruda & Morielle-Versuta (2008)
[frog] *L. podicipinus*	Brazil	GD +50% PB +37%	D_S 0.04	Aruda & Morielle-Versuta (2008)

Table 4.1 (*cont.*)

Species	Region	Change in genetic diversity of urban populations	Divergence of urban and rural populations	Divergence within urban populations	Source
[frog] *L. labyrinthicus*	Brazil	GD +11% PB +6%	D_S 0.06		Aruda & Morielle-Versuta (2008)
Common toad *Bufo bufo*	UK	A −22% H_o −15%		F_{ST} 0.230 (cf. rural F_{ST} 0.066)	Hitchings & Beebee (1998)
Common frog *Rana temporaria*	UK	A −10% H_o −25%		F_{ST} 0.35 (cf. rural F_{ST} 0.11)	Hitchings & Beebee (1997)
Red-backed salamander *Plethodon cinerus*	Canada	A −29% A_R −39% H_o −34%	F_{ST} 0.189	F_{ST} 0.064 (cf. rural $F_{ST} \approx 0$)	Noël *et al.* (2007)
Reptiles					
Caissaca viper *Bothrops moojeni*	Brazil	PL −28%	F_{ST} 0.19		Dutra *et al.* (2008)
Blanding's turtle *Emydoidea blandingii*	USA	PB −18%		DI_S ns	Rubin *et al.* (2001)
Birds					
Dark-eyed junco *Junco hyemalis*	USA	A_R −41% H_o −21%	F_{ST} 0.079		Rasner *et al.* (2004)
Blackbird[a] *Turdus merula*	Europe	A_R −10% H_o −2%	F_{ST} 0.037		Evans *et al.* (2009c)

Species	Location	Genetic diversity	F_{ST}		Reference
Eurasian kestrel *Falco tinnunculus*	Poland	A_R −10% H_o −8%	F_{ST} 0.05		Rutkowski *et al.* (2006)
Mammals					
Red fox *Vulpes vulpes*	Switzerland	A −14% H_o −17%	F_{ST} 0.04	F_{ST} 0.068 (cf. rural F_{ST} 0.009)	Wandeler *et al.* (2003)
Large Japanese fieldmouse *Apodemus speciosus*	Japan	Ha −77% Ha_D −71%	F_{ST} 0.49	F_{ST} 0.77 (cf. rural F_{ST} 0.20)	Hirota *et al.* (2004)
Eurasian badger *Meles meles*	UK	A_R −15%			Huck *et al.* (2008)

Note: [a]Data for the blackbird are means from 12 paired urban and rural populations.

with urban populations that have lost substantial diversity exhibiting greater divergence from rural populations. Limited dispersal of urban individuals will further reduce gene flow and increase divergence. It is thus noteworthy that urban populations of some bird species appear to be more sedentary than those of rural conspecifics (Eurasian kestrel: Plesnik 1990; blackbird: Partecke & Gwinner 2007; K. L. Evans *et al.* unpublished data), and that urbanisation promotes reduced dispersal in some plant species (*Crepis sancta*: Cheptou *et al.* 2008). Divergence between rural and urban populations will also be increased by biased dispersal patterns arising from individuals imprinting on their natal habitat type. This mechanism, often termed natal habitat preference induction, has been documented in some species (Davis & Stamps 2004; Mabry & Stamps 2008), but has not yet been investigated in an urban context. Genetic divergence is also likely to be greater when differential selection pressures act on rural and urban populations, as strong selection pressure can overcome the homogenising influence of gene flow (Garant *et al.* 2005; Senar *et al.* 2006). Evidence for habitat-specific selection pressure is provided by studies indicating that trait divergence between urban and rural populations is a consequence of genetic adaptation. Such traits include heavy metal tolerance in plants (Keane *et al.* 2005), reduced investment in extra-floral nectaries that are associated with ant attraction in the plant *Chamaecrista fasciculate* (Rios *et al.* 2008), and changes in avian stress physiology and migratory behaviour (Partecke *et al.* 2006; Partecke & Gwinner 2007). The impact of time since colonisation on the genetic structure of urban populations has rarely been established, but there is evidence for both decreasing (Wandeler *et al.* 2003) and increasing divergence (Evans *et al.* 2009c) over time, with the latter occurring when selection pressures appear to differ in rural and urban environments.

The final impact of urbanisation on genetic structure concerns genetic differentiation within urban populations. Although data are very limited, this appears to be greater than equivalent differentiation within rural populations (Table 4.1). Genetic structure within urban populations can arise if urban areas are colonised by genetically differentiated rural populations (Hollingsworth & Dickson 1997), but probably arises most frequently from intensive fragmentation of urban habitats reducing gene flow (Vandergast *et al.* 2007). There is a time-lag, however, between fragmentation and impacts on population genetics, such that long-lived species tend to show little genetic structure within urban environments (Rubin *et al.* 2001; van Rossum 2008). Human activities, such as movement of soil between habitat patches, and thus associated faunas and seed banks, can also promote gene flow between isolated populations to a sufficient extent to mitigate the effects of fragmentation (Field *et al.* 2007).

The trends for urban populations to contain less genetic diversity than rural populations, and for the two to be genetically divergent from each other, have important consequences. Theory and empirical data indicate that reductions in genetic diversity will limit the ability of populations undergoing range expansion to adapt to novel environments (Hewitt 2000; Pujol & Pannell 2008). Therefore, reduced evolutionary potential of urban populations may hinder their adaptation to urban environments, possibly contributing to their typically lower species richness relative to rural areas (Chapter 5; Grimm *et al.* 2008). Conversely, genetic differentiation between urban and rural areas is indicative of reduced gene flow, which may facilitate adaptation to towns and cities, as it may buffer locally adapted genotypes from being swamped by genes from rural populations (Hoffmann & Blows 1994; Bridle & Vines 2007). The relative importance of reduced genetic diversity and gene flow in determining evolutionary potential is highly debatable, but urban studies clearly have great potential to contribute to this debate.

Synthesis and directions for future research

Owing to a recent explosion of interest, divergence between conspecific urban and rural populations has been demonstrated in a remarkably wide range of traits. Despite this rapid progress, a number of issues require further attention. First, the studies conducted to date almost certainly concern just a fraction of the traits that are affected by urbanisation, and a broader range of studies is required. Second, studies are currently markedly biased towards northern temperate regions and birds; it is particularly surprising that relatively few studies have looked at intraspecific trait divergence in rural and urban plant populations. These geographic and taxonomic biases must be addressed before general patterns in trait divergence can be ascertained fully. Third, most studies have focused on a single urban population, and simultaneous assessments of divergence across multiple sites and species are required to assess fully the extent to which patterns generalise (but see Figure 4.3). Fourth, the impacts of urbanisation on multiple traits have been assessed in very few species (but see Figure 4.4); more studies of this kind are required to enhance understanding of the interactions and trade-offs that seem likely to influence trait divergence. Fifth, much more attention should be paid to assessing which characteristics of the urban environment drive trait divergence and to assessing the fitness consequences of such divergence in order to establish its adaptive value. Sixth, few studies assess whether divergence arises through phenotypic plasticity or genetic adaptation (but see Partecke *et al.* 2004, 2006; Partecke & Gwinner 2007; Liker *et al.* 2008; Rios *et al.* 2008), and even fewer adequately distinguish between genetic adaptation and epigenetic effects. Seventh, there is very little understanding of the rate at which trait divergence between urban and rural populations has arisen. This is unfortunate, as discussion of evolutionary rates in human-altered environments is currently based on studies of a limited range of selection pressures that do not include the impacts of urbanisation (Hendry *et al.* 2008; Smith & Bernatchez 2008). It would thus be

Figure 4.4 The blackbird *Turdus merula* is emerging as a model species with which to assess trait divergence in urban and rural populations. It provides a very scarce example of a species in which such divergence has been assessed for multiple traits; these include demography, disease risk, migratory behaviour, morphology, population genetic structure, stress physiology and vocalisations (Grégoire *et al.* 2002; Partecke *et al.* 2004, 2006; Partecke & Gwinner 2007; Evans *et al.* 2009a, 2009b, 2009c and unpublished data; Nemeth & Brumm 2009). For some traits the relative contributions of phenotypic plasticity and genetic changes to differentiation have been assessed, and divergence in other traits has been assessed in multiple urban populations. Photograph supplied by Z. G. Davies.

useful to assess the magnitude of trait divergence in populations for which the timing of urbanisation is known. Despite these requirements for additional work, it is clear that selection pressures differ enough in many urban and rural areas to generate trait divergence between conspecific populations occupying these contrasting environments. Moreover, the nature of this trait divergence can often be related to theories derived from investigations of the more natural systems on which ecology has traditionally focused. Urban ecology should thus not be viewed as an isolated discipline; rather, assessing how populations adapt to towns and cities may shed light on many fundamental unresolved ecological and evolutionary issues.

Acknowledgements
This work was funded by the Natural Environment Research Council and the Leverhulme Trust. D. Chamberlain, C. Isaksson and an anonymous reviewer provided assistance.

References

Abs, M. and Bergen, F. (2008). A long term survey of the avifauna in an urban park. In J. M. Marzluff, E. Shulenberger, W. Endlicher et al., eds., *Urban Ecology: An International Perspective on the Interaction Between Humans and Nature*. New York: Springer, pp. 373–6.

Allen, J. A. (1878). The influence of physical conditions in the genesis of species. *Radical Review*, 1, 108–40.

Alves, E. S., Tresmondi, F. and Longui, E. L. (2008). Leaf anatomy of *Eugenia uniflora* L. (Myrtaceae) in urban and rural environments, Sao Paulo State, Brazil. *Acta Botanica Brasílica*, 22, 241–8.

Andrieu, E., Dornier, A., Rouifed, S., Schatz, B. and Cheptou, P. O. (2009). The town *Crepis* and the country *Crepis*: how does fragmentation affect a plant-pollinator interaction? *Acta Oecologica*, 35, 1–7.

Angilletta, M. J., Wilson, R. S., Niehaus, A. C. *et al.* (2007). Urban physiology: city ants possess high heat tolerance. *PLoS One*, 2, e258.

Arizmendi, M. D., Constanza, M. S., Lourdes, F. M., Ivonne, F. M. and Edgar, L. S. (2007). Effect of the presence of nectar feeders on the breeding success of *Salvia mexicana* and *Salvia fulgens* in a suburban park near Mexico City. *Biological Conservation*, 136, 155–8.

Aruda, M. P. and Morielle-Versuta, E. (2008). Cytogenetic and random amplified polymorphic DNA analysis of *Leptodactylus* species from rural and urban environments. *Genetics and Molecular Research*, 7, 161–76.

Auman, H. J., Meathrel, C. E. and Richardson, A. (2008). Supersize me: does anthropogenic food change the body condition of Silver Gulls? A comparison between urbanized and remote, non-urbanized areas. *Waterbirds*, 31, 122–6.

Baycu, G., Tolunay, D., Ozden, H. and Gunebakan, S. (2006). Ecophysiological and seasonal variations in Cd, Pb, Zn, and Ni concentrations in the leaves of urban deciduous trees in Istanbul. *Environmental Pollution*, 143, 545–54.

Beck, N. R. and Heinshon, R. (2006). Group composition and reproductive success of cooperatively breeding white-winged choughs (*Corcorax melanorhamphos*) in urban and non-urban habitat. *Austral Ecology*, 31, 588–96.

Bee, M. A. and Swanson, E. M. (2007). Auditory masking of anuran advertisement calls by road traffic noise. *Animal Behaviour*, 74, 1765–76.

Bergen, F. and Abs, M. (1997). Etho-ecological study of the singing activity of the blue tit (*Parus caeruleus*), great tit (*Parus major*) and chaffinch (*Fringilla coelebs*). *Journal of Ornithology*, 138, 451–67.

Bergmann, C. (1847). Über die verhältnisse der wärmeökonomie der thiere zu ihrer grösse. *Göttinger Studien*, 3, 595–708.

Bermúdez-Cuamatzin, E., Ríos-Chelén, A. A., Gil, D. and Garcia, C. M. (2009). Strategies of song adaptation to urban noise in the house finch: syllable pitch plasticity or differential syllable use? *Behaviour*, 146, 1269–86.

Blair, R. B. (1996). Land use and avian species diversity along an urban gradient. *Ecological Applications*, 6, 506–19.

Bonier, F., Martin, P. R., Sheldon, K. S. *et al.* (2007). Sex-specific consequences of life in the city. *Behavioral Ecology*, 18, 121–9.

Bradley, C. A. and Altizer, S. (2007). Urbanization and the ecology of wildlife diseases. *Trends in Ecology and Evolution*, 22, 95–102.

Bradley, C. A., Gibbs, S. E. J. and Altizer, S. (2008). Urban land use predicts West Nile Virus exposure in songbirds. *Ecological Applications*, 18, 1083–92.

Brasier, C. M. (2008). The biosecurity threat to the UK and global environment from international trade in plants. *Plant Pathology*, 57, 792–808.

Bridle, J. R. and Vines, T. H. (2007). Limits to evolution at range margins: when and why does adaptation fail? *Trends in Ecology and Evolution*, 22, 140–7.

Brittingham, M. C. and Temple, S. A. (1998). Impacts of supplemental feeding on survival rates of black-capped chickadees. *Ecology*, **69**, 581–9.

Brown, J. H. and Lee, A. K. (1969). Bergmann's rule and climatic adaptation in woodrats (*Neotoma*). *Evolution*, **23**, 329–38.

Brumm, H. (2004). The impact of environmental noise on song amplitude in a territorial bird. *Journal of Animal Ecology*, **73**, 434–40.

Brumm, H. and Todt, D. (2002). Noise dependent song amplitude regulation in a territorial songbird. *Animal Behaviour*, **63**, 891–7.

Burger, J. (2001). The behavioral response of basking Northern water (*Nerodia sipedon*) and Eastern garter (*Thamnophis sirtalis*) snakes to pedestrians in a New Jersey park. *Urban Ecosystems*, **5**, 119–29.

Butler, D. (2004). Noise management: sound and vision. *Nature*, **427**, 480–1.

Carey, C., Pessier, A. P. and Peace, A. D. (2003). Pathogens, infectious disease, and immune defenses. In R. D. Semlitsch, ed., *Amphibian Conservation*. Washington, DC: Smithsonian Institution, pp. 127–36.

Chamberlain, D., Cannon, A. R., Toms, M. P. *et al.* (2009). Avian productivity in urban landscapes: a review and meta-analysis. *Ibis*, **151**, 1–18.

Cheptou, P.-O. and Avendaño, L. G. (2006). Pollination processes and the Allee effect in highly fragmented populations: consequences for the mating system in urban environments. *New Phytologist*, **172**, 774–83.

Cheptou, P.-O., Carrue, O., Rouifed, S. and Cantarel, A. (2008). Rapid evolution of seed dispersal in an urban environment in the weed *Crepis sancta*. *Proceedings of the National Academy of Sciences of the USA*, **105**, 3796–9.

Chown, S. L. and Gaston, K. J. (2008). Macrophysiology for a changing world. *Proceedings of the Royal Society of London Series B*, **275**, 1469–78.

Contesse, P., Hegglin, D., Gloor, S., Bontadina, F. and Deplazes, P. (2004). The diet of urban foxes (*Vulpes vulpes*) and the availability of anthropogenic food in the city of Zurich, Switzerland. *Mammalian Biology*, **69**, 81–95.

Cooke, A. S. (1980). Observations on how close certain passerine species will tolerate an approaching human in rural and suburban areas. *Biological Conservation*, **18**, 85–8.

Costantini, D. and Møller, A. P. (2008). Carotenoids are minor antioxidants for birds. *Functional Ecology*, **22**, 367–70.

Costantini, D. and Møller, A. P. (2009). Does immune response cause oxidative stress in birds? A meta-analysis. *Comparative Biochemistry and Physiology A*, **153**, 339–44.

Cresswell, W. (2008). Non-lethal effects of predation in birds. *Ibis*, **150**, 3–17.

Croci, S., Butet, A. and Clergeau, P. (2008). Does urbanization filter birds on the basis of their biological traits? *The Condor*, **110**, 223–40.

Culley, T. M., Sbita, S. J. and Wick, A. (2007). Population genetic effects of urban habitat fragmentation in the perennial herb *Viola pubescens* (Violaceae) using ISSR markers. *Annals of Botany*, **100**, 91–100.

Cynx, J., Bean, N. J. and Rossman, I. (2005). Testosterone implants alter the frequency range of zebra finch songs. *Hormones and Behavior*, **47**, 446–51.

Dabelsteen, T. and Pedersen, S. B. (1985). Correspondence between messages in the full song of the blackbird *Turdus merula* and meanings to territorial males, as inferred from responses to computerized modifications of natural song. *Zeitschrift für Tierpsychologie*, **69**, 149–65.

Davis, J. M. and Stamps, J. A. (2004). The effect of natal experience on habitat preferences. *Trends in Ecology and Evolution*, **19**, 411–16.

Desender, K., Small, E. C., Gaublomme, E. and Verdyck, P. (2005). Rural–urban gradients and the population genetic structure of woodland ground beetles. *Conservation Genetics*, **6**, 51–62.

deSwart, R. L., Ross, P. S., Vos, J. G. and Osterhaus, A. D. M. E. (1996). Impaired immunity in harbour seals (*Phoca vitulina*) exposed to bioaccumulated environmental contaminants: review of a long-term feeding study. *Environmental Health Perspectives*, **104**, (S)823–8.

Dohmen, G. P., McNeil, S. and Bell, J. N. B. (1984). Air pollution increases *Aphis fabae* pest potential. *Nature*, **307**, 52–3.

Dutra, N. C. L., Telles, M. P. C., Dutra, D. L. and Silva, N. J. Jr (2008). Genetic diversity in populations of the viper *Bothrops moojeni* Hoge, 1966 in central Brazil using RAPD markers. *Genetics and Molecular Research*, **7**, 603–13.

Endriss, D. A., Hellgren, E. C., Fox, S. F. and Moody, R. W. (2007). Demography of an urban population of the Texas horned lizard (*Phrynosoma cohnutum*) in central Oklahoma. *Herpetologica*, **63**, 320–31.

Evans, K. L., Chamberlain, D. E., Hatchwell, B. J., Gregory, R. D. and Gaston, K. J. What makes an urban bird? *Global Change Biology*, in press, a. DOI: 10.1111/j.i365-2486.2010.02247.x.

Evans, K. L., Hatchwell, B. J., Parnell, M. and Gaston, K. J. A conceptual framework for the colonisation of urban areas: the blackbird *Turdus merula* as a case study. *Biological Reviews*, in press, b. DOI: 10.1111/j.1469-185X.2010.00121.x.

Evans, K. L., Gaston, K. J., Sharp, S. P., McGowan, A. and Hatchwell, B. J. (2009a). The effect of urbanization on avian morphology and latitudinal gradients in body size. *Oikos*, **118**, 251–9.

Evans, K. L., Gaston, K. J., Sharp, S. P. *et al.* (2009b). Effects of urbanization on disease prevalence and age structure in blackbird *Turdus merula* populations. *Oikos*, **118**, 774–82.

Evans, K. L., Gaston, K. J., Frantz, A. C. *et al.* (2009c). Independent colonization of multiple urban centres by a formerly forest specialist bird species. *Proceedings of the Royal Society of London Series B*, **276**, 2403–10.

Fenoglio, M. S., Saivo, A. and Estallo, E. L. (2009). Effects of urbanization on the parasitoid community of a leafminer. *Acta Oecologica*, **35**, 318–26.

Fernández-Juricic, E., Poston, R., de Collibus, K. *et al.* (2005). Microhabitat selection and

singing behavior patterns of male house finches (*Carpodacus mexicanus*) in urban parks in a heavily urbanized landscape in the western US. *Urban Habitats*, **3**, 49–69.

Field, S. G., Lange, M., Schulenburg, H., Velavan, T. P. and Michiels, N. K. (2007). Genetic diversity and parasite defense in a fragmented urban metapopulation of earthworms. *Animal Conservation*, **10**, 162–75.

Fokidis, H. B., Greiner, E. C. and Deviche, P. (2008). Interspecific variation in avian blood parasites and haematology associated with urbanization in a desert habitat. *Journal of Avian Biology*, **39**, 300–10.

Fokidis, H. B., Orchinik, M. and Deviche, P. (2009). Corticosterone and corticosteroid binding globulin in birds: relation to urbanization in a desert city. *General and Comparative Endocrinology*, **160**, 259–70.

Francis, C. D., Ortega, C. P. and Cruz, A. (2009). Noise pollution changes avian communities and species interactions. *Current Biology*, **19**, 1415–19.

Fuller, R. A., Warren, P. H. and Gaston, K. J. (2007). Daytime noise predicts nocturnal singing in urban robins. *Biology Letters*, **3**, 368–70.

Galloway, T. S. and Depledge, M. H. (2001). Immunotoxicity in invertebrates: measurement and ecotoxicological relevance. *Ecotoxicology*, **10**, 5–23.

Garant, D., Kruuk, L. E. B., Wilkin, T. A., McCleery, R. H. and Sheldon, B. C. (2005). Evolution driven by differential dispersal within a wild bird population. *Nature*, **433**, 60–5.

Gaston, K. J., Chown, S. L. and Evans, K. L. (2008). Ecogeographical rules: elements of a synthesis. *Journal of Biogeography*, **35**, 483–500.

Gazal, R., White, M. A., Gillies, R. *et al.* (2008). GLOBE students, teachers, and scientists demonstrate variable differences between urban and rural leaf phenology. *Global Change Biology*, **14**, 1568–80.

Gerhardt, H. C. and Mudry, K. M. (1980). Temperature effects on frequency preferences and mating call frequencies in the green treefrog, *Hyla cinerea* (Anura: Hylidae). *Journal of Comparative Physiology*, **137**, 1–6.

Gering, J. C. and Blair, R. B. (1999). Predation on artificial bird nests along an urban gradient: predatory risk or relaxation in urban environments? *Ecography*, **22**, 532–41.

Ghalambor, C. K., McKay, J. K., Carroll, S. P. and Reznick, D. N. (2007). Adaptive versus non-adaptive phenotypic plasticity and the potential for contemporary adaptation in new environments. *Functional Ecology*, **21**, 394–407.

Glazier, D. S. (2002). Resource-allocation rules and the heritability of traits. *Evolution*, **56**, 1696–700.

Gliwicz, J., Goszczynski, J. and Luniak, M. (1994). Characteristic features of animal populations under synurbization – the case of the Blackbird and of the Striped Field Mouse. *Memorabilia Zoologica*, **49**, 237–44.

Gloger, C. L. (1883). *Das Abandern der Vogel durch Einfluss des Klimas*. Breslau: A. Schulz.

Gosselink, T. E., Van Deelen, T. R., Warner, R. E. and Mankin, P. C. (2007). Survival and cause-specific mortality of red foxes in agricultural and urban areas of Illinois. *Journal of Wildlife Management*, **71**, 1862–73.

Gratani, L., Crescente, M. F. and Petruzzi, C. (2000). Relationship between leaf life-span and photosynthetic activity of *Quercus ilex* in polluted urban areas (Rome). *Environmental Pollution*, **110**, 19–28.

Gregg, J. W., Jones, C. G. and Dawson, T. E. (2003). Urbanization effects on tree growth in the vicinity of New York City. *Nature*, **424**, 183–7.

Grégoire, A., Faivre, B., Heeb, P. and Cezilly, F. (2002). A comparison of infestation patterns by *Ixodes* ticks in urban and rural populations of the Common Blackbird *Turdus merula*. *Ibis*, **144**, 640–5.

Grimm, N. B., Faeth, S. H., Golubiewsk, N. E. *et al.* (2008). Global change and the ecology of cities. *Science*, **319**, 756–60.

Habib, L., Bayne, E. M. and Boutin, S. (2007). Chronic industrial noise affects pairing success and age structure of ovenbirds *Seiurus aurocapilla*. *Journal of Applied Ecology*, **44**, 176–84.

Hamer, A. J. and McDonnell, M. J. (2008). Amphibian ecology and conservation in the urbanising world: a review. *Biological Conservation*, **141**, 2432–49.

Harris, S. (1984). Ecology of urban badgers *Meles meles*: distribution in Britain and habitat selection, persecution, food and damage in the City of Bristol. *Biological Conservation*, **28**, 349–75.

Hendry, A. P., Farrugia, T. J. and Kinnison, M. T. (2008). Human influences on rates of phenotypic change in wild animal populations. *Molecular Ecology*, **17**, 20–9.

Heuschele, J. and Candolin, U. (2007). An increase in pH boosts olfactory communication in sticklebacks. *Biology Letters*, **3**, 411–13.

Hewitt, G. (2000). The genetic legacy of the quaternary ice ages. *Nature*, **405**, 907–13.

Hirota, T., Hirohata, T., Mashima, H., Satoh, T. and Obara, Y. (2004). Population structure of the large Japanese field mouse, *Apodemus speciosus* (Rodentia: Muridae), in suburban landscape, based on mitochondrial D-loop sequences. *Molecular Ecology*, **13**, 3275–82.

Hitchings, S. P. and Beebee, T. J. C. (1997). Genetic substructuring as a result of barriers to gene flow in urban *Rana temporaria* (common frog) populations: implications for biodiversity conservation. *Heredity*, **79**, 117–27.

Hitchings, S. P. and Beebee, T. J. C. (1998). Loss of genetic diversity and fitness in Common Toad (*Bufo bufo*) populations isolated by inimical habitat. *Journal of Evolutionary Biology*, **11**, 269–83.

Hoffmann, A. A. and Blows, M. W. (1994). Species borders: ecological and evolutionary perspectives. *Trends in Ecology and Evolution*, **9**, 223–7.

Hollingsworth, P. M. and Dickson, J. H. (1997). Genetic variation in rural and urban populations of *Epipactis helleborine* (L) Crantz (Orchidaceae) in Britain. *Botanical Journal of the Linnean Society*, **123**, 321–31.

Hõrak, P., Ots, I., Vellau, H., Spottiswoode, C. and Møller, A. P. (2001). Carotenoid-based plumage coloration reflects hemoparasite infection and local survival in breeding great tits. *Oecologia*, **126**, 166–73.

Huck, M., Frantz, A. C., Dawson, D. A., Burke, T. and Roper, T. J. (2008). Low genetic variability, female-biased dispersal and high movement rates in an urban population of Eurasian badgers *Meles meles*. *Journal of Animal Ecology*, **77**, 905–15.

Hunter, M. L. and Krebs, J. R. (1979). Geographical variation in the song of the great tit (*Parus major*) in relation to ecological factors. *Journal of Animal Ecology*, **48**, 759–85.

Isaksson, C. and Andersson, S. (2007). Carotenoid diet and nestling provisioning in urban and rural great tits *Parus major*. *Journal of Avian Biology*, **38**, 564–72.

Isaksson, C., McLaughlin, P., Monaghan, P. and Andersson, S. (2007). Carotenoid pigmentation does not reflect total non-enzymatic antioxidant activity in plasma of adult and nestling great tits, *Parus major*. *Functional Ecology*, **21**, 1123–9.

Isaksson, C., Sturve, J., Almroth, B. C. and Andersson, S. (2009). The impact of urban environment on oxidative damage (TBARS) and antioxidant systems in lungs and liver of great tits, *Parus major*. *Environmental Research*, **109**, 46–50.

Ishigame, G., Baxter, G. S. and Lisle, A. T. (2006). Effects of artificial foods on the blood chemistry of the Australian magpie. *Austral Ecology*, **31**, 199–207.

Jarrauld, N. (2000). The effects of ambient air pollution on leaf pathogens of rose and sycamore. Unpublished Ph.D. thesis, University of London.

Järvenpää, M. and Lindström, K. (2004). Water turbidity by algal blooms causes mating system breakdown in a shallow-water fish, the sand goby *Pomatoschistus minutes*. *Proceedings of the Royal Society of London Series B*, **271**, 2361–5.

Johnson, P. T. J., Lunde, K. B., Ritchie, E. G. and Launer, A. E. (1999). The effect of trematode infection on amphibian limb development and survivorship. *Science*, **284**, 802–4.

Jokimäki, J., Kaisanlahti-Jokimäki, M. L., Sorace, A. *et al.* (2005). Evaluation of the 'safe nesting zone' hypothesis across an urban gradient: a multi-scale study. *Ecography*, **28**, 59–70.

Jordan, D. S. (1891). *Temperature and Vertebrae: A Study in Evolution*. New York: Wilder-Quarter Century Books.

Joshi, P. C. and Swami, A. (2009). Air pollution induced changes in the photosynthetic pigments of selected plant species. *Journal of Environmental Biology*, **30**, 295–8.

Kandori, I., Hirao, T., Matsunaga, S. and Kurosaki, T. (2009). An invasive dandelion unilaterally reduces the reproduction of a native congener through competition for pollination. *Oecologia*, **159**, 559–69.

Keane, B., Collier, M. H. and Rogstad, S. H. (2005). Pollution and genetic structure of North American populations of the common dandelion (*Taraxacum officinale*). *Environmental Monitoring and Assessment*, **105**, 341–57.

Keyghobadi, N. (2007). The genetic implications of habitat fragmentation for animals. *Canadian Journal of Zoology*, **10**, 1049–64.

Kiat, Y., Perlman, G., Balaban, A. *et al.* (2008). Feeding specialization of urban Long-eared Owls, *Asio otus* (Linnaeus, 1758), in Jerusalem, Israel. *Zoology in the Middle East*, **43**, 49–54.

Kiesecker, J. M. (2002). Synergism between trematode infection and pesticide exposure: a link to amphibian limb deformities in nature? *Proceedings of the National Academy of Sciences of the USA*, **99**, 9900–4.

King, K. C., McLaughlin, J. D., Gendron, A. D. *et al.* (2007). Impacts of agriculture on the parasite communities of northern leopard frogs (*Rana pipiens*) in southern Quebec, Canada. *Parasitology*, **134**, 2063–80.

Kosinski, Z. (2001). Effects of urbanization on nest site selection and nesting success of the Greenfinch *Carduelis chloris* in Krotoszyn, Poland. *Ornis Fennica*, **78**, 175–83.

Kouwenberg, L. L. R., McElwain, J. C., Kurschner, W. M. et al. (2003). Stomatal frequency adjustment of four conifer species to historical changes in atmospheric CO_2. American Journal of Botany, **90**, 610–19.

Kozłowski, J., Czarnoleski, M. and Dańko, M. (2004). Can optimal resource allocation models explain why ectotherms grow larger in cold? Integrative and Comparative Biology, **44**, 480–93.

Kubler, S., Kupko, S. and Zeller, U. (2005). The kestrel (Falco tinnunculus L.) in Berlin: investigation of breeding biology and feeding ecology. Journal of Ornithology, **146**, 271–8.

Lafferty, K. D. (1997). Environmental parasitology: what can parasites tell us about human impacts on the environment? Parasitology Today, **13**, 251–5.

Leader, N., Wright, J. and Yom-Tov, Y. (2005). Acoustic properties of two urban song dialects in the orange-tufted sunbird (Nectarinia osea). Auk, **122**, 231–45.

Lengagne, T. (2008). Traffic noise affects communication behaviour in a breeding anuran, Hyla arborea. Biological Conservation, **141**, 2023–31.

Leonard, M. L. and Horn, A. G. (2008). Does ambient noise affect growth and begging call structure in nestling birds? Behavioral Ecology, **19**, 502–7.

Liker, A., Papp, Z., Bókony, V. and Lendvai, Á. Z. (2008). Lean birds in the city: body size and condition of house sparrows along the urbanization gradient. Journal of Animal Ecology, **77**, 789–95.

Lindstedt, S. L. and Boyce, M. S. (1985). Seasonality, fasting endurance, and body size in mammals. American Naturalist, **125**, 873–8.

Liro, A. (1985). Variation in weights of body and internal organs of the field mouse in a gradient of urban habitats. Acta Theriologica, **30**, 359–77.

Lombard, E. (1911). Le signe de l'elevation de la voix. Annales de Maladies de L'orielle et du Larynx, **37**, 101–19.

Lu, P. L., Yu, Q., Liu, J. D. and Lee, X. H. (2006). Advance of tree-flowering dates in response to urban climate change. Agricultural and Forest Meteorology, **138**, 120–31.

Luiselli, L., Angelici, F. M. and Akani, G. C. (2002). Comparative feeding strategies and dietary plasticity of the sympatric cobras Naja melanoleuca and Naja nigricollis in three diverging Afrotropical habitats. Canadian Journal of Zoology, **80**, 55–63.

Mabry, K. E. and Stamps, J. A. (2008). Dispersing brush mice prefer habitat like home. Proceedings of the Royal Society of London Series B, **275**, 543–8.

Marcogliese, D. J. and Cone, D. K. (1997). Parasite communities as indicators of ecosystem stress. Parasitologia, **39**, 227–32.

Marzluff, J. M. (2001). Worldwide urbanization and its effects on birds. In J. M. Marzluff, R. Bowman and R. Donnelly, eds., Avian Ecology and Conservation in an Urbanising World. Boston: Kluwer Academic, pp. 19–38.

Marzluff, J. M. (2005). Island biogeography for an urbanizing world: how extinction and colonization may determine biological diversity in human-dominated landscapes. Urban Ecosystems, **8**, 157–77.

Marzluff, J. M., Withey, J. C., Whittaker, K. A. et al. (2007). Consequences of habitat utilization by nest predators and breeding songbirds across multiple scales in an urbanizing landscape. The Condor, **109**, 516–34.

McCauley, R. D., Fewtrell, J. and Popper, A. N. (2003). High intensity anthropogenic sound damages fish ears. Journal of the Acoustical Society of America, **113**, 638–42.

McCleery, R. A. (2009). Reproduction, juvenile survival and retention in an urban fox squirrel population. Urban Ecosystems, **12**, 177–84.

McCleery, R. A., Lopez, R. R., Silvy, N. J. and Gallant, D. L. (2008). Fox squirrel survival in urban and rural environments. Journal of Wildlife Management, **72**, 133–7.

McCormack, J. E. and Smith, T. B. (2008). Niche expansion leads to small-scale adaptive divergence along an elevation gradient in a medium-sized passerine bird. *Proceedings of the Royal Society of London Series B*, **275**, 2155–64.

Mennechez, G. and Clergeau, P. (2006). Effect of urbanization on habitat generalists: starlings not so flexible? *Acta Oecologica*, **30**, 182–91.

Mogi, M., Miyagi, I. and Abadi, K. S. (1996). Inter- and intraspecific variation in resistance to desiccation by adult *Aedes* (Stegomyia) spp. (Diptera: Culicidae) from Indonesia. *Journal of Medical Entomology*, **33**, 53–7.

Mockford, E. J. and Marshall, R. C. (2009). Effects of urban noise on song and response behaviour in great tits. *Proceedings of the Royal Society of London Series B*, **276**, 2979–85.

Møller, A. P. (2008). Flight distances of urban birds, predation and selection for urban life. *Behavioral Ecology and Sociobiology*, **63**, 63–75.

Morales, C. L. and Traveset, A. (2009). A meta-analysis of impacts of alien vs native plants on pollinator visitation and reproductive success of co-flowering native plants. *Ecology Letters*, **12**, 716–28.

Morey, P. S., Gese, E. M. and Gehrt, S. (2007). Spatial and temporal variation in the diet of coyotes in the Chicago metropolitan area. *American Midland Naturalist*, **158**, 147–61.

Morton, E. S. (1975). Ecological sources of selection on avian sounds. *American Naturalist*, **109**, 17–34.

NEGTAP (2001). *Transboundary Air Pollution: Acidification, Eutrophication and Ground-level Ozone in the UK*. Edinburgh: CEH.

Neil, K. and Wu, J. (2006). Effects of urbanization on plant flowering phenology: a review. *Urban Ecosystems*, **9**, 243–57.

Nemeth, E. and Brumm, H. (2009). Blackbirds sing higher-pitched songs in cities: adaptation to habitat acoustics or side-effect of urbanization. *Animal Behaviour*, **78**, 637–41.

Newhouse, M. J., Marra, P. P. and Johnson, L. S. (2008). Reproductive success of House Wrens in suburban and rural landscapes. *Wilson Journal of Ornithology*, **120**, 99–104.

Noël, S., Ouellet, M., Galois P. and Lapointe, F. J. (2007). Impact of urban fragmentation on the genetic structure of the eastern red-backed salamander. *Conservation Genetics*, **8**, 599–606.

Nottebohm, F. (1975). Continental patterns of song variability in *Zontrichia capensis*: some possible ecological correlates. *American Naturalist*, **109**, 605–24.

Nowacek, D. P., Thorne, L. H., Johnston, D. W. and Tyack, P. L. (2007). Responses of cetaceans to anthropogenic noise. *Mammal Review*, **37**, 81–115.

Obukhova, N. Y. (2001). Geographic variation of colour in the synanthropic blue rock pigeon. *Russian Journal of Genetics*, **37**, 649–58.

Ord, T. J., Peters, R. A., Clucas, B. and Stamps, J. A. (2007). Lizards speed up visual displays in noisy motion habitats. *Proceedings of the Royal Society of London Series B*, **274**, 1057–62.

Ostfeld, R. S. and Keesing, F. (2000). Biodiversity and disease risk: the case of Lyme disease. *Conservation Biology*, **14**, 722–8.

Ovaska, K. (1991). Reproductive phenology, population-structure, and habitat use of the frog *Eleutherodactylus johnstonei* in Barbados, West Indies. *Journal of Herpetology*, **25**, 424–30.

Page, L. K., Gehrt, S. D. and Robinson, N. P. (2008). Land-use effects on prevalence of raccoon roundworm (*Baylisascaris procyonis*). *Journal of Wildlife Diseases*, **44**, 594–9.

Parris, K. M., Velik-Lord, M. and North, J. M. A. (2009). Frogs call at a higher pitch in traffic noise. *Ecology and Society*, **14**, 25.

Partecke, J. (2005). Underlying physiological control of reproduction in urban and forest-dwelling European blackbirds *Turdus merula*. *Journal of Avian Biology*, **36**, 295–305.

Partecke, J. and Gwinner, E. (2007). Increased sedentariness in European blackbirds following urbanization: a consequence of local adaptation? *Ecology*, **88**, 882–90.

Partecke, J., Schwabl, I. and Gwinner, E. (2006). Stress and the city: urbanization and its effects on the stress physiology in European Blackbirds. *Ecology*, **87**, 1945–52.

Partecke, J., Van't Hof, T. and Gwinner, E. (2004). Differences in the timing of reproduction between urban and forest European Blackbirds (*Turdus merula*): result of phenotypic flexibility or genetic differences? *Proceedings of the Royal Society of London Series B*, **271**, 1995–2001.

Partridge, L., Barrie, B., Fowler, K. and French, V. (1994). Evolution and development of body size and cell size in *Drosophila melanogaster* in response to temperature. *Evolution*, **39**, 1327–34.

Patricelli, G. L. and Blickley, J. L. (2006). Avian communication in urban noise: causes and consequences of vocal adjustment. *The Auk*, **123**, 639–49.

Peters, R. A. (2008). Environmental motion delays the detection of movement-based signals. *Biology Letters*, **4**, 2–5.

Plesnik, J. (1990). Long-term study of some urban and extra-urban populations of the kestrel (*Falco tinnunculus* L.). In K. Stastny and V. Bejcek, eds., *Proceedings of the 11th International Conference on Bird Census and Atlas Work*. Prague: Institute of Systematic and Ecological Biology, pp. 453–8.

Popper, A. N., Fewtrell, J., Smith, M. E. and McCauley, R. D. (2003). Anthropogenic sound: effects on the behavior and physiology of fishes. *Marine Technology Society Journal*, **37**, 35–40.

Postma, E. and van Noordwijk, A. J. (2005). Gene flow maintains a large genetic difference in clutch size at a small spatial scale. *Nature*, **433**, 65–8.

Price, T. D., Yeh, P. J. and Harr, B. (2008). Phenotypic plasticity and the evolution of a socially selected trait following colonization of a novel environment. *American Naturalist*, **172**, S49–S62.

Pujol, B. and Pannell, J. R. (2008). Reduced responses to selection after species range expansion. *Science*, **321**, 96.

Rabin, L. A., McGowan, B., Hooper, S. L. and Owings, D. H. (2003). Anthropogenic noise and its effect on animal communication: an interface between comparative psychology and conservation biology. *International Journal of Comparative Psychology*, **16**, 172–92.

Rasner, C. A., Yeh, P. J., Eggert, L. S. *et al.* (2004). Genetic and morphological evolution following a founder event in the dark-eyed junco, *Junco hyemalis thurberi*. *Molecular Ecology*, **13**, 671–81.

Reisch, C. (2007). Genetic structure of *Saxifrage tridactylites* (Saxifragaceae) from natural and man-made habitats. *Conservation Genetics*, **8**, 893–902.

Reperant, L. A., Hegglin, D., Fischer, C. *et al.* (2007). Influence of urbanization on the epidemiology of intestinal helminths of the red fox (*Vulpes vulpes*) in Geneva, Switzerland. *Parasitology Research*, **101**, 605–11.

Richner, H. (1989). Habitat-specific growth and fitness in carrion crows (*Corvus corone corone*). *Journal of Animal Ecology*, **58**, 427–40.

Rios, R. S., Marquis, R. J. and Flunker, J. C. (2008). Population variation in plant traits associated with ant attraction and herbivory in *Chamaecrista fasciculata* (Fabaceae). *Oecologia*, **156**, 577–88.

Robb, G. N., McDonald, R. A., Chamberlain, D. E. *et al.* (2008). Winter feeding of birds increases productivity in the subsequent breeding season. *Biology Letters*, **4**, 220–3.

Roberts, D. G., Ayre, D. J. and Whelan, R. J. (2007). Urban plants as genetic reservoirs or threats to the integrity of bushland plant populations. *Conservation Biology*, **21**, 842–52.

Rodewald, A. D. and Shustack, D. P. (2008). Urban flight: understanding individual and population-level responses of Nearctic–Neotropical migratory birds to urbanization. *Journal of Animal Ecology*, **77**, 83–91.

Roetzer, T., Wittenzeller, M., Haeckel, H. and Nekovar, J. (2000). Phenology in central Europe – differences and trends of spring phenophases in urban and rural areas. *International Journal of Biometeorology*, **44**, 60–6.

Roff, D. (1980). Optimizing development time in a seasonal environment: the 'ups and downs' of clinal variation. *Oecologia*, **45**, 202–8.

Romanova, E. B. and Egorikhina, M. N. (2006). Changes in hematological parameters of *Rana* frogs in a transformed urban environment. *Russian Journal of Ecology*, **37**, 188–92.

Rubin, C. S., Warner, R. E., Bouzat, J. L. and Paige, K. N. (2001). Population genetic structure of Blanding's turtles (*Emydoidea blandingii*) in an urban landscape. *Biological Conservation*, **99**, 323–30.

Rubin, C. S., Warner, R. E., Ludwig, D. R. and Thiel, R. (2004). Survival and population structure of Blanding's turtles (*Emydoidea blandingii*) in two suburban Chicago forest preserves. *Natural Areas Journal*, **24**, 44–8.

Ruiz, G., Rosenmann, M., Novoa, F. F. and Sabat, P. (2002). Haematological parameters and stress index in rufous-collared sparrows dwelling in urban environments. *The Condor*, **104**, 162–6.

Rutkowski, R., Rejt, L. and Szczuka, A. (2006). Analysis of microsatellite polymorphism and genetic differentiation in urban and rural kestrels *Falco tinnunculus* (L.). *Polish Journal of Ecology*, **54**, 473–80.

Ryan, M. J., Cocroft, R. B. and Wilczynski, W. (1990). The role of environmental selection in intraspecific selection in intraspecific divergence of mate recognition signals in the cricket frog, *Acris crepitans*. *Evolution*, **44**, 1869–72.

Saunders, P. J. W. (1966). The toxicity of sulphur dioxide to *Diplocarpon rosae* Wolf causing black spot of roses. *Annals of Applied Biology*, **58**, 103–14.

Savidge, J. A. (1991). Population characteristics of the introduced brown tree snake (*Boiga irregularis*) on Guam. *Biotropica*, **23**, 294–300.

Schoech, S. J. and Bowman, R. (2003). Does differential access to protein influence differences in timing of breeding of Florida scrub-jays (*Aphelocoma coerulescens*) in suburban and wildland habitats? *Auk*, **120**, 1114–27.

Schoech, S. J., Bowman, R. and Bridge, E. S. (2007). Baseline and acute levels of corticosterone in Florida Scrub-Jays (*Aphelocoma coerulescens*): effects of food supplementation, suburban habitat, and year. *General and Comparative Endocrinology*, **154**, 150–60.

Scott, L. K., Hume, I. D. and Dickman, C. R. (1999). Ecology and population biology of long-nosed bandicoots (*Perameles nasuta*) at North Head, Sydney Harbour National Park. *Wildlife Research*, **26**, 805–21.

Seehausen, O., van Alphen, J. J. M. and Witte, F. (1997). Cichlid fish diversity threatened by eutrophication that curbs sexual selection. *Science*, **277**, 1808–11.

Senar, J. C., Borras, B., Cabrera, J., Cabrera, T. and Björklund, M. (2006). Local differentiation in the presence of gene flow in the citril finch *Serinus citronella*. *Biology Letters*, **2**, 85–7.

Severtsova, E. A., Kornilova, M. B., Severtsov, A. S. and Kabardina, Y. A. (2002). Comparative analysis of *Rana temporaria* and *Rana arvalis* fecundity in populations of Moscow City and Moscow Oblast. *Zoologichesky Zhurnal*, **81**, 82–90.

Shargo, E. S. (1988). Home range, movements, and activity patterns of coyotes (*Canis latrans*) in Los Angeles suburbs. Unpublished Ph.D. thesis, University of California, Los Angeles.

Shaw, A. J. (1999). *Heavy Metal Tolerance in Plants: Evolutionary Aspects*. Boca Raton, FL: CRC Press.

Sims, V., Evans, K. L., Newson, S. E., Tratalos, J. and Gaston, K. J. (2008). Avian assemblage structure and domestic cat densities in urban environments. *Diversity and Distributions*, **14**, 387–99.

Slabbekoorn, H. and den Boer-Visser, A. (2006). Cities change the songs of birds. *Current Biology*, **16**, 2326–31.

Slabbekoorn, H. and Ripmeester, E. A. P. (2008). Birdsong and anthropogenic noise:

implications and applications for conservation. *Molecular Ecology*, **17**, 72–83.

Smith, H. and Engeman, R. (2002). An extraordinary raccoon, *Procyon lotor*, density at an urban park. *Canadian Field Naturalist*, **116**, 636–9.

Smith, T. B. and Bernatchez, L. (2008). Evolutionary change in human-altered environments. *Molecular Ecology*, **17**, 1–8.

Spagnuolo, V., Muscariello, L., Terracciano, S. and Giordano, S. (2007). Molecular biodiversity in the moss *Leptodon smithii* (Neckeraceae) in relation to habitat disturbance and fragmentation. *Journal of Plant Research*, **120**, 595–604.

Sun, J. W. C. and Narins, P. M. (2005). Anthropogenic sounds differentially affect amphibian call rate. *Biological Conservation*, **121**, 419–27.

Swei, A., Brylski, P. V., Spencer, W. D., Dodd, S. C. and Patton, J. L. (2003). Hierarchical genetic structure in fragmented populations of the Little Pocket Mouse (*Perognathus longimembris*) in Southern California. *Conservation Genetics*, **4**, 501–14.

Takami, Y., Koshio, C., Ishii, M. *et al.* (2004). Genetic diversity and structure of urban populations of *Pieris* butterflies assessed using amplified fragment length polymorphism. *Molecular Ecology*, **13**, 245–58.

Thompson, K. and McCarthy, M. A. (2008). Traits of British alien and native urban plants. *Journal of Ecology*, **96**, 853–9.

Thorington, K. K. and Bowman, R. (2003). Predation rate on artificial nests increases with human housing density in suburban habitats. *Ecography*, **26**, 188–96.

Tratalos, J., Fuller, R. A., Evans, K. L. *et al.* (2007). Bird densities are associated with household densities. *Global Change Biology*, **13**, 1685–95.

van Rossum, F. (2008). Conservation of long-lived perennial forest herbs in an urban context: *Primula elatior* as study case. *Conservation Genetics*, **9**, 119–28.

Vandergast, A. G., Bohonak, A. J., Weissman, D. B. and Fisher, R. N. (2007). Understanding the genetic effects of recent habitat fragmentation in the context of evolutionary history: phylogeography and landscape genetics of a southern California endemic Jerusalem cricket (Orthoptera: Stenopelmatidae: Stenopelmatus). *Molecular Ecology*, **16**, 977–92.

Vogel, R. L. and Tuomenpuro, J. (1997). Dunnock *Prunella modularis*. In W. J. M. Hagemeijer and M. J. Blair, eds., *The EBCC Atlas of European Breeding Birds: Their Distribution and Abundance*. London: T & A. D. Poyser, pp. 506–7.

Walters, R. J. and Hassall, M. (2006). The temperature-size rule in ectotherms: may a general explanation exist after all? *American Naturalist*, **167**, 510–23.

Wandeler, P., Funk, S. M., Largiadér, C. R., Gloor, S. and Breitenmoser, U. (2003). The city-fox phenomenon: genetic consequences of a recent colonization of urban habitat. *Molecular Ecology*, **12**, 647–56.

Warren, P. S., Katti, M., Ermann, M. and Brazel, A. (2006). Urban bioacoustics: it's not just noise. *Animal Behaviour*, **71**, 491–502.

Weilgart, L. S. (2007). The impacts of anthropogenic ocean noise on cetaceans and implications for management. *Canadian Journal of Zoology*, **85**, 1091–116.

Weller, B. and Ganzhorn, J. U. (2004). Carabid beetle community composition, body size, and fluctuating asymmetry along an urban-rural gradient. *Basic and Applied Ecology*, **5**, 193–201.

Wilson, G. B. and Bell, J. N. B. (1985). Studies on the tolerance to SO_2 of grass populations in polluted areas. III. Investigations on the rate of development of tolerance. *New Phytologist*, **100**, 63–77.

Witter, M. S. and Cuthill, I. C. (1993). The ecological costs of avian fat storage. *Philosophical Transactions of the Royal Society of London Series B*, **340**, 73–92.

Wong, B. B. M., Candolin, U. and Lindström, K. (2007). Environmental deterioration compromises socially enforced signals of male quality in three-spined sticklebacks. *American Naturalist*, **170**, 184–9.

Wood, W. E. and Yezerinac, S. M. (2006). Song Sparrow (*Melospiza melodia*) song varies with urban noise. *Auk*, **123**, 650–9.

Yeh, P. J. (2004). Rapid evolution of a sexually selected trait following population establishment in a novel habitat. *Evolution*, **58**, 166–74.

Yom-Tov, T. (2003). Body sizes of carnivores commensal with humans have increased over the past 50 years. *Functional Ecology*, **17**, 323–7.

Zeveloff, S. I. (2002). *Raccoons: A Natural History*. Washington DC: Smithsonian Books.

Zeveloff, S. I. and Boyce, M. S. (1988). Body size patterns in North American mammal faunas. In M. S. Boyce, ed., *Evolution of Life Histories of Mammals*. New Haven: Yale University Press, pp. 123–46.

Zhang, X., Friedl, M. A., Schaaf, C. B., Strahler, A. H. and Schneider, A. (2004). The footprint of urban climates on vegetation phenology. *Geophysical Research Letters*, **31**, 12209.

Zorenko, T. and Leontyeva, T. (2003). Species diversity and distribution of mammals in Riga. *Acta Zoologica Lituanica*, **13**, 78–86.

CHAPTER FIVE

Species diversity and urbanisation: patterns, drivers and implications

GARY W. LUCK AND LISA T. SMALLBONE

Spatiotemporal patterns in species diversity have intrigued ecologists for many decades (MacArthur 1965; Brown 1981; Gaston 2000). These patterns occur at a variety of scales, for example the latitudinal gradient of increasing species richness from the poles to the Equator (Willig et al. 2003; Hillebrand 2004) and altitudinal gradients across regions (Sanders 2002; McCain 2004). Their appropriate identification and explanation have generated enormous interest and debate (Rahbek & Graves 2001; Willig et al. 2003; Field et al. 2009). It is only logical to extend this interest to urban systems and examine species richness patterns, and drivers of these patterns, across gradients of urbanisation at multiple spatial and temporal scales.

What sets urban studies apart is the implicit or explicit recognition of the major influence of human landscape modification on species distribution. Indeed, the effects of anthropogenic activity are virtually impossible to ignore in urban systems. Moreover, integrating social, cultural, economic, human demographic and ecological data, and exploring the role of human activities in moderating patterns in species diversity, are exciting developments that are forging new ground in interdisciplinary research.

In this chapter, we explore patterns in species diversity in urban systems across space and time, and briefly review some of the key drivers of these patterns and their implications. The chapter's focus is on patterns in species richness, particularly native species, as variation in species abundance and density are covered in Chapter 4. Much more attention is given to spatial patterns in richness, and this reflects the bias in the literature. We group patterns under the headings 'broad-scale', 'comparative', 'gradient' and 'urban-centric'. Broad-scale refers to patterns that occur over large spatial extents (e.g. national, continental or global) with a large grain size (the size of the sampling unit used; e.g. 1° grid cells) where researchers have examined covariation in species richness and some measure of urbanisation (e.g. human

Urban Ecology, ed. Kevin J. Gaston. Published by Cambridge University Press.
© British Ecological Society 2010.

population density). Comparative studies are those that compare species richness in native habitats or rural landscapes with urban areas. A particular type of comparative study is the rural–urban or native–urban gradient, and we treat these studies separately since they are prominent in the literature. In comparative and gradient studies, spatial extent is usually much less than broad-scale studies, and grain size is almost always much smaller. Urban-centric refers to patterns that occur when sampling locations are primarily nested within a single urban centre or across centres. Researchers generally compare species assemblages across different levels of urbanisation, but may occasionally include direct comparisons with native or rural landscapes.

We use the terms species diversity and species richness interchangeably to describe the number of species in an area, as this is consistent with their use in the literature. However, we note that, technically, 'diversity' is a function of both species richness and the abundance of each species and is often encapsulated in diversity indices such as the Shannon index. Also, 'urban' is a term that is poorly defined across studies and is occasionally used qualitatively rather than quantitatively. Pickett and Cadenasso (2006) suggested that definitions of urban will necessarily be flexible and case-specific, but nevertheless the term needs to be defined in each study. This could be achieved by quantifying key measures such as housing density, road density or percentage cover of impervious surfaces (Chapter 2). We try to avoid confusion by referring to patterns in diversity with increasing urbanisation or between urban and non-urban areas (as indicated by authors). Hence, urban is used in a relative rather than absolute sense. We make no attempt to compare levels of urbanisation across studies. While this might be highly desirable, suitable data often do not exist in the literature.

Patterns in space
Broad-scale

At very broad scales (national to continental), human population density (HPD) is a useful surrogate for the level of urbanisation, particularly in developed countries. A growing number of studies have examined correlations between HPD and species richness for a range of taxonomic groups across various spatial extents and using a diversity of grain sizes (e.g. Balmford et al. 2001; Araújo 2003; Gaston & Evans 2004; Luck et al. 2004; Fjeldså & Burgess 2008). A somewhat surprising result from these studies is the consistent reporting of positive correlations between HPD and species richness; surprising because of the undoubted negative impacts of human landscape modification on the persistence of many species.

Luck (2007a) conducted a major review of the relationships between HPD and biodiversity, including a meta-analysis of studies correlating HPD with species richness. He found strong, positive population effect sizes (a population effect size is a single value combining correlation coefficients across studies using

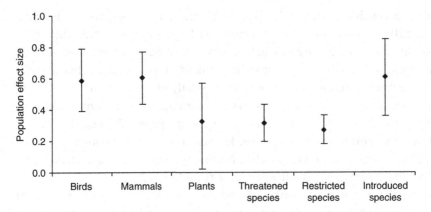

Figure 5.1 Population effect sizes (combining correlation coefficients across studies) between human population density and the species richness of birds, mammals and plants, and threatened, restricted and introduced species. Error bars are 95% confidence intervals. Data from Luck (2007a).

meta-analytical techniques) between HPD and the species richness of plants, birds and mammals (Figure 5.1). However, there was substantial variability across studies owing to, among other things, different spatial extents, grain sizes, sample sizes, habitats and approaches.

A key variable in these studies is grain size, which ranged from 100 m^2 plots up to countries in the research reviewed by Luck (2007a). As grain size increases, so does the strength of the positive correlation between HPD and species richness (also see Pautasso 2007), although this can be complicated by the fact that higher population densities may be characteristic of smaller sampling areas. At small grain sizes the correlation can be, not unexpectedly, negative rather than positive (e.g. Drake & Pereira 2002; Urquiza-Haas *et al.* 2009). Moreover, broad-scale studies using alternative measures of level of urbanisation may yield results that differ from those relying on HPD. Clergeau *et al.* (2001) conducted a biogeographical meta-analysis of 18 studies from towns in temperate and boreal climatic regions where degree of urbanisation was measured as peri-urban, suburban and town centre. They found that bird species richness was negatively correlated with latitude and increasing urbanisation. Lepczyk *et al.* (2008) found that, across the mid-western United States, native bird richness was highest where anthropogenic land cover was lowest and housing density was intermediate.

The general conclusion from the above studies is that people co-occur with a diversity of species at the regional level ($\geq$2500 km^2), but only a few species persist with humans at the local level, particularly in highly urbanised locations. 'Regions' are broad enough to encompass human settlements and relatively undisturbed areas, and co-occurrence across regions seems to be driven by people and other species responding similarly to energy and/or productivity gradients moderated by human land-use (see 'Drivers' below, and Figure 5.7).

Lower species diversity at the local level results from substantial habitat loss in highly urbanised locations. This realisation helps to reconcile the apparently contradictory findings among broad-scale and local-scale studies.

Across regions, Luck (2007a) found that HPD was also positively correlated with the number of threatened species, geographically restricted species (endemic to a localised region) and introduced species (especially for plants; Figure 5.1), although the last of these has been poorly explored at broad scales (Chapter 6). While the correlation with threatened species infers that increasing urbanisation adversely affects species persistence, such spatial correlations are weak causal explanations. Stronger evidence of the negative impacts of urbanisation on species persistence comes from studies that track persistence over time (e.g. Thompson & Jones 1999; Parks & Harcourt 2002; Tait *et al.* 2005). Yet there are only a handful of these studies in the literature and there is a desperate need for long-term monitoring of species persistence in urban areas as development proceeds.

Geographically restricted species are those with small geographic ranges. The positive correlation between HPD and these species suggests their ranges occur, more often than not, close to human settlements. This presents a conservation challenge since there may be few locations within their range that are distant from human settlements and associated negative impacts. It also underscores the dire need to implement effective conservation strategies in urban areas (Schwartz *et al.* 2006; Lawson *et al.* 2008).

Patterns of HPD and species richness in Australia match those from other countries and continents across a range of taxonomic groups. This is true even if the effects of biased sampling effort and spatial autocorrelation are controlled for (Luck *et al.*, in press; Figure 5.2). However, reptiles represent an important exception to the general pattern, as there is a slightly negative correlation between HPD and reptile species richness across Australia at a grain size of 1° (although strong positive correlations between HPD and richness are recorded elsewhere, e.g. North America; Luck *et al.* 2004). This is because diversity patterns in reptiles, which are ectothermic, are probably more closely related to spatiotemporal variation in solar radiation or temperature than other Australian species (including humans) which are responding to gradients in rainfall and productivity. This results in many reptiles occurring in sparsely populated arid and semi-arid regions.

The broad-scale trends described above mask more complex patterns between species richness and urbanisation occurring at local levels. Nevertheless, these studies yield important information that may guide regional management strategies. They also encourage a hierarchical approach to examining species–people relationships and demonstrate that different patterns and drivers manifest themselves at different spatial scales. A multi-scaled approach to ecological studies has gained substantial support in recent decades

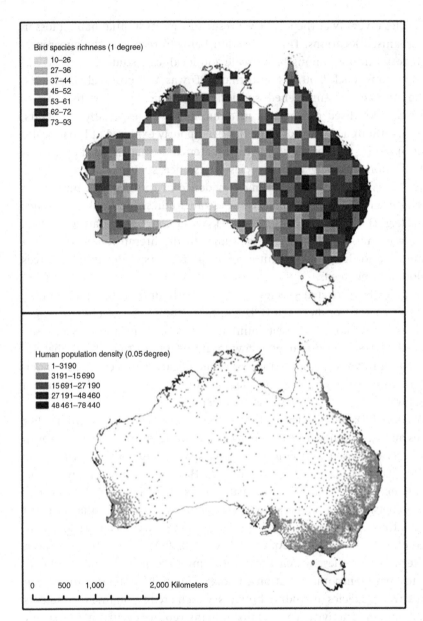

Figure 5.2 The distribution of bird species richness and human population density across mainland Australia. Bird data are at a resolution of 1° grid cells with equal survey effort per cell (10 random surveys). Blank (white) cells represent insufficient survey effort (fewer than 10 surveys). Human population density is number of people per 0.05° grid cell. Reproduced with permission from Luck *et al.* (in press).

(Lindenmayer & Franklin 2002), and urban ecology should continue to promote this approach.

Comparative

Comparisons of species richness between urban areas and those dominated by agriculture or native vegetation generally find that richness is lower in urban areas, while total species abundance is higher with a handful of species contributing the majority of individuals. There are important exceptions to this general pattern identified from studies across different taxonomic groups and subsets of species from a particular group (e.g. cavity-nesting birds). It is also important to note that some researchers confine their sampling entirely to patches of native vegetation (e.g. surrounded by agriculture or urbanised areas), while others sample the full range of available habitats. Taxonomic bias inevitably exists in these studies, with much work conducted on birds and to a lesser extent plants, and fewer studies on other taxonomic groups. This reflects a combination of ease of sampling and the fact that a range of bird and plant species have adapted relatively well to urbanisation.

Lower species diversity with increasing urbanisation has been found for birds (Cam et al. 2000; Sandström et al. 2006), bats (Kurta & Teramino 1992; De Cornulier & Clergeau 2001), terrestrial mammals (Tait et al. 2005) and amphibians (Gagné & Fahrig 2007; see Hamer & McDonnell 2008 for a review of urbanisation effects on amphibians), but the occasional study does not fit this trend. For example, Palomino and Carrascal (2007) found that bird species richness in native vegetation patches was not adversely affected with decreasing distance to small cities (<15 000 people) in Central Spain, although roads had generally negative impacts. Fewer studies have been conducted on invertebrates (see McIntyre 2000 for a review), but Rickman and Connor (2003) found that the species richness and total abundance of leaf-mining Lepidoptera in habitat remnants around San Francisco Bay was not influenced by the extent of urbanisation occurring in a 500-m radius around each remnant. However, increasing urban density in Melbourne, Australia had substantial negative impacts on macro-invertebrate communities of small streams, whereby metropolitan communities were dominated by a few abundant species (Walsh et al. 2001).

Variation in the results among studies can be explained in part by the varied responses of particular species groups. A number of studies have examined changes in the diversity of subsets of bird species with increasing urbanisation, and Chace and Walsh (2006) conducted a comprehensive review of the effects of urbanisation on avifaunal assemblages. Results across studies are not always consistent, but some general trends can be identified (Table 5.1). For example, housing density was negatively correlated with the richness of territorial species, forest interior species and neotropical migrants in some regions of the USA (Mills et al. 1989; Friesen et al. 1995; Green & Baker 2003), while

Table 5.1 *Generalised trends in species richness and abundance of subsets of avian species in response to increasing urbanisation.*

Exotic species ↑	Sedentary species ↑?
Habitat generalists ↑	Short-distance migrants ↓?
Habitat specialists ↓	Widely distributed species ↑
Ground-nesters ↓	Narrowly distributed species ↓
Cavity-nesters ?	Omnivores ↑
Shrub-nesters ↓?	Insectivores ↓?
Tree-nesters ↑?	Granivores ↑?

Note: A question mark indicates a degree of uncertainty in likely responses, while strong conflicting trends have been reported for cavity-nesters.

omnivores and exotic species have been shown to increase in urban areas in the USA and Europe (Jokimäki & Suhonen 1998; Allen & O'Connor 2000). Pidgeon *et al.* (2007) found that forest-dependent birds were mostly negatively affected by increasing house density across the USA, although results varied for different species groups and ecoregions. Cavity-nesters, short-distance migrants and narrowly distributed species were some of the groups to show negative responses.

For bird communities in Singapore, insectivores, carnivores, shrub-nesters and primary cavity excavators were adversely affected by increasing urbanisation, whereas frugivores prospered in areas of low-density housing (Lim & Sodhi 2004). Other studies support the trend of insectivore decline with increasing urbanisation, while also showing that granivores may adapt well to urban environments (e.g. Allen & O'Connor 2000; Lindsay *et al.* 2002). In Australia, urbanisation may favour behaviourally aggressive and medium- to large-bodied bird species (Garden *et al.* 2006). Many of these species are responding to resource availability in the urban landscape. For example, large honeyeaters are attracted to streetscapes with a high density of flowering native trees and shrubs.

In contrast to the results of Pidgeon *et al.* (2007) and Sandström *et al.* (2006), Chace and Walsh (2006) suggested that cavity-nesters are favoured by urbanisation (also see Miller *et al.* 2003). Such apparently contradictory findings highlight the importance of acknowledging contextual and ecological differences across studies. For example, Australia has many cavity-[hollow-]nesting birds, but no species that can excavate cavities, which are formed from insect and fungal attack. Hollow-bearing trees are rare in Australia's urban environments, since most are cleared during development, and hollows take many decades to form in native trees so are mostly absent from plantings post development. Hence, urban areas in Australia do not support a high diversity of hollow-nesting species (Garden *et al.* 2006). It is also important to acknowledge that few studies that examine how species groups with particular traits vary in richness with urbanisation simultaneously assess multiple traits or address species phylogeny.

A number of studies classify bird species into 'urban-adapted' or 'urban-sensitive' (e.g. Blair 1996; Lim & Sodhi 2004). Classifications based purely on variation in population abundance across gradients of urbanisation are not particularly informative without exploration of the ecological traits that allow species to adapt to urban environments. Thankfully, a handful of studies have begun to explore this issue (e.g. Kark *et al.* 2007; Croci *et al.* 2008). Splitting species into urban adapters or urban-sensitive can result in different patterns in richness. For example, Tratalos *et al.* (2007) found that, across Britain, total bird species richness and that of urban-adapted species increased from low to moderate housing density then declined at higher density (following the hump-shaped pattern often reported in rural–urban gradient studies; see below), while the richness of urban-sensitive species consistently declined with increasing density. Total species abundance and the abundance of urban-adapted species increased with housing density, declining only at the highest densities.

Richness patterns for plants often differ from those of faunal species. Urban areas have been shown to harbour more plant species than surrounding non-urban areas (Kühn *et al.* 2004), and species richness is positively correlated with city size (Pyšek 1998), although this may reflect a simple species–area relationship. Positive correlations still exist even when plants are split between native and introduced species. For example, Pyšek (1998) found that, across 54 cities in central Europe, the richness of native plants, aliens, neophytes (introduced to central Europe after AD 1500) and archaeophytes (introduced before AD 1500) increased with city area, number of inhabitants and HPD. The increase was strongest for neophytes, the most recent introductions. An important result from this study was that the proportional representation of aliens was much higher in cities than in the regional species pool (Figure 5.3). This is indicative of urban areas being highly modified landscapes, although still supporting a rich diversity of plant species.

Kühn *et al.* (2004) demonstrated that native and naturalised alien plant species richness were significantly higher in city grid cells than in non-city grid cells in Germany. Native plant species richness was largely explained by the number of geological types per grid cell (130 km^2), and cities were mostly settled in areas of high geological diversity. Hence, human settlers were attracted to inherently diverse locations. Similar to Pyšek (1998), Kühn *et al.* (2004) found that the proportional representation of alien (neophytes and archaeophytes) versus native species increased in city (26%) compared with non-city (rural; 19%) grid cells. Of course, contrary patterns can generally be found. For example, Roy *et al.* (1999) showed that total plant species richness and the richness of native plants did not increase with urbanisation in Britain,

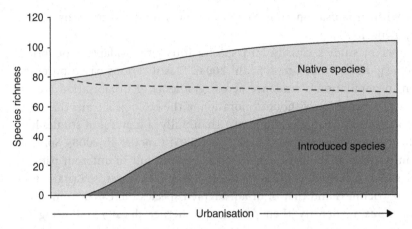

Figure 5.3 A stylized representation of broad trends in plant species richness with increasing urbanisation. Introduced species generally increase in richness and in their proportional contribution to total species richness. Native species richness may either increase slightly or decline with increasing urbanisation (dashed line). The relationships are asymptotic, and total species richness may decline at very high levels of urbanisation.

although alien species richness did. Importantly, they used a much smaller grain size (2 km²) than the above studies.

In support of the results found for bird species, work on plants has also determined that species with particular traits are more likely to persist in urban environments. For example, Thompson and McCarthy (2008) demonstrated that in Sheffield and Birmingham in England, urbanisation favoured plant species that were larger (based on plant height) and that preferred base-rich habitats and, to some extent, dry, unshaded and moderately fertile habitats. Their results were largely consistent for both native and introduced species. Plant height also dictated the likelihood of extinction for plants in Middlesex, England, whereby short plants (mostly natives) were more likely to suffer extinction (Preston 2000; a similar result was recorded in Auckland, New Zealand by Duncan and Young 2000).

In an important follow-up to the study by Kühn *et al.* (2004), Knapp *et al.* (2008) showed that plant species richness in German cities was not related to phylogenetic diversity. That is, high species richness mostly resulted from a greater number of more closely related, functionally similar species adapted to urban areas. This suggests that functional diversity is reduced in urban versus non-urban areas. Such a conclusion is well supported in the growing literature on biotic homogenisation, whereby increasing urbanisation reduces species diversity across sites and leads to greater community similarity among urban areas than among non-urban areas (Rahel 2000; Olden & Poff 2003; McKinney 2006; Olden *et al.* 2006; Devictor *et al.* 2007). Homogenisation of species assemblages in urban areas appears to be driven by invasions or intro-ductions of generalist, urban-adapted species, local extinctions of specialist species (McKinney & Lockwood 1999; Olden *et al.* 2004) and the promulgation of similar habitat components across urban localities. The process of homogen-isation raises substantial challenges for protecting diverse species assemblages within human settlements.

Gradient studies

One way to look at patterns of urbanisation is to sample along a gradient of increasing settlement intensity from native habitats or rural areas to city centres. Sampling can occur at non-contiguous points in different land-use/urbanisation categories along a transect running from city centre to rural/native zone, or at points located randomly throughout the entire study area. Studies of this type have been undertaken across a range of settlement sizes from major cities (population >1 000 000) to regional towns (population <250 000) (e.g. Sewell & Catterall 1998; Melles *et al.* 2003; Caula *et al.* 2008). Grain sizes varied from <10 ha (Sewell & Catterall 1998; Smith & Wachob 2006; Caula *et al.* 2008) to much larger plots (>2000 ha; Weng 2007; Gaublomme *et al.* 2008). Native–urban or rural–urban gradients are generally defined by some index of urbanisation using physical (e.g. building density, land-use types, vegetation cover, density of roads or distance to the central business district; Melles *et al.* 2003; Clergeau *et al.* 2006; Pillsbury & Miller 2008), demographic (e.g. population density or demographic indices; Rubbo & Kiesecker 2005; Hahs & McDonnell 2006) and/or landscape metrics (e.g. patch fragmentation, patch size, edge or land-use heterogeneity; Luck & Wu 2002; Weng 2007).

Studies across gradients often report species diversity peaking at moderate levels of development with reduced species richness occurring at high levels (urban) and low levels (rural and native) of development (Sewell & Catterall 1998; Blair 1999, 2004; Smith & Wachob 2006). Rural–urban gradients in particular often display a hump-shaped pattern in species richness (Figure 5.4). This is consistent with the intermediate disturbance hypothesis, which predicts that species richness will be highest at intermediate levels of disturbance (Connell 1978). Many studies have observed this relationship for a variety of taxonomic groups including birds (Sewell & Catterall 1998; Blair 2004; Chace & Walsh 2006), bats (Gehrt & Chelsvig 2003; Duchamp *et al.* 2004) and lizards (Germaine & Wakeling 2001). This response may be the result of increased resources at moderate levels of urbanisation, with gardens, parks, reserves and other land-uses characterising these areas and providing habitat for particular groups of species (French *et al.* 2005; Gaston *et al.* 2007; Young *et al.* 2007).

McKinney (2008) conducted a comprehensive review of studies on plants, invertebrates and non-avian vertebrates (i.e. mammals, reptiles and amphibians), comparing species richness along a gradient from low (rural) to

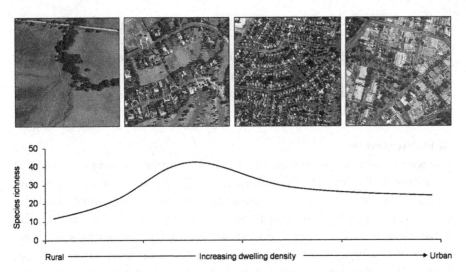

Figure 5.4 Stylised trend of species richness (synthesising results of gradient studies) along a rural–urban gradient defined by increasing dwelling density with corresponding aerial photos of landscape types. Each figure represents an area of 600 by 600 m.

moderate (suburban) to high (urban core) levels of urbanisation. He found that 11 of 17 plant studies recorded an increase in species richness from low to moderate urbanisation, while studies of invertebrates and non-avian vertebrates were more likely to record a decrease in richness (30 of 47 studies and 14 of 17 studies, respectively; although 14 of the invertebrate studies documented an increase in species richness from low to moderate levels of urbanisation). Conversely, in the transition from moderate to high urbanisation, no study on plants or vertebrates and only one study on invertebrates recorded an increase in species richness. Most studies recorded a decrease in richness for vertebrates and invertebrates, while the number of plant studies recording a decrease in richness or no change was evenly split. McKinney (2008) concluded that plant species richness tended to peak at moderate levels of urbanisation, while the richness of invertebrates and non-avian vertebrates peaked at low levels of urbanisation.

Species display different tolerances to urban disturbance and have been grouped as 'exploiters', 'adapters' or 'avoiders' of urbanisation (Blair 1996; Kark et al. 2007; Croci et al. 2008). Species classified as adapted to moderate levels of disturbance will tend to be dominant at the rural–urban interface where landscapes are the most heterogenous (Sewell & Catterall 1998; Melles et al. 2003; Caula et al. 2008). This is a distinct group of species able to use resources typical in this landscape, but the assemblage is very different from the species that occurred in the area prior to development (see Tait et al. 2005; McKinney 2006).

Another recurring pattern along rural–urban gradients is the greater abundance of species at the urban centre, with a few urban exploiters contributing the majority of individuals. This has been noted for birds and micro-bats (Kurta & Teramino 1992; Lesinski *et al.* 2000; Blair 2004; Palomino & Carrascal 2007). Other taxonomic groups show a different pattern even when a few species are dominant. For example, abundance has been observed to decrease at the urban centre for amphibians (Rubbo & Kiesecker 2005; Pillsbury & Miller 2008), lizards (Germaine & Wakeling 2001), butterflies (Blair 1999) and ground beetles (Gaublomme *et al.* 2008).

The difference in responses of certain taxa may be related to their ability to disperse through the landscape, their habitat preferences and the location of habitat along the rural–urban gradient (Niemelä *et al.* 2000; Atauri & De Lucio 2001). For example, frogs are highly sensitive to habitat fragmentation and road density (Rubbo & Kiesecker 2005; Parris 2006; Pillsbury & Miller 2008), which is highest at the city centre, and they often rely on ephemeral ponds for breeding, which are more likely to be found in the peri-urban and rural zone (Hamer & McDonnell 2008). The level of urban disturbance a species is able to tolerate depends on its ecological traits, and these will determine a species' capacity to occupy niches along the gradient (see 'Urban-centric' below).

Some studies have observed seasonal variation in species richness patterns along rural–urban gradients, whereby richness generally declines with increasing building intensity during favourable seasons when resources are abundant (e.g. spring), whereas in the harshest seasons with greater resource restrictions (winter/summer) more species are recorded using urban environments of intermediate disturbance. For example, in Brisbane, Australia, Sewell and Catterall (1998) found more bird species in moderately disturbed suburban environments than large vegetated remnants (>100 ha) during summer compared with winter (summer is a harsher season for this location). In Montpellier, Southern France, Caula *et al.* (2008) found more bird species using suburban environments compared with croplands and woodlands during winter than in spring. Hence, some species may be exploiting resources within moderately disturbed urban environments when harsh climatic conditions are affecting resource availability in the surrounding landscape. This suggests that urbanisation may limit seasonal variation in important resources (e.g. water), and this is a crucial question for future research. Understanding the patterns of species distribution along rural–urban gradients can help to focus conservation and restoration efforts where they will be the most effective.

Urban-centric

Comparative studies generally show that species richness declines with increasing urbanisation or that richness peaks at some intermediate level of development especially across rural–urban gradients. However, patterns vary across taxonomic groups and for species with specific ecological traits, and some species can readily adapt to or exploit urban habitats. Simple comparisons of native versus urban landscapes sometimes treat 'urban' as a homogeneous unit, ignoring the substantial heterogeneity that can occur within an urban environment. This heterogeneity leads to changes in species diversity and composition across neighbourhoods within towns and cities.

The species richness of fauna across neighbourhoods is often positively correlated with the cover of native and/or exotic vegetation. Studies on birds consistently report on the importance of retaining native tree canopy cover, vegetation structure and streetscape vegetation to support a higher number of species (e.g. Mills *et al.* 1989; Fernández-Juricic 2000; Hennings & Edge 2003; White *et al.* 2005; MacGregor-Fors 2008). White *et al.* (2005) found that parks and streetscapes with native vegetation supported more bird species than recently developed streetscapes or those with exotic vegetation in Melbourne suburbs. Hennings and Edge (2003) found that urban canopy cover promoted native bird richness in Portland, Oregon, and Green (1984), also working in Melbourne, reported that total native vegetation cover was positively correlated with native bird richness, but negatively correlated with exotic bird richness. Studies also report the importance of native remnants and riparian zones within urban environments as potential conservation reserves (Bush *et al.* 2003; Hennings & Edge 2003; Hodgson *et al.* 2006; Pennington *et al.* 2008).

Surveys confined entirely within green space surrounded by urbanisation (e.g. remnant vegetation or recreation parks) generally show that species richness is highest in the areas that most closely resemble the previous land cover (e.g. forest reserves), and the size of remaining vegetation patches is positively correlated with species richness, consistent with well-established species–area relationships (e.g. Mörtberg 1998; Koh & Sodhi 2004; Palmer *et al.* 2008). Vegetation area seems to be particularly important for those species that cannot readily use the surrounding urban landscape, although Antos *et al.* (2006) found that vegetation patch size was also positively correlated with introduced bird species richness in urban areas in Melbourne. Moreover, area might not be the most critical factor driving patterns in species richness in some taxonomic groups. Garden *et al.* (2007) found that habitat structural elements were the most important determinants of reptile and mammal assemblages in urban vegetation patches in Brisbane.

In sum, these results show, unsurprisingly, that native faunal species respond positively to the retention of native vegetation and that richness increases with increasing vegetation cover. Hence, they demonstrate the importance of retaining vegetation in urban landscapes. Not only can this promote species conservation, but vegetation offers human residents a range of additional benefits including microclimate regulation (Harlan *et al.* 2006; Jenerette *et al.* 2007), control of air and water pollution (Randolph 2004), carbon storage (Nowak 1994), recreational opportunities (Miller 2006), and cultural and health benefits (Fuller *et al.* 2007; Tzoulas *et al.* 2007).

Some of the more interesting urban-centric work links household or neighbourhood socioeconomic characteristics with vegetation cover and species diversity. This begins to tease apart some of the driving forces of neighbourhood variation in urban ecosystems and also highlights demographic groups that may be disadvantaged through lack of direct access to the benefits of nature. For example, Hope *et al.* (2003) found that plant species richness was positively related to, and primarily a factor of, family income and housing age across neighbourhoods in Phoenix, Arizona. Income is a key variable in studies of vegetation cover (e.g. Grove & Burch 1997), and this suggests that people with greater financial security are either attracted to neighbourhoods with a higher level of 'naturalness' or promote vegetation cover and diversity through their activities. The flipside of this trend is that residents with a 'lower' socioeconomic status often live in neighbourhoods with few species and little vegetation cover (Iverson & Cook 2000; Pauleit *et al.* 2005).

Urban-centric studies are now beginning to focus on species richness patterns in household gardens and how householder behaviour may influence these patterns. For example, Thompson *et al.* (2003) found that a third of all plant species in gardens in Sheffield, England, were native and, collectively, garden quadrats contained twice as many species as semi-natural habitats (although mean species richness per quadrat was lower in gardens). Moreover, plant species richness increased with garden area, with the relationship being potentially stronger for natives than exotics (Smith *et al.* 2006). Householder activities such as watering and fertilising can play an important role in maintaining small plant populations and increasing the diversity of gardens, and some potential exists for coordinating householder activities to achieve broader conservation objectives in line with incentive schemes currently used in rural landscapes.

The structure and composition of household gardens may also have a stronger influence on bird assemblages than broader landscape factors (Daniels & Kirkpatrick 2006). Garden area has been shown to be positively related to the species richness of garden birds, and native birds may be particularly attracted to native plants (Thompson *et al.* 1993; Chamberlain *et al.* 2004). For example, French *et al.* (2005) found that native nectarivores, a substantial component of Australian urban bird communities, preferred to feed on native *Banksia* spp. and *Grevillea* spp. than on introduced *Camellia* spp. and *Hibiscus* spp. in Sydney, Australia.

Provision of supplementary food will generally increase bird species richness and abundance in gardens (Savard *et al.* 2000; Daniels & Kirkpatrick 2006; Parsons *et al.* 2006). Fuller *et al.* (2008) documented evidence that bird assemblages had important links with bird feeding stations and with the proportion

Socioeconomic status

Low ——————————————————————————————→ High

Species richness

Figure 5.5 Suburban streetscapes in southeastern Australia along a gradient of socioeconomic 'status' based on income and the proportion of residents with a tertiary education. Bird and plant species richness tend to be highest in neighbourhoods with higher status (i.e. higher income and education levels; G. W. Luck and L. T. Smallbone, unpublished data). Photographs used with permission of L. T. Smallbone.

of households that provide supplementary food in the neighbourhoods of Sheffield. Interestingly, they also found that species richness was higher in middle- and high-income neighbourhoods, mirroring the results of Hope *et al.* (2003) for plants in Arizona. We have found similar results in towns (17 000 – 80 000 people) across southeastern Australia showing that vegetation cover, bird species richness and bat activity are all higher in high-income suburbs where a greater proportion of residents have completed tertiary education (Luck *et al.* 2009, and unpublished data) (Figure 5.5).

While there is much variation in the results of urban-centric studies, some general patterns are still evident, such as the decline in species diversity from the edge to the interior of metropolitan areas (Green 1984; Clergeau *et al.* 1998; Cam *et al.* 2000; Melles *et al.* 2003), although it is unclear how such a pattern might be influenced by town size. The drivers of these patterns reflect complex interactions among environmental gradients, neighbourhood socioeconomic characteristics, householder behaviour and species ecological traits. Potential drivers of species richness patterns in urban landscapes across scales are further explored below.

Patterns in time

Although there are many studies on spatial patterns in diversity across urban areas, there are relatively few on patterns over time. This is not unusual in ecology owing to the rarity of accurate, long-term databases. The most comprehensive work relates to temporal changes in vegetation cover in urban areas, which is relatively easy to map with satellite images or aerial photographs (e.g. Morawitz *et al.* 2006; DiBari 2007). Moreover, some studies have tracked changes in urban land cover or housing development in urban and peri-urban areas and matched this with loss of vegetation (e.g. Hammer *et al.* 2004; Tian *et al.* 2005; Gonzalez-Abraham *et al.* 2007). Hence, we have reasonably good information on temporal patterns in urban development and changes in vegetation cover, but few data exist on changes in species diversity over time (and there are even fewer studies on changes in population size, although see, for example, Brichetti *et al.* 2008).

Nevertheless, a key study in this area is that of Tait *et al.* (2005). These authors tracked changes in the species richness of native and introduced plants, mammals, birds, reptiles and amphibians in Adelaide, Australia, from 1836 to 2002 (Adelaide is currently a city with over one million people). Some results were predictable, while others were surprising. Native species richness declined in most groups, but only marginally for reptiles (3.6% of 56 species) and not at all for amphibians. Native mammals suffered the most, losing 50% of species ($n = 40$; a similar result was reported by van der Ree and McCarthy (2005) for mammal species loss in Melbourne since European settlement). Interestingly, total species richness remained practically the same for birds and reptiles. That is, while 21 native bird species were lost, 20 introduced species established themselves in the city. Similarly, two native reptiles were replaced by two introduced species. The same was not true for mammals (a loss of 20 species, but a gain of only 9), while there was an overall increase of 46% in the total number of plant species.

Other than for mammals, these patterns do not represent the devastating loss of species that is normally associated with urban development. However, it is important to note that Tait *et al.* (2005) included the entire metropolitan area in their study, and this incorporates patches of native and other vegetation. Moreover, species richness is not necessarily the best measure of human impact on nature since many populations can be lost, with dire implications for species persistence and ecosystem functioning, even though a species may remain extant (Hughes *et al.* 1997; Luck *et al.* 2003).

Puth and Burns (2009) also reviewed studies that examined changes in species richness over time for the metropolitan region of New York. They found that of the 26 studies documenting temporal change, 65% recorded declines in species richness (increasing to 77% when only trends in native species were considered). Only six studies found that species richness increased over time, and most of these examined recolonisation after disturbance. Decreasing species richness was recorded across a variety of taxonomic groups including plants, fish, amphibians, reptiles and birds. Other studies of temporal trends in New York flora concur with these findings, showing that the area has lost over 40% of its native plant species in the past century or so (Robinson *et al.* 1994; DeCandido *et al.* 2004). These latter studies also found that exotic plant species richness increased over the same time period.

A decline in native plants and corresponding increase in introduced species over time, sometimes leading to a net gain in plant species richness, is consistent with studies conducted over broader areas (e.g. McKinney 2002;

Pyšek *et al.* 2005), although site-specific results may vary from this trend (e.g. Chocholouskova & Pyšek 2003). Parody *et al.* (2001) reported no change in bird species richness in moderately populated areas over a 50-year period in Michigan, while Jones and Wieneke (2000) recorded the same result for a single suburban bird community studied over 16 years in Townsville, Australia. Yet some studies report a substantial decline in bird species richness over time for parks surrounded by urban development (e.g. Diamond *et al.* 1987; Corlett 1988; Recher & Serventy 1991), although it is unclear how much this species loss is a factor of park size and isolation (or habitat changes) and how much can be attributed to increasing urbanisation.

Lane *et al.* (2006) estimated that, since 1819, between 33% and 72% of bat species have been lost from Singapore, an island that has been substantially developed (50% 'built-up'). The upper bound is based on a species assemblage inferred from a neighbouring mainland area and may be an overestimate. The authors make the important point that many species are probably lost during initial habitat clearance prior to extensive urban development and prior to the advent of comprehensive and systematic species monitoring. Therefore, it is likely that the species richness of original assemblages used as baselines for comparing historical changes is almost always underestimated, and this is problematic for detailing changes over time with increasing urbanisation.

Matching spatial variation in settlement or neighbourhood age with species richness can also allude to likely temporal changes. For example, broad-scale studies by McKinney (2001, 2002) showed positive correlations between introduced plant richness or net gain in the number of plant species (i.e. introductions minus local extinctions) and HPD and time since settlement in the United States. The same was not true for fish species analysed in the same studies. Other research found that bird species richness was higher in older suburbs, probably reflecting the development of urban vegetation over time (e.g. Jones 1981; Munyenyembe *et al.* 1989). This probably occurs when vegetation is cleared at the time of settlement or settlement develops on previously cleared land (e.g. rural areas) and the establishment of domestic gardens increases vegetation cover and species diversity.

In sum, changes over time appear to follow these general trends (Figure 5.6). For plants, early human settlers clear native vegetation, but introduce many new species (for agricultural or domestic purposes) such that the number of introductions outpaces the number of local extinctions leading to a net gain in richness over time. The number of species introductions is likely to decline over time and the same is probably true for local extinctions as most urban-sensitive species have already been lost from long-established settlements. This should lead to a plateau in total species richness (although the number of exotic species will probably continue to increase slowly over time, especially for plants). A similar plateau would be reached for animal (i.e. vertebrate)

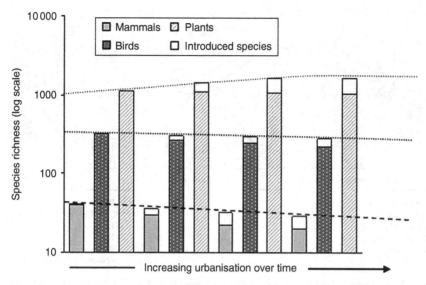

Figure 5.6 General patterns in total species richness and the richness of native and introduced species with increasing urbanisation over time. From data in Tait *et al.* (2005).

species richness, but possibly through different processes. These could be a replacement of native species with a similar number of introduced species (e.g. birds), no change in native species (e.g. reptiles) or a substantial decline in natives resulting in a new, and much lower asymptote in total species richness (e.g. mammals).

The trends in Figure 5.6 are, of course, simplifications and ignore the fact that rapid gains and losses in species richness can occur over short periods (Tait *et al.* 2005). More importantly, native species richness almost always declines with increasing urbanisation and highly developed locations have substantially different species assemblages from those that existed prior to development. Spatiotemporal patterns in species diversity in urban landscapes result from a myriad of complex, interacting processes occurring over various spatial and temporal scales (Kent *et al.* 1999).

Drivers

The drivers of patterns in urban diversity vary across spatial scales (Figure 5.7). What drives the broad-scale congruence between species richness and HPD across regions is subject to some speculation, but little comprehensive assessment. People and other species may co-occur either because both are responding to underlying abiotic or biotic conditions (e.g. climate) or human landscape modification leads to increased species richness near human settlements (e.g. through creation of more heterogeneous landscapes). Moreover, humans are

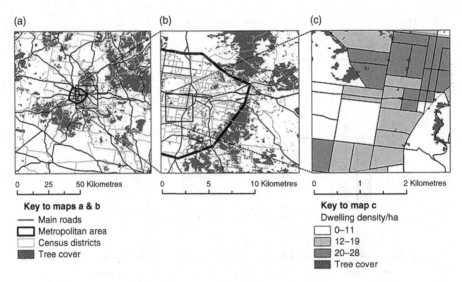

Key to maps a & b
— Main roads
☐ Metropolitan area
☐ Census districts
■ Tree cover

Key to map c
Dwelling density/ha
☐ 0–11
▨ 12–19
■ 20–28
■ Tree cover

Figure 5.7 Summary of drivers across scales. (a) Across broad extents, using large grain sizes (e.g. 1°), human population density and species richness positively covary, probably owing to common drivers such as available energy and primary productivity. This relationship is moderated by human land-use policy (e.g. area of conservation land near human settlements). (b) Within a region, rural–urban gradients often show peaks in species richness in fringe suburbs of metropolitan areas (intermediate disturbance) owing to greater vegetation cover and landscape heterogeneity in these locations compared with rural areas and city centres. (c) Within urban centres, the socioeconomic characteristics of neighbourhoods are linked to vegetation cover and diversity, and this in turn affects faunal assemblages. Neighbourhoods characterised by relatively low housing density, high-income and high education levels support a greater diversity of species.

often responsible for deliberately introducing or promoting colonisation of exotic species in urban environments. This is especially true for plants, but these introductions do not explain the positive correlation between HPD and species richness across regions, because this correlation still exists when examining only native species (Luck 2007a).

A useful starting point is to assess whether well-established drivers of biogeographic gradients in species richness also explain patterns in human distribution. Drivers in common may include potential energy and/or primary productivity, habitat heterogeneity and evolutionary time/long-term climatic stability (Rahbek & Graves 2001; Willig *et al.* 2003). Potential energy and primary productivity are positively correlated with HPD and species richness in some locations and offer possible explanations for their co-occurrence across regions (Evans & Gaston 2005; Luck 2007b). Fjeldså and Rahbek (1998) suggested that historical climatic stability may have led to the spatial congruence between people and species richness.

Alternatively, human land-use policy may be solely responsible for the regional co-occurrence of human settlements and species richness through anthropogenic landscape heterogeneity (Fairbanks 2004), increased productivity (Hugo & van Rensburg 2008) or conservation policy (Gaston 2005). The first two explanations suggest that humans increase heterogeneity (e.g. through establishing a diversity of land-uses) or productivity (e.g. through irrigated agriculture) near human settlements, leading to greater species richness. The third predicts that HPD and species richness are correlated because both positively covary with conservation activities, or that conservation policy mediates the relationship between people and richness. The former is highly unlikely given reports of strong negative correlations between HPD and the size of conservation reserves. This relationship is likely to increase the probability of species extinction inside reserves (Brashares *et al.* 2001; Harcourt *et al.* 2001; Luck 2007b).

Hugo and van Rensburg (2008) tested each of the human land-use hypotheses for HPD and bird species correlations across South Africa at a quarter degree resolution. They found support for the anthropogenic heterogeneity and productivity hypotheses, and concluded that the co-occurrence of humans and other species is strongest in regions with greater cover of conservation land-use. Extending this work, we tested the explanatory power of a range of hypotheses for biotic drivers (primary productivity and inherent habitat heterogeneity) and for human land-use drivers (anthropogenic landscape heterogeneity, increased productivity and conservation policy) of HPD and bird species richness co-occurrence across Australia at a 1° resolution (Luck *et al.* in press). We found that potential energy (annual rainfall) and net primary productivity appeared to be the key drivers of the people–richness correlation after controlling for spatial autocorrelation and biased sampling effort. Inherent habitat heterogeneity was more strongly, positively related to species richness in low-energy than high-energy regions, contrary to previous findings (Kerr & Packer 1997). Similar to Hugo and van Rensburg (2008), we found that conservation policy appears to mediate the relationship between humans and other species, whereby the HPD–diversity correlation was strongest in areas with a greater percentage cover of conservation land-use and declined steadily with a reduction in conservation land-use. This means that in regions where conservation land-use is low, species are restricted to a few small conservation reserves that are mostly distant from human settlements. In contrast to Hugo and van Rensburg (2008), we found little support for the anthropogenic or increased productivity hypotheses.

Another potential driver of the HPD–species richness correlation that has not received adequate attention is biased sampling effort. This is particularly problematic when species lists are derived from atlas data collected by volunteer observers. Sampling effort has been shown to be greatest in regions of high

HPD using such data (Luck *et al.* 2004), and this may lead to more species being recorded in these regions, confounding relationships with other potential drivers. However, recent studies have now demonstrated that the HPD–richness correlation still exists after controlling for biased sampling effort (e.g. Evans *et al.* 2007; Pautasso & McKinney 2007).

At smaller scales, the correlation between species richness and urbanisation is generally negative. Therefore, the examination of potential drivers is usually focused on the question of why more species occur in some urban areas than others. Landscape variables change considerably when moving from highly urbanised centres to peri-urban and rural areas. Vegetation patch size and density, land-use category and fragmentation show varying patterns across degrees of urbanisation (Luck & Wu 2002; Hahs & McDonnell 2006; Weng 2007). As urbanisation intensifies, patch size generally becomes smaller and fragmentation increases (Luck & Wu 2002; Smith & Wachob 2006; Weng 2007).

Across rural–urban gradients, faunal species richness may peak at intermediate locations (e.g. outer suburbs) owing to greater vegetation cover and the diversity of land-uses in these locations compared with rural areas (which are largely cleared) and urban centres (Luck & Wu 2002; Weng 2007). Moreover, intermediate locations are areas where species that respond positively to urbanisation may coexist with those more dependent on native vegetation. Landscape/habitat heterogeneity (i.e. the variety of landscape elements) plays an important role in driving faunal species richness patterns (e.g. Kerr & Packer 1997; Atauri & de Lucio 2001; Tratalos *et al.* 2007), although promotion of heterogeneity per se is unwarranted without some understanding of the key landscape elements required by particular taxonomic groups.

At various scales, bird species have been shown to respond positively to increasing vegetation cover, composition and structure (Mills *et al.* 1989; Munyenyembe *et al.* 1989; Pidgeon *et al.* 2007; Bino *et al.* 2008). Vegetation cover is often negatively correlated with housing density and the cover of impervious surfaces (e.g. roads and footpaths). The availability of anthropogenic food may also have a strong influence on bird assemblages in urban environments (e.g. Jokimäki & Suhonen 1998; Fuller *et al.* 2008), although this may increase the abundance of a few readily adaptable species rather than species richness per se. At the scale of individual streets and gardens, plant species composition has an important influence on bird assemblages. For example, Young *et al.* (2007) found that the presence of flowering native trees was more important in driving nectarivore assemblages in urban streets in Adelaide than the area surrounding the tree.

In sum, faunal species will respond to the availability of critical resources (e.g. for feeding and breeding) across urban landscapes, and use of these resources will be dictated by their capacity to move around the landscape, interspecific interactions (e.g. competition) and species' ecological traits. Arguably, a more interesting question then is what dictates the distribution of resources. Vegetation is an important resource for many faunal species and has been shown to vary greatly across urban areas. In contrast to fauna, plant species richness (natives and exotics) tends to be higher in urban than rural areas and increases over time with increasing urbanisation (although the relationship is likely to be asymptotic). This largely reflects the introduction of new plant species for agricultural or domestic purposes (Pyšek 1998). Moreover, humans are able to maintain very small plant populations (e.g. one to two individuals) and ensure the persistence of non-endemic species through watering, fertilisation and other maintenance activities, and this leads to an 'artificially' high diversity (i.e. many of these populations or species would become 'extinct' without human intervention). Also, in some regions, home gardeners have access to a huge species pool through local nurseries (Thompson et al. 2003).

While underlying heterogeneity (e.g. topographic or geological) may have an influence on plant species diversity in urban areas (Pyšek 1998; Kühn et al. 2004), an increasing number of studies demonstrate a relationship between the socioeconomic profile of neighbourhood residents or householder behaviour and vegetation cover and plant species diversity (Hope et al. 2003; Grove et al. 2006). These studies often show that income or some other measure of social 'status' (e.g. education) is positively correlated with species richness or vegetation cover. Teasing apart the causal direction of this relationship will take some detailed interdisciplinary work. For example, do people with high-incomes prefer to live in more vegetated suburbs (and have the financial capacity to do so) or do they increase vegetation cover through their behaviour? Some of our own work has shown that level of education may be a better predictor of vegetation cover than income (although the two are often correlated), whereby cover is positively correlated with the proportion of residents with a tertiary education (Luck et al. 2009).

If resident behaviour is important in determining levels of vegetation cover, then improving social status is a potential mechanism for increasing vegetation and subsequently plant and fauna richness in neighbourhoods. Of course, not everyone can be rich or highly educated (although income and education levels can always be improved), and local government planning is still key to improving the nature of urban environments. Moreover, more easily managed factors like housing density also have strong relationships with vegetation cover and species richness (e.g. Hope et al. 2003; Tratalos et al. 2007; Luck et al. 2009), and a lot more research is required to determine how interactions between landscape planning, socioeconomic status, resident behaviour and species' ecological traits influence flora and fauna in human settlements.

Implications and future research

There are two primary implications of the broad-scale congruence between people and species richness. The first is the substantial conservation challenges raised by this congruence given the negative impacts of human activities on nature. Alleviating spatial conflict between people and biodiversity has been explored in a number of studies through conservation planning strategies (e.g. Balmford *et al.* 2001; Araújo *et al.* 2002; Luck *et al.* 2004; O'Dea *et al.* 2006). These studies suggest that opportunities still exist for promoting conservation in the midst of human development, but these opportunities are more restricted when conservation objectives become more demanding (e.g. protecting 20% of a species range rather than 10%).

The second implication is that there is substantial opportunity for people to experience species-rich locations near where they live if conservation close to human settlements can be improved (Schwartz *et al.* 2006; Lawson *et al.* 2008). Maintaining high diversity can be important for the production of various ecosystem services, many of which are used locally (Díaz *et al.* 2006). Future research should continue to focus on key drivers of spatial congruence and the development of intra- and inter-regional management strategies that ensure species persistence through appropriate settlement planning. More work is needed to assess the ecological, social and environmental (e.g. energy use) trade-offs between concentrated, high-density living in increasingly large cities versus more dispersed, smaller regional centres. We know little, for example, of how rural–urban gradients vary with city size or if neighbourhoods in small towns are characterised by the same spatial variation in species richness as those in large cities. At a smaller scale, clustered development, although promoted as a tool to reduce the impacts of human settlement on nature, offered no greater conservation benefits than dispersed housing in a study in Boulder County, Colorado (Lenth *et al.* 2006).

Within human settlements, people living on the rural–urban fringe, in low-density housing or 'wealthier' neighbourhoods have the greatest opportunities to conserve and interact with a diverse nature. However, the stark reality for most people, particularly in large cities, is that their neighbourhood is likely to be species-poor and dominated by a handful of abundant urban exploiters (Turner *et al.* 2004). A reduction in the opportunities for people to interact with nature can impact negatively on human health and psychological wellbeing (Harlan *et al.* 2006; Tzoulas *et al.* 2007), and there is some evidence to suggest that species diversity per se is important in promoting the latter (Fuller *et al.* 2007). For an increasing number of people, their entire experience with nature may be dominated by interactions in urban environments. Interactions with nature can influence people's perception of and desire to protect nature, and may link directly with environmental activism (Dunn *et al.* 2006). Since most of the Earth's population now live in urban areas (United Nations 2008), and these people will have a substantial influence on conservation policy through voter numbers, improving human–nature interactions in human settlements is not only important for meeting local conservation objectives, but could have major implications for global conservation (Dunn *et al.* 2006).

In the future, researchers must build on the ground-breaking interdisciplinary studies that link ecological, socioeconomic and environmental data (e.g. Hope *et al.* 2003). Only through integrated research can we begin to understand the patterns of diversity that occur in urban landscapes and tease apart the complex, interconnected drivers of these patterns.

Acknowledgements

This work was funded by an Australian Research Council Discovery Grant (DP0770261) to G.L. Thanks to S. McDonald (Spatial Data Analysis Network, Charles Sturt University) for assistance with remote sensing and Geographical Information System analyses. K. J. Gaston and two anonymous reviewers provided thoughtful comments on a draft manuscript.

References

Allen, A.P. and O'Connor, R.J. (2000). Hierarchical correlates of bird assemblage structure on north-eastern USA lakes. *Environmental Monitoring and Assessment*, **62**, 15–35.

Antos, M.J., Fitzsimons, J.A., Palmer, G.C. and White, J.G. (2006). Introduced birds in urban remnant vegetation: does remnant size really matter? *Austral Ecology*, **31**, 254–61.

Araújo, M.B. (2003). The coincidence of people and biodiversity in Europe. *Global Ecology and Biogeography*, **12**, 5–12.

Araújo, M.B., Williams, P.H. and Turner, A. (2002). A sequential approach to minimise threats within selected conservation areas. *Biodiversity and Conservation*, **11**, 1011–24.

Atauri, J.A. and De Lucio, J.V. (2001). The role of landscape structure in species richness distribution of birds, amphibians, reptiles and Lepidopterans in Mediterranean landscapes. *Landscape Ecology*, **16**, 147–59.

Balmford, A., Moore, J.L., Brooks, T. *et al.* (2001). Conservation conflicts across Africa. *Science*, **291**, 2616–19.

Bino, G., Levin, N., Darawshi, S. *et al.* (2008). Accurate prediction of bird species richness patterns in an urban environment using Landsat-derived NDVI and spectral unmixing. *International Journal of Remote Sensing*, **29**, 3675–700.

Blair, R.B. (1996). Land use and avian species diversity along an urban gradient. *Ecological Applications*, **6**, 506–19.

Blair, R.B. (1999). Birds and butterflies along an urban gradient: surrogate taxa for assessing biodiversity? *Ecological Applications*, **9**, 164–70.

Blair, R.B. (2004). The effects of urban sprawl on birds at multiple levels of biological organization. *Ecology and Society*, **9**, 5. Available at http://www. ecologyandsociety.org/vol9/iss5/art2

Brashares, J.S., Arcese, P. and Sam, M.K. (2001). Human demography and reserve size predict wildlife extinction in West Africa. *Proceedings of the Royal Society of London Series B*, **268**, 2473–8.

Brichetti, P., Rubolini, D., Galeotti, P. and Fasola, M. (2008). Recent declines in urban Italian Sparrow *Passer* (*domesticus*) *italiae* populations in northern Italy. *Ibis*, **150**, 177–81.

Brown, J. H. (1981). Two decades of homage to Santa Rosalia: toward a general theory of diversity. *American Zoologist*, **21**, 877–88.

Bush, J., Miles, B. and Bainbridge, B. (2003). Merri Creek: managing an urban waterway for people and nature. *Ecological Management and Restoration*, **4**, 170–9.

Cam, E., Nichols, J. D., Sauer, J. R., Hines, J. E. and Flather, C. H. (2000). Relative species richness and community completeness: birds and urbanization in the Mid-Atlantic states. *Ecological Applications*, **10**, 1196–210.

Caula, S., Marty, P. and Martin, J. (2008). Seasonal variation in species composition of an urban bird community in Mediterranean France. *Landscape and Urban Planning*, **87**, 1–9.

Chace, J. F. and Walsh, J. J. (2006). Urban effects on native avifauna: a review. *Landscape and Urban Planning*, **74**, 46–69.

Chamberlain, D. E., Cannon, A. R. and Toms, M. P. (2004). Associations of garden birds with gradients in garden habitat and local habitat. *Ecography*, **27**, 589–600.

Chocholouskova, Z. and Pyšek, P. (2003). Changes in composition and structure of urban flora over 120 years: a case study of the city of Plzen. *Flora*, **198**, 366–76.

Clergeau, P., Croci, S., Jokimaki, J., Kaisanlahti-Jokimaki, M. L. and Dinetti, M. (2006). Avifauna homogenisation by urbanization: analysis at different European latitudes. *Biological Conservation*, **127**, 336–44.

Clergeau, P., Jokimäki, J. and Savard, J.-P. L. (2001). Are urban bird communities influenced by the bird diversity of adjacent landscapes? *Journal of Applied Ecology*, **38**, 1122–34.

Clergeau, P., Savard, J.-P. L., Mennechez, G. and Falardeau, G. (1998). Bird abundance and diversity along an urban–rural gradient: a comparative study between two cities on different continents. *The Condor*, **100**, 413–25.

Connell, J. H. (1978). Diversity in tropical rainforests and coral reefs. *Science*, **199**, 1302–10.

Corlett, R. T. (1988). Bukit Timah: the history and significance of a small rain forest reserve. *Environmental Conservation*, **15**, 37–44.

Croci, S., Butet, A. and Clergeau, P. (2008). Does urbanization filter birds on the basis of their biological traits? *The Condor*, **110**, 223–40.

Daniels, G. D. and Kirkpatrick, J. B. (2006). Does variation in garden characteristics influence the conservation of birds in suburbia? *Biological Conservation*, **133**, 326–35.

DeCandido, R., Muir, A. A. and Gargiullo, M. B. (2004). A first approximation of the historical and extant vascular flora of New York City: implications for native plant species conservation. *Journal of the Torrey Botanical Society*, **131**, 243–51.

De Cornuiler, T. and Clergeau, P. (2001). Bat diversity in French urban areas. *Mammalia*, **65**, 540–3.

Devictor, V., Romain, J., Couvet, D., Lee, A. and Jiguet, F. (2007). Functional homogenization effect of urbanization on bird communities. *Conservation Biology*, **21**, 741–51.

Diamond, J. M., Bishop, K. D. and van Balen, S. (1987). Bird survival in an isolated Javan woodland: island or mirror? *Conservation Biology*, **1**, 132–42.

Díaz, S., Fargione, J., Chapin, F. S. and Tilman, D. (2006). Biodiversity loss threatens human well-being. *PLoS Biology*, **4**(8), e277. Available at DOI:10.1371/journal.pbio.0040277

DiBari, J. N. (2007). Evaluation of five landscape-level metrics for measuring effects of urbanization on landscape structure: the case of Tucson, Arizona, USA. *Landscape and Urban Planning*, **79**, 308–13.

Drake, M. T. and Pereira, D. L. (2002). Development of a fish-based index of biotic integrity for small inland lakes in central Minnesota. *North American Journal of Fisheries Management*, **22**, 1105–23.

Duchamp, J.E., Sparks, D.W. and Whitaker, J.O.
(2004). Foraging-habitat selection by bats at
an urban–rural interface: comparison
between a successful and a less successful
species. *Canadian Journal of Zoology-Revue
Canadienne De Zoologie*, **82**, 1157–64.

Duncan, R.P. and Young, J.R. (2000).
Determinants of plant extinction and rarity
145 years after European settlement of
Auckland, New Zealand. *Ecology*, **81**,
3048–61.

Dunn, R.R., Gavin, M.C., Sanchez, M.C. and
Solomon, J.N. (2006). The pigeon paradox:
dependence of global conservation on
urban nature. *Conservation Biology*, **20**,
1814–16.

Evans, K.L. and Gaston, K.J. (2005). People,
energy and avian species richness. *Global
Ecology and Biogeography*, **14**, 187–96.

Evans, K.L., Greenwood, J.J.D. and Gaston, K.J.
(2007). The positive correlation between
avian species richness and human
population density in Britain is not
attributable to sampling bias. *Global Ecology
and Biogeography*, **16**, 300–4.

Fairbanks, D.H.K. (2004). Regional land-use
impacts affecting avian richness patterns in
southern Africa – insights from historical
avian atlas data. *Agriculture Ecosystems and
Environment*, **101**, 269–88.

Fernández-Juricic, E. (2000). Avifaunal use of
wooded streets in an urban landscape.
Conservation Biology, **14**, 513–21.

Field, R., Hawkins, B.A., Cornell, H.V. et al.
(2009). Spatial species-richness gradients
across scales: a meta-analysis. *Journal of
Biogeography*, **36**, 132–47.

Fjeldså, J. and Burgess, N.D. (2008). The
coincidence of biodiversity patterns and
human settlement in Africa. *African Journal
of Ecology*, **46**, 33–42.

Fjeldså, J. and Rahbek, C. (1998). Continent-wide
conservation priorities and diversification
processes. In G.M. Mace, A. Balmford and
J.R. Ginsberg, eds., *Conservation in a Changing
World*. Cambridge: Cambridge University
Press, pp. 139–60.

French, K., Major, R. and Hely, K. (2005). Use of
native and exotic garden plants by
suburban nectarivorous birds. *Biological
Conservation*, **121**, 545–59.

Friesen, P.F., Eagles, J. and MacKay, R.J. (1995).
Effects of residential development on forest
dwelling Neotropical migrant songbirds.
Conservation Biology, **9**, 1408–14.

Fuller, R.A., Irvine, K.N., Devine-Wright, P.,
Warren, P.H. and Gaston, K.J. (2007).
Psychological benefits of greenspace
increase with biodiversity. *Biology Letters*, **3**,
390–4.

Fuller, R.A., Warren, P.H., Armsworth, P.R.,
Barbosa, O. and Gaston, K.J. (2008). Garden
bird feeding predicts the structure of urban
avian assemblages. *Diversity and Distributions*,
14, 131–7.

Gagné, S.A. and Fahrig, L. (2007). Effect of
landscape context on anuran communities
in breeding ponds in the National Capital
Region, Canada. *Landscape Ecology*, **22**,
205–15.

Garden, J., McAlpine, C., Peterson, A., Jones, D.
and Possingham, H. (2006). Review of the
ecology of Australian urban fauna: a focus
on spatially explicit processes. *Austral
Ecology*, **31**, 126–48.

Garden, J., McAlpine, C., Possingham, H.P. and
Jones, D.N. (2007). Habitat structure is more
important than vegetation composition for
local-level management of native terrestrial
reptile and small mammal species living in
urban remnants: a case study from
Brisbane, Australia. *Austral Ecology*, **32**,
669–85.

Gaston, K.J. (2000). Global patterns in
biodiversity. *Nature*, **405**, 220–7.

Gaston, K.J. (2005). Biodiversity and extinction:
species and people. *Progress in Physical
Geography*, **29**, 239–47.

Gaston, K.J. and Evans, K.L. (2004). Birds
and people in Europe. *Proceedings of the
Royal Society of London Series B*,
271, 1649–55.

Gaston, K.J., Fuller, R.A., Loram, A. et al. (2007).
Urban domestic gardens (XI): variation in

urban wildlife gardening in the UK. *Biodiversity and Conservation*, **16**, 3227–38.

Gaublomme, E., Hendrickx, F., Dhuyvetter, H. and Desender, K. (2008). The effects of forest patch size and matrix type on changes in carabid beetle assemblages in an urbanized landscape. *Biological Conservation*, **141**, 2585–96.

Gehrt, S. D. and Chelsvig, J. E. (2003). Bat activity in an urban landscape: patterns at the landscape and microhabitat scale. *Ecological Applications*, **13**, 939–50.

Germaine, S. S. and Wakeling, B. F. (2001). Lizard species distributions and habitat occupation along an urban gradient in Tucson, Arizona, USA. *Biological Conservation*, **97**, 229–37.

Gonzalez-Abraham, C. E., Radeloff, V. C., Hawbaker, T. J. *et al.* (2007). Patterns of houses and habitat loss from 1937 to 1999 in northern Wisconsin, USA. *Ecological Applications*, **17**, 2011–23.

Green, D. A. and Baker, M. G. (2003). Urbanization impacts on habitat and bird communities in a Sonoran Desert ecosystem. *Landscape and Urban Planning*, **63**, 225–39.

Green, R. J. (1984). Native and exotic birds in a suburban habitat. *Australian Wildlife Research*, **11**, 181–90.

Grove, J. M. and Burch, W. R. (1997). A social ecology approach and applications of urban ecosystem and landscape analyses: a case study of Baltimore, Maryland. *Urban Ecosystems*, **1**, 259–79.

Grove, J. M., Troy, A. R., O'Neil-Dunne, J. P. M. *et al.* (2006). Characterization of households and its implications for the vegetation of urban ecosystems. *Ecosystems*, **9**, 578–97.

Hahs, A. K. and McDonnell, M. J. (2006). Selecting independent measures to quantify Melbourne's urban–rural gradient. *Landscape and Urban Planning*, **78**, 435–48.

Hamer, A. J. and McDonnell, M. J. (2008). Amphibian ecology and conservation in the urbanising world: a review. *Biological Conservation*, **141**, 2432–49.

Hammer, R. B., Stewart, S. I., Winkler, R. L., Radeloff, V. C. and Voss, P. R. (2004). Characterizing dynamic spatial and temporal residential density patterns from 1940–1990 across the north central United States. *Landscape and Urban Planning*, **69**, 183–99.

Harcourt, A. H., Parks, S. A. and Woodroffe, R. (2001). Human density as an influence on species/area relationships: double jeopardy for small African reserves? *Biodiversity and Conservation*, **10**, 1011–26.

Harlan, S. L., Brazel, A. J., Prashad, L., Stefanov, W. L. and Larsen, L. (2006). Neighbourhood microclimates and vulnerability to heat stress. *Social Science and Medicine*, **63**, 2847–63.

Hennings, L. A. and Edge, W. D. (2003). Riparian bird community structure in Portland, Oregon: habitat, urbanization, and spatial scale patterns. *The Condor*, **105**, 288–302.

Hillebrand, H. (2004). On the generality of the latitudinal diversity gradient. *American Naturalist*, **163**, 192–211.

Hodgson, P., French, K. and Major, R. E. (2006). Comparison of foraging behaviour of small, urban-sensitive insectivores in continuous woodland and woodland remnants in a suburban landscape. *Wildlife Research*, **33**, 591–603.

Hope, D., Gries, C., Zhu, W. *et al.* (2003). Socioeconomics drives urban plant diversity. *Proceedings of the National Academy of Sciences of the USA*, **100**, 8788–92.

Hughes, J. B., Daily, G. C. and Ehrlich, P. R. (1997). Population diversity: its extent and extinction. *Science*, **278**, 689–92.

Hugo, S. and van Rensburg, B. J. (2008). The maintenance of a positive spatial correlation between South African bird species richness and human population density. *Global Ecology and Biogeography*, **17**, 611–21.

Iverson, L. R. and Cook, E. A. (2000). Urban forest cover of the Chicago region and its relation to household density and income. *Urban Ecosystems*, **4**, 105–24.

Jenerette, G. D., Harlan, S. L., Brazel, A. *et al.* (2007). Regional relationships between surface temperature, vegetation, and human settlement in a rapidly urbanizing ecosystem. *Landscape Ecology*, **22**, 353–65.

Jokimäki, J. and Suhonen, J. (1998). Distribution and habitat selection of wintering birds in urban environments. *Landscape and Urban Planning*, **39**, 253–63.

Jones, D. N. (1981). Temporal changes in the suburban avifauna of an inland city. *Australian Wildlife Research*, **8**, 109–19.

Jones, D. N. and Wieneke, J. (2000). The suburban bird community of Townsville revisited: changes over 16 years. *Corella*, **24**, 53–60.

Kark, S., Iwaniuk, A., Schalimtzek, A. and Banker, E. (2007). Living in the city: can anyone become an 'urban exploiter'? *Journal of Biogeography*, **34**, 638–51.

Kerr, J. T. and Packer, L. (1997). Habitat heterogeneity as a determinant of mammal species richness in high-energy regions. *Nature*, **385**, 252–4.

Kent, M., Stevens, R. A. and Zhang, L. (1999). Urban plant ecology patterns and processes: a case study of the flora of the city of Plymouth, Devon, U.K. *Journal of Biogeography*, **26**, 1281–98.

Knapp, S., Kühn, I., Schweiger, O. and Klotz, S. (2008). Challenging urban species diversity: contrasting phylogenetic patterns across plant functional groups in Germany. *Ecology Letters*, **11**, 1054–64.

Koh, L. P. and Sodhi, N. S. (2004). Importance of reserves, fragments, and parks for butterfly conservation in a tropical urban landscape. *Ecological Applications*, **14**, 1695–708.

Kühn, I., Brandl, R. and Klotz, S. (2004). The flora of German cities is naturally species rich. *Evolutionary Ecology Research*, **6**, 749–64.

Kurta, A. and Teramino, J. A. (1992). Bat community structure in an urban park. *Ecography*, **15**, 257–61.

Lane, D. J. W., Kingston, T. and Lee, B. P. H. (2006). Dramatic decline in bat species richness in Singapore, with implications for Southeast Asia. *Biological Conservation*, **131**, 584–93.

Lawson, D. M., Lamar, C. K. and Schwartz, M. W. (2008). Quantifying plant population persistence in human-dominated landscapes. *Conservation Biology*, **4**, 922–8.

Lenth, B. A., Knight, R. L. and Gilgert, W. C. (2006). Conservation value of clustered housing developments. *Conservation Biology*, **20**, 1445–56.

Lepczyk, C. A., Flather, C. H., Radeloff, V. C. *et al.* (2008). Human impacts on regional avian diversity and abundance. *Conservation Biology*, **22**, 405–16.

Lesinski, G., Fuszara, E. and Kowalski, M. (2000). Foraging areas and relative density of bats (Chiroptera) in differently human transformed landscapes. *Zeitschrift für Saugetierkunde – International Journal of Mammalian Biology*, **65**, 129–37.

Lim, H. C. and Sodhi, N. S. (2004). Responses of avian guilds to urbanisation in a tropical city. *Landscape and Urban Planning*, **66**, 199–215.

Lindenmayer, D. B. and Franklin, J. F. (2002). *Conserving Forest Biodiversity: A Comprehensive Multiscaled Approach*. Washington, DC: Island Press.

Lindsay, A. R., Gillum, S. S. and Meyer, M. W. (2002). Influence of lakeshore development on breeding bird communities in a mixed northern forest. *Biological Conservation*, **107**, 1–11.

Luck, G. W. (2007a). A review of the relationships between human population density and biodiversity. *Biological Reviews*, **82**, 607–45.

Luck, G. W. (2007b). The relationships between net primary productivity, human population density and species conservation. *Journal of Biogeography*, **34**, 201–12.

Luck, M. and Wu, J. G. (2002). A gradient analysis of urban landscape pattern: a case study from the Phoenix metropolitan region, Arizona, USA. *Landscape Ecology*, **17**, 327–39.

Luck, G. W., Daily, G. C. and Ehrlich, P. R. (2003). Population diversity and ecosystem services. *Trends in Ecology and Evolution*, **18**, 331–6.

Luck, G. W., Ricketts, T. H., Daily, G. C. and Imhoff, M. (2004). Alleviating spatial conflict between people and biodiversity. *Proceedings of the National Academy of Sciences of the USA*, **101**, 182–6.

Luck, G. W., Smallbone, L., McDonald, S. and Duffy, D. (in press). What drives the positive correlation between human population density and bird species richness in Australia? *Global Ecology and Biogeography*, in press.

Luck, G. W., Smallbone, L. T. and O'Brien, R. (2009). Socio-economics and vegetation change in urban ecosystems: patterns in space and time. *Ecosystems*. Available at DOI:10.1007/s10021-009-9244-6.

MacArthur, R. H. (1965). Patterns of species diversity. *Biological Review*, **40**, 510–33.

MacGregor-Fors, I. (2008). Relation between habitat attributes and bird richness in a western Mexico suburb. *Landscape and Urban Planning*, **84**, 92–8.

McCain, C. M. (2004). The mid-domain effect applied to elevational gradients: species richness of small mammals in Costa Rica. *Journal of Biogeography*, **31**, 19–31.

McIntyre, N. E. (2000). Ecology of urban arthropods: a review and a call to action. *Annals of the Entomological Society of America*, **93**, 825–35.

McKinney, M. L. (2001). Effects of human population, area, and time on non-native plant and fish diversity in the United States. *Biological Conservation*, **100**, 243–52.

McKinney, M. L. (2002). Do human activities raise species richness? Contrasting patterns in United States plants and fishes. *Global Ecology and Biogeography*, **11**, 343–8.

McKinney, M. L. (2006). Urbanization as a major cause of biotic homogenization. *Biological Conservation*, **127**, 247–60.

McKinney, M. L. (2008). Effects of urbanization on species richness: a review of plants and animals. *Urban Ecosystems*, **11**, 161–76.

McKinney, M. L. and Lockwood, J. L. (1999). Biotic homogenization: a few winners replacing many losers in the next mass extinction. *Trends in Ecology and Evolution*, **14**, 450–3.

Melles, S., Glenn, S. and Martin, K. (2003). Urban bird diversity and landscape complexity: species–environment associations along a multiscale habitat gradient. *Conservation Ecology*, **7**, 5. Available at http://www.consecol.org/vol7/iss1/art5/

Miller, J. R. (2006). Restoration, reconciliation, and reconnecting with nature. *Biological Conservation*, **127**, 356–61.

Miller, J. R., Wiens, J. A., Hobbs, N. T. and Theobald, D. M. (2003). Effects of human settlement on bird communities in lowland riparian areas of Colorado (USA). *Ecological Applications*, **13**, 1041–59.

Mills, G. S., Dunning, J. B. Jr and Bates, J. M. (1989). Effects of urbanization on breeding bird community structure in southwestern desert habitats. *The Condor*, **91**, 416–28.

Morawitz, D. F., Blewett, T. M., Cohen, A. and Alberti, M. (2006). Using NDVI to assess vegetative land cover change in central Puget Sound. *Environmental Monitoring and Assessment*, **114**, 85–106.

Mörtberg, U. (1998). Bird species diversity in urban forest remnants: landscape pattern and habitat quality. In J. W. Dover and R. G. H. Bunce, eds., *Key Concepts in Landscape Ecology*. Preston, UK: IALE, pp. 239–44.

Munyenyembe, F., Harris, J. and Hone, J. (1989). Determinants of bird populations in an urban area. *Australian Journal of Ecology*, **14**, 549–57.

Niemelä, J., Kotze, J., Ashworth, A. *et al.* (2000). The search for common anthropogenic impacts on biodiversity: a global network. *Journal of Insect Conservation*, **4**, 3–9.

Nowak, D. J. (1994). Atmospheric carbon dioxide reduction by Chicago's urban forest. In E. C. McPherson, ed., *Chicago's Urban Forest Ecosystem: Results of the Chicago Urban Climate Project*. Radnor, PA: Northeastern Forest Experiment Station, pp. 83–94.

O'Dea, N., Araújo, M. B. and Whittaker, R. J. (2006). How well do important bird areas represent species and minimize conservation conflict in the tropical Andes? *Diversity and Distributions*, **12**, 205–14.

Olden, J.D. and Poff, N.L. (2003). Toward a
mechanistic understanding and prediction
of biotic homogenization. *American
Naturalist*, **162**, 442–60.

Olden, J.D., Poff, N.L., Douglas, M.R.,
Douglas, M.E. and Fausch, K.D. (2004).
Ecological and evolutionary consequences
of biotic homogenization. *Trends in Ecology
and Evolution*, **19**, 18–24.

Olden, J.D., Poff, N.L. and McKinney, M.L. (2006).
Forecasting faunal and floral
homogenization associated with human
population geography in North America.
Biological Conservation, **127**, 261–71.

Palmer, G.C., Fitzsimons, J.A., Antos, M.J. and
White, J.G. (2008). Determinants of native
avian richness in suburban remnant
vegetation: implications for conservation
planning. *Biological Conservation*, **141**,
2329–41.

Palomino, D. and Carrascal, L.M. (2007).
Threshold distances to nearby cities and
roads influence the bird community of a
mosaic landscape. *Biological Conservation*,
140, 100–9.

Parks, S.A. and Harcourt, A.H. (2002). Reserve
size, local human density, and mammalian
extinctions in U.S. protected areas.
Conservation Biology, **16**, 800–8.

Parody, J.M., Cuthbert, F.J. and Decker, E.H.
(2001). The effect of 50 years of landscape
change on species richness and community
composition. *Global Ecology and Biogeography*,
10, 305–13.

Parris, K.M. (2006). Urban amphibian
assemblages as metacommunities. *Journal of
Animal Ecology*, **75**, 757–64.

Parsons, H., Major, R.E. and French, K. (2006).
Species interactions and habitat
associations of birds inhabiting urban areas
of Sydney, Australia. *Austral Ecology*, **31**,
217–27.

Pauleit, S., Ennos, R. and Golding, Y. (2005).
Modeling the environmental impacts of
urban land use and land cover change: a
study in Merseyside, UK. *Landscape and Urban
Planning*, **71**, 295–310.

Pautasso, M. (2007). Scale dependence of the
correlation between human population
presence and vertebrate and plant species
richness. *Ecology Letters*, **10**, 16–24.

Pautasso, M. and McKinney, M.L. (2007). The
botanist effect revisited: plant species
richness, county area, and human
population size in the United States.
Conservation Biology, **21**, 1333–40.

Pennington, D.N., Hansel, J. and Blair, R.B.
(2008). The conservation value of urban
riparian areas for landbirds during spring
migration: land cover, scale, and vegetation
effects. *Biological Conservation*, **141**, 1235–48.

Pidgeon, A.M., Radeloff, V.C., Flather, C.H. *et al.*
(2007). Associations of forest bird species
richness with housing and landscape
patterns across the USA. *Ecological
Applications*, **17**, 1989–2010.

Pickett, S.T.A. and Cadenasso, M.L. (2006).
Advancing urban ecological studies:
frameworks, concepts, and results from the
Baltimore Ecosystem Study. *Austral Ecology*,
31, 114–25.

Pillsbury, F.C. and Miller, J.R. (2008). Habitat
and landscape characteristics underlying
anuran community structure along an
urban–rural gradient. *Ecological Applications*,
18, 1107–18.

Preston, C.D. (2000). Engulfed by suburbia or
destroyed by the plough: the ecology of
extinction in Middlesex and
Cambridgeshire. *Watsonia*, **23**, 59–81.

Puth, L.M. and Burns C.E. (2009). New York's
nature: a review of the status and trends in
species richness across the metropolitan
region. *Diversity and Distributions*, **15**, 12–21.

Pyšek, P. (1998). Alien and native species in
central European urban floras: a
quantitative comparison. *Journal of
Biogeography*, **25**, 155–63.

Pyšek, P., Jarošík, V., Chytrý, M. *et al.* (2005). Alien
plants in temperate weed communities:
prehistoric and recent invaders occupy
different habitats. *Ecology*, **86**, 772–85.

Rahbek, C. and Graves, G.R. (2001). Multiscale
assessment of patterns of avian species

richness. *Proceedings of the National Academy of Sciences of the USA*, **98**, 4534–9.

Rahel, F. J. (2000). Homogenization of fish faunas across the United States. *Science*, **288**, 854–6.

Randolph, J. (2004). *Environmental Land Use Planning*. Washington, DC: Island Press.

Recher, H. F. and Serventy, D. L. (1991). Long term changes in the relative abundances of birds in Kings Park, Perth, Western Australia. *Conservation Biology*, **5**, 90–102.

Rickman, J. K. and Connor, E. F. (2003). The effect of urbanization on the quality of remnant habitats for leaf-mining Lepidoptera on *Quercus agrifolia*. *Ecography*, **26**, 777–87.

Robinson, G. R., Yurlina, M. E. and Handel, S. N. (1994). A century of change in the Staten-island flora – ecological correlates of species losses and invasions. *Bulletin of the Torrey Botanical Club*, **121**, 119–29.

Roy, D. B., Hill, M. O. and Rothery, P. (1999). Effects of urban land cover on the local species pool in Britain. *Ecography*, **22**, 507–15.

Rubbo, M. J. and Kiesecker, J. M. (2005). Amphibian breeding distribution in an urbanized landscape. *Conservation Biology*, **19**, 504–11.

Sanders, N. J. (2002). Elevational gradients in ant species richness: area, geometry, and Rapoport's rule. *Ecography*, **25**, 25–32.

Sandström, U. G., Angelstam, P. and Mikusiński, G. (2006). Ecological diversity of birds in relation to the structure of urban green space. *Landscape and Urban Planning*, **77**, 39–53.

Savard, J.-P. L., Clergeau, P. and Mennechez, G. (2000). Biodiversity concepts and urban ecosystems. *Landscape and Urban Planning*, **48**, 131–42.

Schwartz, M. W., Thorne, J. H. and Viers, J. H. (2006). Biotic homogenization of the California flora in urban and urbanizing regions. *Biological Conservation*, **127**, 282–91.

Sewell, S. R. and Catterall, C. P. (1998). Bushland modification and styles of urban development: their effects on birds in south-east Queensland. *Wildlife Research*, **25**, 41–63.

Smith, C. M. and Wachob, D. G. (2006). Trends associated with residential development in riparian breeding bird habitat along the Snake River in Jackson Hole, WY, USA: implications for conservation planning. *Biological Conservation*, **128**, 431–46.

Smith, R. M., Thompson, K., Hodgson, J. G., Warren, P. H. and Gaston, K. J. (2006). Urban domestic gardens (IX): composition and richness of the vascular plant flora, and implications for native biodiversity. *Biological Conservation*, **129**, 312–22.

Tait, C. J., Daniels, C. B. and Hill, R. S. (2005). Changes in species assemblages within the Adelaide metropolitan area, Australia, 1836–2002. *Ecological Applications*, **15**, 346–59.

Thompson, K. and Jones, A. (1999). Human population density and prediction of local plant extinction in Britain. *Conservation Biology*, **13**, 185–9.

Thompson, K. and McCarthy, M. A. (2008). Traits of British alien and native urban plants. *Journal of Ecology*, **96**, 853–9.

Thompson, K., Austin, K. C., Smith, R. M. et al. (2003). Urban domestic gardens (I): putting small-scale plant diversity in context. *Journal of Vegetation Science*, **14**, 71–8.

Thompson, P. S., Greenwood, J. J. D. and Greenaway, K. (1993). Birds in European gardens in the winter and spring of 1988–89. *Bird Study*, **40**, 120–34.

Tian, G., Liu, J., Xie, Y. et al. (2005). Analysis of spatio-temporal dynamic pattern and driving forces of urban land in China in 1990s using TM images and GIS. *Cities*, **22**, 400–10.

Tratalos, J., Fuller, R. A., Evans, K. L. et al. (2007). Bird densities are associated with household densities. *Global Change Biology*, **13**, 1685–95.

Turner, W. R., Nakamura, T. and Dinetti, M. (2004). Global urbanization and the separation of humans from nature. *BioScience*, **54**, 585–90.

Tzoulas, K., Korpela, K., Venn, S. et al. (2007). Promoting ecosystem and human health in urban areas using green infrastructure: a literature review. *Landscape and Urban Planning*, **81**, 167–78.

United Nations (Department of Economic and Social Affairs) (2008). *World Urbanization Prospects: The 2007 Revision*. New York: United Nations.

Urquiza-Haas, T., Peres, C. A. and Dolman, P. M. (2009). Regional scale effects of human density and forest disturbance on large-bodied vertebrates throughout the Yucatán Peninsula, Mexico. *Biological Conservation*, **142**, 134–48.

van der Ree, R. and McCarthy, M. A. (2005). Inferring persistence of indigenous mammals in response to urbanisation. *Animal Conservation*, **8**, 309–19.

Walsh, C. J., Sharpe, A. K., Breen, P. F. and Sonneman, J. A. (2001). Effects of urbanization on streams of the Melbourne region, Victoria, Australia. I. Benthic macroinvertebrate communities. *Freshwater Biology*, **46**, 535–51.

Weng, Y.-C. (2007). Spatiotemporal changes of landscape pattern in response to urbanization. *Landscape and Urban Planning*, **81**, 341–53.

White, J. G., Antos, M. J., Fitzsimons, J. A. and Palmer, G. C. (2005). Non-uniform bird assemblages in urban environments: the influence of streetscape vegetation. *Landscape and Urban Planning*, **71**, 123–35.

Willig, M. R., Kaufman, D. M. and Stevens, R. D. (2003). Latitudinal gradients of biodiversity: pattern, process, scale, and synthesis. *Annual Review of Ecology, Evolution and Systematics*, **34**, 273–309.

Young, K. M., Daniels, C. B. and Johnston, G. (2007). Species of street tree is important for southern hemisphere bird trophic guilds. *Austral Ecology*, **32**, 541–50.

CHAPTER SIX

Urbanisation and alien invasion

STEFAN KLOTZ AND INGOLF KÜHN

Introduction

Urbanisation is a rapidly developing process of global change. The world's human population in 1900 was around 1.6 billion, of which 13% lived in cities (UN-HABITAT 2003, 2006). Within 100 years the world population increased to 6.5 billion and the urban population increased to 50% (UN-HABITAT 2006). This increase will continue rapidly. In Europe and other highly industrialised regions the percentage of urban population is already much higher (>70%). This urban population growth is strongly connected with an increase in the size and intensity of urban land-use. As a consequence of land-use change, more and more plant and animal habitats have been lost. Urbanisation is a major driver of plant and animal extinction on a regional scale (Fuller & Gaston 2009). The species most affected have been those of wet and very pristine habitats, as well as those of extensive agricultural lands like meadow and pasture, all of which have become rarer in heavily used landscapes such as those of Germany (Klotz 1989).

The concentration of the human population has resulted in an increase in traffic and transport of food, raw materials etc., and growing needs for open space recreation activities such as gardening and walking. Additionally, landscaping and gardening, as well as vegetable and ornamental plant production, are concentrated in urbanised regions. Therefore we have two of the general prerequisites for species introductions into cities. First, there is the concentration of intentional introductions of plant and animal species. Second, resulting from the increasing traffic and transportation of different materials (Hulme et al. 2008), urban areas are hotspots of unintentional introductions. Cities are hence not only rich in native plant species (Kühn et al. 2004a) but even more so in alien plant species (Pyšek 1998; Kühn et al. 2004a; for a general overview of species richness in urban areas see Chapter 5). This can easily be seen, for example, in Germany (Figure 6.1). Hotspots of neophyte

Urban Ecology, ed. Kevin J. Gaston. Published by Cambridge University Press.

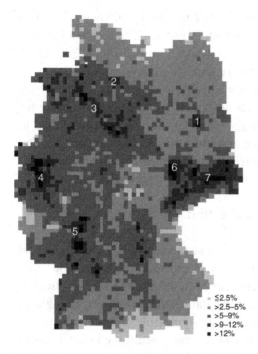

Figure 6.1 Hotspots of neophyte species (alien plant species introduced after the discovery of the Americas) in Germany (percentage of total plant species richness). 1: Berlin; 2: Hamburg; 3: Bremen and the Weser valley; 4: Ruhr area; 5: Rhine–Main (greater Frankfurt) area; 6: Leipzig–Halle; 7: Saxonian industrial region.

≤2.5%
>2.5–5%
>5–9%
>9–12%
>12%

distribution (i.e. alien plant species introduced after the discovery of the Americas; Pyšek *et al.* 2004) are the major urban areas of Berlin, Hamburg, the Ruhr and Rhine–Main areas, Leipzig–Halle and the Saxonian industrial region. Another striking feature is the richness in neophytes of the Weser river valley. This river is characterised by high salinity due to potash mining in the upper reaches.

The urban environment has a small grain size and is very patchy, being made up of very different land-use structures and habitats such as residential areas with gardens, lawns and shrubby vegetation, as well as parks and cemeteries and areas of industrial production and traffic (roads, railways, ports, airports etc.). Former mining lands as well as solid waste dumps and landfills may also be parts of urban conglomerations. Additionally, remnants of pristine habitats as well as forests, woodlands and agricultural habitats remain in urban areas (see Gilbert 1989). However, owing to the small spatial sizes of the different habitats, all biological communities are influenced by urbanisation in general. Totally new urban habitats as well as the degraded pristine remnants are target habitats for new alien species arriving in cities (Wania *et al.* 2006).

Thus far, most studies of alien species in urban areas have been carried out in Europe. Far fewer are from other parts of the world, and these have often mainly been descriptive or have covered single species. Most studies have analysed plant species as focal organisms of urbanisation. General studies using animal groups are much rarer and often restricted to selected habitats within the city. Birds are the best-studied group of animals (e.g. Fuller *et al.* 2009; Loss *et al.* 2009). Knapp *et al.* (2008a) compared different species groups in selected protected habitats within a city and in the surrounding agricultural landscape of Halle, Germany. They found that the total species number for carabid beetles and butterflies was higher in the rural protected areas than in the urban protected areas. Large-scale and multi-species comparisons of alien animals between habitats of different degrees of urbanisation are, however, mostly lacking. In this chapter we will focus on plant species in particular.

Ecological characteristics of urbanisation and their importance for invasion

Urbanised areas are different from other landscapes in many environmental characteristics (Chapter 3; Sukopp 1998). First, micro-climates and even regional climates are clearly distinct from those of surrounding landscapes. Urban climate studies in cities of the temperate zone have shown that global radiation is decreased by air pollution, and sunshine duration decreased because of increased development of clouds. The greenhouse effect leads to an increase in temperatures (Landsberg 1981; Oke 1982). Frost events are rarer than or not as strong as in the surrounding landscapes. At higher latitudes, cities can thus be seen as outposts of warmer and milder climates. Wind speed and storm events are diminished in cities with consequences for biotic processes such as species dispersal and pollination. The high coverage of non-vegetated surfaces (including areas paved by asphalt and concrete) increases the water runoff after rainfall and decreases evapotranspiration. The main consequence is a reduction in air humidity. Precipitation is increased in cities owing to the heat island effect and the higher cloud density. Snow fall is reduced by the generally higher temperatures.

Besides climatic changes, soils are heavily affected by urban land-use, including by construction activities, pollution and sedimentation of dusts, as well as by frequent disturbances. Typical processes in urban soils are the destruction of the native soil layer structure by soil intermixing, covering of surface soil layers, and the addition of natural (sand, gravel etc.) and/or artificial substrates (asphalt, demolished bricks etc.). The covering by impermeable surfaces results in fossilisation of soils over a longer period. Gardening activities lead to nutrient-rich hortisols. Pollution increases the nutrient content and the pH level of soils in general. The use of de-icing salt in winter leads to higher salt concentrations locally along roads and on dumps.

The hydrology of urban ecosystems mainly depends on the canalisation of the original waters as well as on the water treatment systems. Generally, wet sites are ameliorated and wet habitats become rarer. Conversely, modern gardens often include artificial waters such as small ornamental ponds with swamp and mire plants. Vegetation strips along waterbodies are often heavily disturbed and open enough for the immigration of new species.

Typical urban species: natives and aliens

Despite the heterogeneity of urban areas, specific urban species can be found. Wittig *et al.* (1985), using grid mapping techniques, identified species which are more frequent or less frequent in urban areas. They thus distinguished urbanophilic, urbanoneutral and urbanphobic plants. Wittig (2002) presented a list of most of the urbanophilic species in Germany. Most of the species on that list are alien. That means that in the native species pool, plants which are well adapted to urban conditions are rare. Urbanophilic aliens are concentrated in the inner part of the cities where the climatic differences and the general environmental changes from non-urban regions are greatest. Using gradient analyses, Kunick (1974) found a clear zonation in the relative numbers of alien species, decreasing from the centre of urban agglomerations to the margin of Berlin, Germany. This pattern highlights that inner urban habitats are targets for alien species (mostly thermophilous species and those adapted to a high frequency of disturbances). These thermophilous alien species are mostly restricted to urban areas, and it is reported that they invade the countryside only at lower latitudes. One such example is the Tree of Heaven *Ailanthus altissima* (Mill.) Swingle in several European cities (e.g. London, Paris, Berlin, Halle, Leipzig). In central and northern German cities, the species is restricted to the urban heat islands. In southern Germany and in the neighbouring southern countries, the tree can invade the open countryside (Gutte *et al.* 1987).

An analysis of the levels of invasion of urban and non-urban habitats across Europe (Pyšek *et al.* 2009) shows that intentionally and unintentionally introduced species occur in highest numbers in urbanised habitats (Table 6.1). The process of acclimatisation and further spread starts here. Urban areas can thus be the sources of further immigration of plants and animals.

Former intentionally introduced and widely used plants colonise new habitats and new landscapes. Common chicory *Cichorium intybus*, formerly widely sown in Central Europe, was used as a surrogate for coffee, but lost importance because of the increase in coffee production and the decrease in coffee prices. Now, this species is only rarely planted but is present to some degree in many city meadows and lawns. A similar example in subtropical and tropical cities worldwide is tree tobacco *Nicotiana glauca*, currently one of the most important invaders. It was formerly used as a surrogate for tobacco. Another example is Canada golden-rod *Solidago canadensis*. This species was mainly introduced to Central Europe as an ornamental plant and widely grown in gardens. It is well adapted to grow on fallow land, urban

Table 6.1 *Levels of invasion of European EUNIS (European Nature Information System) habitats by alien plant species in Europe and those to Europe (after Pyšek et al. 2009).*

Category	No. of aliens in Europe	No. of aliens to Europe
Number of species classified	2122	1059
% classified of the total	56.6	57.7
A. Marine habitats	12	7
D. Mires, bogs & fens	220	118
B. Coastal habitats	343	170
C. Inland surface waters	444	260
F. Heathland & scrub	462	206
H. Inland sparsely vegetated habitats	497	211
G. Woodland & forests	668	310
E. Grasslands	793	276
I. Arable land, gardens & parks	1240	533
J. Industrial habitats	1360	658

Note: Aliens in Europe are species which are native to some regions in Europe and alien to others; aliens to Europe are species with the area of origin outside Europe. The letters A to J are those used in the European Nature Information System (EUNIS) classification.

brownfields and along railways. In late summer and autumn the species flowers intensively. By escaping and flowering in ruderal habitats the species lost its former importance as an ornamental.

In analyses of the flora of the city of Halle in central Germany and the rural surroundings, Wania *et al.* (2006) identified on a small scale (grid cells with 250-m resolution) those species which are typical for the city and for the rural region. Of 86 species identified as being significantly more frequent in cities than the surroundings, 27 (31.4%) were alien and 16 (18.6%) were neophytes. Of seven species characteristic of the more rural surroundings, only one (14.3%) was alien and none were neophytic.

To survive, grow and reproduce in particular urban habitats, plant and animal species need specific adaptations to these environments. Urban plant species in Germany, for example, more often are wind-pollinated, have sclero-morphic leaves and are dispersed by animals, and less often are insect-pollinated, have hygromorphic leaves and are dispersed by wind (Knapp *et al.* 2008c). In the Czech Republic, Lososová *et al.* (2006) found that, compared with weed vegetation of arable land, in the ruderal vegetation of settlements species were often stronger competitors, wind-pollinated, flowered in mid-summer, reproduced both by seeds and vegetatively, dispersed by wind or humans, and had high demands for light and nutrients and more continental distribution ranges. Hence they are often adapted to grow in warm, dry and nutrient-rich habitats. Species which have several reproductive strategies and are adapted to human dispersal therefore have an advantage in urban regions. Interestingly, although in general the phylogenetic diversity of species in urban areas is reduced (Knapp *et al.* 2008b), species sharing traits which are better adapted to urban environments showed a higher phylogenetic diversity than those better adapted to rural environments. Many rare species (alien as well as native), especially those preferring cool or acidic habitats, might already have disappeared from urban areas (Knapp *et al.* 2009).

Evolutionary adaptability might play some role in urban areas, but is not well studied. New 'anthropogenic species', such as hybrids between native and alien species, or between alien species from different biogeographical regions and escaped cultivars of ornamental plants, could play an important role. A first nationwide list of such plant species is available within the BIOLFLOR database of biological and ecological traits of the flora of Germany (Kühn & Klotz 2002; Kühn *et al.* 2004b).

Immigration pathways and urbanisation

There are several different pathways by which alien species arrive in new regions which might eventually become their non-native range. Understanding the processes involved is crucial for successful management but also for improved projections of future invasions. Cities here are just a special, albeit probably a most important, system. As stated previously, new species may be introduced intentionally or unintentionally. This, however, is no clear dichotomy but rather represents the two extremes in a gradual change. Hulme *et al.* (2008) present a simplified framework to categorise six major pathways of initial introduction, namely release, escape, contaminant, stowaway, corridor and unaided. Releases are intentional introductions of a commodity, such as biocontrol agents, game animals or plants for erosion control. Escapes are also intentionally introduced as a commodity but escaped unintentionally, such as feral crops and livestock, pets, garden plants and live baits. Contaminants are unintentionally introduced with a specific (intentionally introduced) commodity, such as parasites, pests and commensals of traded plants and animals. Stowaways are unintentionally introduced while being attached to or within a transport vector. Typical examples are hull fouling, and ballast water/soil/sediment organisms. Corridors are defined as human infrastructures linking previously unconnected regions. This can result in unintentional introductions of, for example, marine organisms moving through the Suez Canal or Ponto-Caspian organisms migrating along rivers and newly established canals to the Baltic Sea. Unaided introduction is of course also unintentional through natural dispersal of species from a region

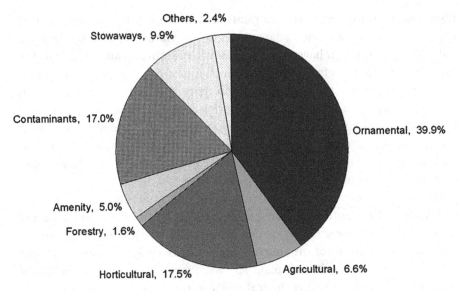

Figure 6.2 Relative contributions of pathways of introduction shown for naturalised aliens to Europe, i.e. plant species with the area of origin outside Europe (modified from Pyšek *et al.* 2009; based on 1983 naturalised aliens; from data in Lambdon *et al.* 2008).

where they are already alien across political borders. Most of the European alien plant species were intentionally introduced into this region and were released deliberately or escaped unintentionally (Figure 6.2; Pyšek *et al.* 2009). Of these, the majority were introduced for ornamental, agricultural, horticultural or forestry reasons or as an amenity. Contaminants also play a considerable role but fewer species arrived unintentionally attached to or within a transport vector.

The relative importance of invasion pathways specifically to cities is not as well analysed. Still, one can assume that the proportion of species introduced for ornamental or horticultural reasons is even higher than for the complete European species set and makes up a high proportion of those species introduced into Europe. Several of the ornamental species are typical for urban parks, gardens or even historic garden cultures (Kowarik 2003). Nonetheless, agricultural seeds contributed considerably to seed rain along road tunnels in Berlin (von der Lippe & Kowarik 2007a, 2007b).

Cities as sources of invasions

There is considerable concern that cities, being so rich in alien species, can act as foci for the spread of alien species into the wider environment (e.g. Sullivan *et al.* 2005; Houlahan *et al.* 2006). Road verges are known to be rich in non-native plant species (Parendes & Jones 2000; Gelbard & Belnap 2003). Alien species such as the South African ragwort *Senecio inaequidens* DC. in Europe

(Ernst 1998; Heger & Böhmer 2005) or common milkweed *Asclepias syriaca* L. and purple loosestrife *Lythrum salicaria* L. in North America (Wilcox 1989; Wyatt 1996) are reported to spread along roadsides. Therefore, several studies have explored to what extent, for example, traffic and traffic infrastructure in cities might be able to facilitate biological invasions from cities into the surroundings. Von der Lippe and Kowarik (2007b) attempted to describe the spatial and quantitative effectiveness of traffic as a dispersal vector by addressing the role of vehicles in long-distance dispersal and by quantifying seed deposition at roadsides in Berlin. They analysed seed rain on the roadsides of five different lanes within three tunnels along a single urban motorway. They found that half of the plant species were alien, with 33.8% of all species in their seed traps being neophytes. Non-native species were therefore slightly overrepresented, compared with the local species pool. Approximately one-third of the species found in the tunnel traps were not present in any of the tunnel surroundings, indicating long-distance dispersal, with alien species (on average 38.5%) showing significantly higher proportions of such dispersal than native species (4.1%). In a second analysis, von der Lippe and Kowarik (2008) differentiated between seed deposition along the inbound and outbound tunnel lanes. Native and non-native species richness of the tunnel samples was significantly higher along lanes leading out of the city compared to the inbound lanes. The same pattern was true for seed deposition, but one order of magnitude higher. The export of seeds from non-native species out of the city by vehicles was not only higher in numbers than the import, it was also higher in relation to their overall proportion in the flora. Cities can therefore act as foci for the spread of alien species into the wider environment.

Brunzel *et al.* (2009) found that in a region north of Frankfurt, Germany, many species of neophytes increased their abundance between a first survey (1974–81) and a second one (2003). Furthermore, neophytes were more abundant in urbanised settlements closer to Frankfurt than those further away. The occurrence of neophytes was better predicted by environmental variables, using a multiple regression framework, than was that of native plant species. Distance to Frankfurt (negative), number of inhabitants (positive) and connectivity to Frankfurt (assessed as the quality of the transport infrastructure; positive) were the three most important significant predictors explaining neophyte plant species richness. In another analysis of the same dataset, dispersal kernels based on patterns of human movement between settlements led to a better match with the observed distribution pattern than a null model simulating pure distance-dependent dispersal for all species (Niggemann *et al.* 2009). Interestingly, alien species seemed to benefit more from human dispersal than native species. Nonetheless, not only vehicles and transport infrastructure facilitate species dispersal. Even walkers can contribute considerably to the long-distance dispersal of plant species by carrying seeds underneath and on their shoes (Wichmann et al. 2009).

Different patterns have been found for the United Kingdom (Botham et al. 2009). Comparing two mapping schemes (1987–88 versus 2003–04), the authors reported a strong association of neophytes with urban land cover but no apparent spread into the wider countryside. Neophytes as a group showed some reduction in the strength of their association with urban land cover over time. This was true even for the species for which strong positive urban associations were identified. However, very few species changed in their individual urban association over time. There are several factors which could potentially mask the evidence of species spreading into the wider countryside: (i) urban sprawl may outweigh species dispersal; (ii) the study is based on presence/absence only and not on abundances; and (iii) the time period of 16 years may not be long enough. Alternatively, contrasting patterns may genuinely exist, perhaps as a consequence of the situation in Britain being different in several respects from the one in Central Europe. Unlike many other studies (e.g. Araújo 2003; Hope et al. 2003; Kühn et al. 2004a; Wania et al. 2006), Britain did not show an increase in species richness in urban areas compared with rural ones (Roy et al. 1999).

Alien species and biotic homogenisation in urban regions

The spread of alien species into the wider countryside as well as the introduction of the same species into cities around the globe may lead to a phenomenon called biotic homogenisation (McKinney & Lockwood 1999; Lockwood et al. 2000). This is often interpreted as formerly distinct biota becoming more similar by the replacement of rare native species with widely spread alien species. There are, however, two mechanisms involved, the introduction of species and the extinction of species, which can lead to increased, decreased or unchanged similarity among biota (Olden & Poff 2003). Increasing similarity is generally called homogenisation, decreasing similarity called differentiation. This is not restricted to species but can also occur at a phylogenetic or functional level (Olden & Poff 2003; Winter et al. 2008).

The process of species introduction can, however, have different effects on different scales. Especially on small scales and at short distances, differentiation can occur (e.g. Qian et al. 2008; Winter et al. 2008, 2010), whereas on larger scales and across larger distances homogenisation is frequently found (e.g. McKinney 2005; Qian & Ricklefs 2006; Winter et al. 2010).

Considering cities, McKinney (2006) discusses the special case of urbanisation and homogenisation. The expansion of towns and cities promotes the loss of native species and their replacement by alien species. And not only can the same plants be found in cities all over the globe (plants: Tree of Heaven, goldenrain tree *Koelreuteria paniculata*, common fig *Ficus carica*; animals: house sparrow *Passer domesticus*) but urban habitats also tend to be very similar. In this paper, McKinney showed that plant communities of state parks and national wildlands are less similar than those of urban habitats.

For California, Schwartz *et al.* (2006) found that in densely populated counties there is a low similarity of plant species that have been extirpated but a high similarity of noxious weeds. This will lead to homogenisation. For Germany, the pattern is slightly more complex. Kühn and Klotz (2006) compared 60 urbanised grid cells of the German floristic mapping scheme with more rural cells, and found that urban areas are more similar to each other in plant composition than rural areas (i.e. homogenised). Disentangling the contributions of the different groups of species, assemblages of native species and archaeophytes (i.e. alien plant species introduced before the discovery of the Americas; Pyšek *et al.* 2004) are more homogenous in urban areas compared with rural ones while assemblages of neophytes are more differentiated. Hence, urbanisation can lead to homogenisation, but this is not caused by neophytes. Similarly, La Sorte *et al.* (2008) found for 22 cities across Europe that archaeophytes were associated with a higher compositional similarity while neophytes were associated with a lower compositional similarity. The former can be interpreted as leading to homogenisation, the latter to differentiation. A comparable pattern was also found in a cross-continental analysis of eight US cities and seven European cities (La Sorte *et al.* 2007).

Interestingly, while all plant species (natives and aliens) were more species-rich in urbanised areas in Germany than in rural areas, neophytes were disproportionately more frequent (Kühn & Klotz 2006). While high species richness is usually considered beneficial for biodiversity conservation, this can often be associated with homogenisation. Of course, the identity of species is important and whether they are generalists or specialists. Still, this leads to a conservation challenge: cities are species-rich despite the effect of homogenisation (McKinney 2006). Appropriate management strategies therefore need to be developed to maintain a high native diversity but counteract the homogenising processes.

Summary

Because of their specific environmental conditions as well as a high heterogeneity of habitats, urbanised areas support very many plant species. Some are native species but alien (i.e. non-native) species are especially prevalent. Typical alien plant species in Central Europe are, for example, Tree of Heaven or Canadian golden-rod. The former is especially adapted to the warmer climates of cities, which may anticipate more general climate warming. Typical urban plant species are often better adapted to dry or nutrient-rich soils and are more often wind-pollinated than insect-pollinated. Most alien plant species were intentionally introduced, often as ornamentals or for horticultural reasons; fewer were unintentionally introduced, often as contaminants of commodities (seeds, wool, minerals etc.), especially as a result of transport infrastructures to and within cities. But cities act also as sources of biological invasions into the wider environment. Here, traffic infrastructure especially facilitates spread into surrounding landscapes. This may potentially result in a homogenisation process (i.e. formerly more distinct biotas become more similar). However, currently urbanisation has mostly been observed to result in homogenisation (among urban floras) of native species, whereas alien species show differentiation (i.e. becoming more dissimilar). Therefore, specific management strategies need to be developed to meet the specific demands of conserving a high biodiversity with many specialised or rare species and concurrently preventing homogenisation to preserve distinct regional species compositions.

References

Araújo, M. B. (2003). The coincidence of people and biodiversity in Europe. *Global Ecology and Biogeography*, **12**, 5–12.

Botham, M. S., Rothery, P., Hulme, P. E. *et al.* (2009). Do urban areas act as foci for the spread of alien plant species? An assessment of temporal trends in the UK. *Diversity and Distributions*, **15**, 338–45.

Brunzel, S., Fischer, S. F., Schneider, J., Jetzkowitz, J. and Brandl, R. (2009). Neo- and archaeophytes respond more strongly than natives to socio-economic mobility and disturbance patterns along an urban–rural gradient. *Journal of Biogeography*, **36**, 835–44.

Ernst, W. H. O. (1998). Invasion, dispersal and ecology of the South African neophyte *Senecio inaequidens* in the Netherlands: from wool alien to railway and road alien. *Acta Botanica Neerlandica*, **47**, 131–51.

Fuller, R. A. and Gaston, K. J. (2009). The scaling of green space coverage in European cities. *Biology Letters*, **5**, 352–5.

Fuller, R. A., Tratalos, J. and Gaston, K. J. (2009). How many birds are there in a city of half a million people? *Diversity and Distributions*, **15**, 328–37.

Gelbard, J. L. and Belnap, J. (2003). Roads as conduits for exotic plant invasions in a semiarid landscape. *Conservation Biology*, **17**, 420–32.

Gilbert, O. L. (1989). *Ecology of Urban Habitats*. London: Chapman and Hall.

Gutte, P., Klotz, S., Lahr, C. and Trefflich, A. (1987). *Ailanthus altissima* (Mill.) Swingle – eine vergleichende pflanzengeographische Studie. *Folia Geobotanica et Phytotaxonomica*, **22**, 241–62.

Heger, T. and Böhmer, H. J. (2005). The invasion of Central Europe by *Senecio inaequidens* DC. – a complex biogeographical problem. *Erdkunde*, **59**, 34–49.

Hope, D., Gries, C., Zhu, W. X. *et al.* (2003). Socioeconomics drive urban plant diversity. *Proceedings of the National Academy of Sciences of the USA*, **100**, 8788–92.

Houlahan, J. E., Keddy, P. A., Makkay, K. and Findlay, C. S. (2006). The effects of adjacent land use on wetland species richness and community composition. *Wetlands*, **26**, 79–96.

Hulme, P. E., Bacher, S., Kenis, M. *et al.* (2008). Grasping at the routes of biological invasions: a framework for integrating pathways into policy. *Journal of Applied Ecology*, **45**, 403–14.

Klotz, S. (1989). Flora und Vegetation in der Stadt – ihre Spezifik und Indikatorfunktion. *Bauforschung-Baupraxis*, **244**, 29–33.

Knapp, S., Kühn, I., Bakker, J. P. et al. (2009). How species traits and affinity to urban land use control large-scale species frequency. *Diversity and Distributions*, **15**, 533–46.

Knapp, S., Kühn, I., Mosbrugger, V. and Klotz, S. (2008a). Do protected areas in urban and rural landscapes differ in species diversity? *Biodiversity and Conservation*, **17**, 1595–612.

Knapp, S., Kühn, I., Schweiger, O. and Klotz, S. (2008b). Challenging urban species diversity: contrasting phylogenetic patterns across plant functional groups in Germany. *Ecology Letters*, **11**, 1054–64.

Knapp, S., Kühn, I., Wittig, R. et al. (2008c). Urbanization causes shifts in species' trait state frequencies. *Preslia*, **80**, 375–88.

Kowarik, I. (2003). *Biologische Invasionen. Neophyten und Neozoen in Mitteleuropa.* Stuttgart: Ulmer.

Kühn, I. and Klotz, S. (2002). Floristischer Status und gebietsfremde Arten. In S. Klotz, I. Kühn and W. Durka, eds. BIOLFLOR – *Eine Datenbank zu biologisch-ökologischen Merkmalen der Gefäßpflanzen in Deutschland.* Bonn: Bundesamt für Naturschutz, Schriftenreihe für Vegetationskunde Vol. **38**, pp. 47–56.

Kühn, I. and Klotz, S. (2006). Urbanisation and homogenization – comparing the floras of urban and rural areas in Germany. *Biological Conservation*, **127**, 292–300.

Kühn, I., Brandl, R. and Klotz, S. (2004a). The flora of German cities is naturally species rich. *Evolutionary Ecology Research*, **6**, 749–64.

Kühn, I., Durka, W. and Klotz, S. (2004b). BiolFlor – a new plant-trait database as a tool for plant invasion ecology. *Diversity and Distributions*, **10**, 363–5.

Kunick, W. (1974). Veränderungen von Flora und Vegetation einer Großstadt, dargestellt am Beispiel von Berlin (West). Dissertation, Technische Universität Berlin.

La Sorte, F. A., McKinney, M. L. and Pyšek, P. (2007). Compositional similarity among urban floras within and across continents:

biogeographical consequences of human-mediated biotic interchange. *Global Change Biology*, **13**, 913–21.

La Sorte, F. A., McKinney, M. L., Pyšek, P. et al. (2008). Distance decay of similarity among European urban floras: the impact of anthropogenic activities on beta diversity. *Global Ecology and Biogeography*, **17**, 363–71.

Lambdon, P. W., Pyšek, P., Basnou, C. et al. (2008). Alien flora of Europe: species diversity, temporal trends, geographical patterns and research needs. *Preslia*, **80**, 101–49.

Landsberg, H. (1981). *The Urban Climate.* New York: Academic Press.

Lockwood, J. L., Brooks, T. M. and McKinney, M. L. (2000). Taxonomic homogenization of the global avifauna. *Animal Conservation*, **3**, 27–35.

Lososová, Z., Chytrý, M., Kühn, I. et al. (2006). Patterns of plant traits in annual vegetation of man-made habitats in central Europe. *Perspectives in Plant Ecology, Evolution and Systematics*, **8**, 69–81.

Loss, S. R., Ruiz, M. O. and Brawn, J. D. (2009). Relationships between avian diversity, neighborhood age, income, and environmental characteristics of an urban landscape. *Biological Conservation*, **142**, 2578–85.

McKinney, M. L. (2005). Species introduced from nearby sources have a more homogenizing effect than species from distant sources: evidence from plants and fishes in the USA. *Diversity and Distributions*, **11**, 367–74.

McKinney, M. L. (2006). Urbanization as a major cause of biotic homogenization. *Biological Conservation*, **127**, 247–60.

McKinney, M. L. and Lockwood, J. L. (1999). Biotic homogenization: a few winners replacing many losers in the next mass extinction. *Trends in Ecology and Evolution*, **14**, 450–3.

Niggemann, M., Jetzkowitz, J., Brunzel, S., Wichmann, M. C. and Bialozyt, R. (2009). Distribution patterns of plants explained by human movement behaviour. *Ecological Modelling*, **220**, 1339–46.

Oke, T. R. (1982). The energetic basis of the urban heat island. *Quarterly Journal of the Royal Meteorological Society*, **108**, 1–24.

Olden, J. D. and Poff, N. L. (2003). Toward a mechanistic understanding and prediction of biotic homogenization. *American Naturalist*, **162**, 442–60.

Parendes, L. A. and Jones, J. A. (2000). Role of light availability and dispersal in exotic plant invasion along roads and streams in the H. J. Andrews Experimental Forest, Oregon. *Conservation Biology*, **14**, 64–75.

Pyšek, P. (1998). Alien and native species in Central European urban floras: a quantitative comparison. *Journal of Biogeography*, **25**, 155–63.

Pyšek, P., Lambdon, P. W., Arianoutsou, M. *et al.* (2009). Alien vascular plants of Europe. In DAISIE, ed., *The Handbook of Alien Species in Europe*. Dordrecht: Springer, pp. 43–61.

Pyšek, P., Richardson, D. M., Rejmánek, M. *et al.* (2004). Alien plants in checklists and floras: towards better communication between taxonomists and ecologists. *Taxon*, **53**, 131–43.

Qian, H. and Ricklefs, R. E. (2006). The role of exotic species in homogenizing the North American flora. *Ecology Letters*, **9**, 1293–8.

Qian, H., McKinney, M. L. and Kühn, I. (2008). Effects of introduced species on floristic similarity: comparing two US states. *Basic and Applied Ecology*, **9**, 617–25.

Roy, D. B., Hill, M. O. and Rothery, P. (1999). Effects of urban land cover on the local species pool in Britain. *Ecography*, **22**, 507–15.

Schwartz, M. W., Thorne, J. H. and Viers, J. H. (2006). Biotic homogenization of the California flora in urban and urbanizing regions. *Biological Conservation*, **127**, 282–91.

Sukopp, H. (1998). Urban ecology: scientific and practical aspects. In J. Breuste, H. Feldmann and O. Uhlmann, eds., *Urban Ecology*. Berlin and Heidelberg: Springer, pp. 3–16.

Sullivan, J. J., Timmins, S. M. and Williams, P. A. (2005). Movement of exotic plants into coastal native forests from gardens in northern New Zealand. *New Zealand Journal of Ecology*, **29**, 1–10.

UN-HABITAT (2003). *Population, Education and Development. The Concise Report*. New York: United Nations.

UN-HABITAT (2006). *World Urbanization Prospects. The 2005 Revision. Executive Summary, Fact Sheets, Data Tables*. New York: United Nations.

von der Lippe, M. and Kowarik, I. (2007a). Crop seed spillage along roads: a factor of uncertainty in the containment of GMO. *Ecography*, **30**, 483–90.

von der Lippe, M. and Kowarik, I. (2007b). Long-distance dispersal of plants by vehicles as a driver of plant invasions. *Conservation Biology*, **21**, 986–96.

von der Lippe, M. and Kowarik, I. (2008). Do cities export biodiversity? Traffic as dispersal vector across urban–rural gradients. *Diversity and Distributions*, **14**, 18–25.

Wania, A., Kühn, I. and Klotz, S. (2006). Biodiversity patterns of plants in agricultural and urban landscapes in Central Germany – spatial gradients of species richness. *Landscape and Urban Planning*, **75**, 97–110.

Wichmann, M. C., Alexander, M. J., Soons, M. B. et al. (2009). Human-mediated dispersal of seeds over long distances. *Proceedings of the Royal Society B – Biological Sciences*, **276**, 523–32.

Wilcox, D. A. (1989). Migration and control of purple loosestrife (*Lythrum salicaria* L.) along highway corridors. *Environmental Management*, **13**, 365–70.

Winter, M., Kühn, I., La Sorte, F. A., *et al.* (2010). The role of non-native plants and vertebrates in defining patterns of compositional dissimilarity within and across continents. *Global Ecology and Biogeography*, **19**, 332–42.

Winter, M., Kühn, I., Nentwig, W. and Klotz, S. (2008). Spatial aspects of trait homogenization within the German flora. *Journal of Biogeography*, 35, 2289–97.

Wittig, R. (2002). *Siedlungsvegetation*. Stuttgart: Ulmer.

Wittig, R., Diesing, D. and Godde, M. (1985). Urbanophob-Urbanoneutral-Urbanophil – behavior of species concerning the urban habitat. *Flora*, **177**, 265–82.

Wyatt, R. (1996). More on the southward spread of common milkweed, *Asclepias syriaca* L. *Bulletin of the Torrey Botanical Club*, **123**, 68–9.

CHAPTER SEVEN

Interactions between people and nature in urban environments

RICHARD A. FULLER AND KATHERINE N. IRVINE

Since the dawn of human civilisation, people have interacted with nature, most notably to harness the resources that have fuelled the human enterprise (Vitousek et al. 1997). The sheer rate and scale of human appropriation of natural resources has precipitated a biodiversity crisis currently being manifested in rapid rates of species extinctions, extensive transformation of the structure and function of ecosystems, and rapid alterations to the Earth's climate (Vitousek et al. 1986; Pimm & Raven 2000). The biodiversity crisis is a result of human activity, so the solutions to it will depend largely on human actions, on understanding and enhancing the way that we all interact with nature (Collins et al. 2000; Ehrlich 2002). Because most people on the planet live in towns and cities, the majority of our daily interactions with nature take place in urban environments, and this has led to a recent upsurge of interest in the dynamics of these relationships (Bradshaw & Bekoff 2000; Miller & Hobbs 2002; Pyle 2003; Saunders et al. 2006). Despite the manifest impoverishment of the natural environment in urban areas, or perhaps because of it, many urban dwellers seek out interaction with nature in some form, for example by visiting a local green space, or feeding backyard birds. Yet our understanding of these interactions is nascent, principally because their study requires work across several disciplinary boundaries (Alberti et al. 2003; Braun 2005). For example, ecologists often lack the interest or the tools to study people (Collins et al. 2000).

Interactions between people and urban nature are widespread across those societies in which they have been studied. Small-scale urban agriculture for the sale of produce in local markets is prevalent across much of the developing world (Silk 1986), and gardening of private plots for subsistence is common across many African cities (Obudho & Foeken 1999). Fifty-nine per cent of respondents to a survey in Guangzhou, China, reported visiting urban parks often or very often (Jim & Chen 2006a), and 92% of participants in a recent UK survey reported visiting urban green spaces (GreenSpace 2007). In the USA,

Urban Ecology, ed. Kevin J. Gaston. Published by Cambridge University Press.

about 68 million people engage in wildlife-watching activity in or within a mile of their home, spending about US $33 billion in the process (US Department of the Interior, Fish and Wildlife Service 2006). In parts of Europe, North America and Australia, between one-fifth and one-third of households provide supplementary food for wild birds (Clergeau et al. 1997; Rollinson et al. 2003; Lepczyk et al. 2004). Total annual expenditure on outdoor feeding of birds in the UK has recently been estimated at £200 million (British Trust for Ornithology 2006), while in the USA $23.2 billion is spent annually on wildlife-watching equipment (US Department of the Interior, Fish and Wildlife Service 2006).

Despite this, there is concern that urbanisation is progressively isolating people from the natural environment, and that the problem is getting worse as cities continue to grow in size and density (Turner et al. 2004; Miller 2005; Louv 2008). Even the term 'outside' suggests a degree of separation from nature (Irvine & Warber 2002a). Modern humans spend most of their lives inside, and urbanisation exacerbates this problem by adding substantial geographic distance from natural, or even sometimes any vegetated, landscapes. About one-third of Hispanic children in the USA are routinely kept indoors because their neighbourhoods are perceived as dangerous (Dye & Johnson 2007). In short, we are seeing an 'extinction of experience' (Pyle 1978). This is of profound concern, because interactions between people and nature lead to a variety of measurable benefits, at both individual and societal levels (Irvine & Warber 2002b; Bird 2004, 2007; Brown & Grant 2005). It is also concerning from a conservation viewpoint because demand for venues that allow interaction with nature imply provision of a green space network that will also benefit biodiversity, and access to nature in turn predicts active engagement (Schultz 2000; Williams & Cary 2002; Lohr & Pearson-Mims 2005). A simple example illustrates this. Surveys in the USA have discovered that people living in large cities are less likely to be birdwatchers than residents of small cities and rural areas (La Rouche 2003; Sali et al. 2008). The explanation for this is no doubt partly a practical one – suitable birdwatching areas might be a considerable distance away in a large city – but there is also the possibility that levels of daily interaction with nature partly determine engagement in birdwatching in a kind of positive feedback loop. It is thus perhaps ironic that much of the philosophy of conservation is built around the idea of separating biodiversity from people and the threats associated with them, a thought to which we shall return in closing.

However, the story is more complicated than this. Urbanisation embodies a shift from an economy based on directly harvesting nature to one based on industrial and service sectors, a shift correlated with the growth of protective environmental values at least in the post-war USA (Mertig et al. 2002). This is manifested in a change from direct utilitarian interactions with nature, to a mutualistic relationship with 'fabricated or disrupted' remnants of nature

(Bell 1976). While there are obvious benefits of concentrating people into cities, it is critical that we identify the potential costs to human quality of life, and how we can manage them.

While urbanisation can be viewed as a manifestation of economic growth and developmental progress, from a biological viewpoint it is profoundly disrupting, causing (i) the loss and degradation of natural habitats; (ii) reductions in ecosystem services such as climate and water regulation; (iii) increased levels of pollution and disturbance; and (iv) major changes to the structure of biological assemblages including invasion by non-native species (Chapters 3–6; for reviews see McKinney 2002, 2005; Chace & Walsh 2006). Although urban areas are clearly biologically disrupted systems, they can support significant levels of biodiversity (Fuller et al. 2009) and have some regional conservation value (Mason 2000; Bland et al. 2004; Vähä-Piikkiö et al. 2004). This is particularly so in developed regions such as Europe and North America, where intensive agriculture has led to substantial population declines across the wider landscape, and increased the relative importance of urban areas to sustaining species' overall abundances (Gregory & Baillie 1998; Mason 2000). For example, the density of birds in urban Sheffield, UK, is estimated at more than six times the national average, although most of the abundant species in the city's avifauna are also those common across the UK at large (Fuller et al. 2009).

Resource fluxes, geochemical cycles and land cover composition are under continuous control by humans in urban environments, and thus the pattern of daily human activity will shape the biodiversity value of towns and cities. Large-scale patterns in human activity result ultimately from a collection of individual decisions by urban citizens. For example, the establishment of a new pond in just 10% of gardens within urban Sheffield (a city in central UK with a population of c. 500 000) would result in the formation of 17 500 new habitat patches (Gaston et al. 2007a). Widespread engagement in citizen science can reveal much about the dynamics of urban biodiversity (Greenwood 2007), and provide alerts to changes in populations, for example of garden birds (Cannon et al. 2005).

There is enormous variation in the extent of urban green space networks across 386 cities in Europe (Fuller & Gaston 2009), which leads not only to variation in the biological quality of those urban environments, but also to disparities across human society in access to experiences of nature (Barbosa et al. 2007). Likewise, reaping the human benefits of such interactions with nature depends on the presence, but also the qualities, of accessible nature in the local environment. These reciprocal effects lead to a complex interplay between human activity and biodiversity. Consequently, management for urban biodiversity should not, and probably cannot, be separated from programmes to improve human quality of life in urban environments. Provision of green spaces within urban areas historically focused almost exclusively on the area provided for each inhabitant (Turner 2006), although attention has recently switched to an emphasis on green space quality (CABE Space 2004, 2005a). Indeed, emerging evidence of a positive relationship between biodiversity value and environmental benefits to human wellbeing (e.g. Fuller *et al.* 2007) suggests that management to enhance biodiversity value could also benefit the human population in a win–win scenario (Irvine *et al.* 2010).

What are people–nature interactions?

Dictionary definitions of the term 'interaction' emphasise its mutual or reciprocal quality; two elements engaging and each influencing the other. As such, we consider interactions between people and nature to be those actions that result in measurable changes both to people and to nature; in other words, interactions are two-way relationships (Irvine *et al.* 2010). This is distinct from studying, for example, the effects of exposure to nature on people, or human impacts on ecological systems in isolation. Examples of interactions in urban environments include visiting urban public green spaces, maintaining a private garden and feeding wildlife. In each of these cases, human activities result in measurable changes to the natural environment (e.g. providing habitat for plants and animals, enhancing survival of garden birds, disturbing wildlife). Simultaneously, being in the natural environment, or the very act of engaging with nature, has measurable impacts on various aspects of human quality of life (e.g. physical health, psychological wellbeing, economic gain). In all cases, the outcomes of interactions can be positive or negative. Research on positive experiences has predominated in the literature thus far, although negative experiences are doubtless important (Bixler & Floyd 1997).

Interactions occur via the full range of human senses, from the visual appreciation of a brightly coloured bird to the smell of fragrant flowers, the taste of wild blackberries or garden-grown tomatoes, the sound of birdsong, wind in the trees, or intermittent silence, the feel of the wind on one's skin or kicking through autumn leaves (Irvine & Warber 2002a). There is growing interest in the 'sensescape' of urban environments that challenges the dominance of visual senses in the research agenda thus far (Rodaway 1994).

Human interactions with nature may usefully be thought of as falling along a continuum from passive to active (Irvine & Warber 2002b; Brown & Grant 2005). Merely knowing that a garden or urban green space is nearby represents one of the most passive forms of interaction (Kaplan *et al.* 1996; Bell *et al.* 2004), increasing through watching wildlife documentaries on television, having pictures of nature on the wall or having a view through a window. More active forms of interaction progressively include, for example, sitting in or walking through a local green space, tending a garden, regularly feeding garden birds, campaigning for local environmental change, volunteering for local conservation management groups or designing a garden around its suitability for biodiversity. These interactions can take place in private, in public or within institutions (e.g. workplaces, prisons, hospitals, schools).

People–nature interactions are many and varied, and each will have particular dynamics. However, most are poorly studied, and it remains unclear whether they stem from a single phenomenon, such as an underlying need for human interaction with the natural world. One such unifying explanation is biophilia, the idea advanced by Wilson (1984) that humans have a fundamental connection with the natural world, and an innate need to be exposed to its complexity of form and function. Rapid global urbanisation has lent urgency to discovering the importance of people–nature interactions, because opportunities for them are declining in both quality and quantity (Pyle 2003; Miller 2005). Research in this field is particularly critical because (i) much of the focus thus far has been on interactions in the countryside or wilderness, and (ii) there is evidence for substantial cultural variation in people–nature interactions (Teel et al. 2007).

In response to cultural bias in research into people–nature interactions, the Wildlife Values Globally project is documenting variation in wildlife values across cultures in the USA, China, Thailand, Estonia, the Netherlands, Mongolia, Malaysia, Uganda and Kenya (Teel et al. 2007). The results of the project thus far support the possibility of a consistent shift in wildlife value orientations towards a mutualistic rather than consumptive or utilitarian view of nature across all of these diverse cultures. This said, there is ample evidence for enormous cultural variation in values, attitudes and forms of people–nature interactions. History suggests that cultural change can proceed rapidly, and so a clearer understanding of culturally specific attitudes toward nature conservation will help in devising ways to shape environmental attitudes in the future (Ehrlich 2002). For example, in the USA traditional utilitarian values of wildlife long held by those consumptively using nature (hunting, fishing) have given way to a dominance by mutualistic and protective values (wildlife appreciation), and the degree of this shift is positively correlated with urbanisation (Manfredo et al. 2003). However, the situation in China appears to be very different. Perhaps partially driven by lower income levels, materialistic dimensions of valuing wildlife emerged as dominant in a small interview-based survey of urban and rural residents (Zinn & Shen 2007).

Other aspects of people–nature interactions are less culturally variable. For example, evidence for environmental altruism (a negative relationship between value of self-enhancement and pro-environmental behaviour) was consistent across respondents in Mexico, Nicaragua, Peru, Spain and the USA (Schultz & Zelezny 1998). On balance, though, the evidence thus far suggests that many interactions are culturally specific, and much remains to be done to document global variation in the motivations for, and the form, prevalence and outcomes of, interactions between people and urban nature.

Examining three people–nature interactions that are reasonably well understood, namely visiting urban public green spaces, maintaining a private garden and feeding urban wildlife, we will consider how people–nature interactions manifest themselves, their prevalence across human society, motivations for engaging in them and their consequences for both people and ecosystems. We conclude by drawing out common themes, and suggesting ways in which the research agenda can be moved forward in this emerging field.

Visiting urban green spaces

Urban green spaces have been the arena for much of the work on the relationship between people and the natural world (Speirs 2003; Balram & Dragicevic 2005; Pincetl & Gearin 2005; Jim & Chen 2006b). More than half of the respondents to a survey in Guangzhou used urban green spaces more than once per week (Jim & Chen 2006a), and the equivalent figure was 46% in an English survey (Dunnett et al. 2002), equating to about 33 million people making over 2.5 billion visits annually to urban green spaces. A visit to an urban green space might involve no more than indirect exposure to its natural, semi-natural or managed vegetation while engaging in another activity such as sport or walking the dog. More intimate interactions might involve sitting and relaxing, or actively watching and engaging with nature by birdwatching or feeding the wildlife. Whatever form the interaction might take, access to urban green spaces is an important aspect of quality of life for urban dwellers, given that it leads to a wide range of personal and societal benefits including those to physical and psychological health, urban regeneration, economic gain, social cohesion, crime reduction, sense of community and environmental awareness (Chapter 9; Bird 2004, 2007; Department of Transport, Local Government and the Regions 2006; Maller et al. 2006). As a result of this, access to green space is emphasised in numerous policy documents across the world, and is a key dimension of the sustainable city (Bengston et al. 2004; Irvine et al. 2010).

While there is more or less complete consensus that green spaces are beneficial to human quality of life, rather less is known about how the presence of nature delivers these benefits (Stilgoe 2001). To answer this kind of question, ecological datasets must be brought directly alongside those from different disciplines, including, but not limited to, psychology, sociology and geography. This approach was recently taken in a study of urban green space users in Sheffield, where the degree of psychological benefit gained by visitors to 15 urban parks was correlated positively with plant species richness in the sites, suggesting that the biological complexity of the environment is important in determining the outcome of interactions between people and nature (Figure 7.1; Fuller et al. 2007).

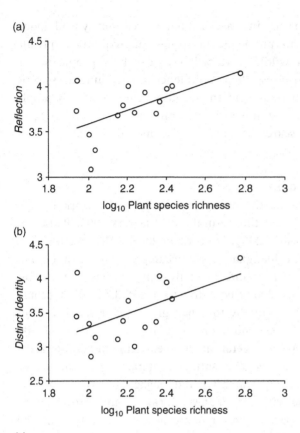

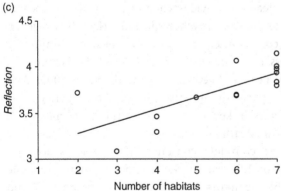

Figure 7.1 Two measures of psychological wellbeing among visitors to 15 urban green spaces in Sheffield, UK, increased with the richness of plant species in those spaces. The two measures are *Reflection* (the ability to think and gain perspective) and *Distinct Identity* (the degree of feeling unique or different through association with a particular place). *Reflection* also increased with the number of habitats present in the green space. Reproduced with permission from Fuller *et al.* (2007).

This said, the proportion of respondents who mention direct contact with nature as a reason for visiting green spaces varies markedly, with reported values ranging from 22% (GreenSpace 2007) to 54% (Chiesura 2004) and 55% (Irvine *et al.* 2010). A study in Sheffield that asked green space visitors to give specific reasons for visiting discovered that explicit mentions of flora and fauna were rare, but that broader constructions of nature such as fresh air, being outside, peace and quiet, open space and topography were commonly mentioned (Irvine *et al.* 2010). This suggests that specific elements of biodiversity value (e.g. species richness, habitat heterogeneity) may not be directly perceived as important, yet the combination of these components into a natural scene is part of the reason why green spaces are used. Visitors to the Vondelpark in Amsterdam, the Netherlands, indicated that relaxation was their primary motivation for visiting (Chiesura 2004), yet without further analysis it is hard to ascertain how much of this relates to the importance of the natural environment. Thus, while contact with nature might not be cited explicitly by park visitors, frequent contact with nature via visits to green spaces features in the lives of many urban dwellers.

Preferences for particular features within urban green spaces often reveal ambivalence. This is typified by residents of Guangzhou, who give low explicit value to the presence of nature and wildlife, but state a preference on the one hand for large, well-vegetated green spaces with lots of mature trees, and on the other hand for those with a wide range of recreational facilities (Jim & Chen 2006a). This paradox is also evident in Sheffield, where people avoided biodiverse sectors within urban green spaces, spending most time in areas of low-density vegetation and on paved surfaces, yet highly endorsed the presence of trees when asked about their preferences for park design (Irvine *et al.* 2010). In addition, park visitors in the city were able accurately to assess relative biodiversity levels within urban green spaces, suggesting that urban residents can be highly attuned to the presence of natural features within urban green spaces (Fuller *et al.* 2007). Hearing mechanical sounds is consistently rated less pleasant than natural sounds both within green spaces and in public squares (Yang & Kang 2005). Irvine *et al.* (2009) reported that the sounds most frequently mentioned by visitors to green spaces in Sheffield were those of natural origin (e.g. bird song, dogs barking, wind in the trees), and vegetation buffers are widely used to reduce the prevalence of mechanical sounds (for review see Bucur 2006).

Human interactions with nature in urban green spaces influence biodiversity both directly and indirectly. First, although many urban green spaces are provided chiefly for human benefit, they directly enhance the ability of an urban landscape to support biodiversity. Green spaces provide habitat networks for plants and animals, including those of regional and national conservation priority (Vähä-Piikkiö *et al.* 2004; Angold *et al.* 2006; Bryant 2006;

Mason *et al.* 2007). Indeed, nature conservation was explicitly considered in UK green space policy from 1983 onwards, latterly emphasising ecological quality and connectivity in the green space network (Turner 1992, 2006), concepts directly applied from landscape ecology. The ways in which green spaces are managed will also directly affect their biological quality and ability to deliver ecosystem services such as temperature regulation and flood alleviation (Cilliers & Bredenkamp 2000; Cornelis & Hermy 2004; Jim & Chen 2006b). For example, clearing of dead wood and removal of dense vegetation can reduce habitat quality for birds (Sandström *et al.* 2006); availability of certain exotic plants can benefit urban butterflies (Shapiro 2002); and extensive mowing of grasses reduces invertebrate populations and the richness of native plants (Helden & Leather 2004; Kirkpatrick 2004).

Second, because interacting with nature in urban green spaces has been empirically linked to positive attitudes toward nature (Lohr 2007), it is possible that facilitating such interactions will encourage interest in conservation issues beyond the arena of the interaction itself (Miller 2005). The intriguing possibility of this indirect feedback has received much recent attention and has been mooted as a possible long-term outcome of urban environmental education programmes (Hale 1993).

The history of urban green space provision, and thus the social and geographical distribution of opportunities to interact with nature, varies markedly across the world. The availability of public green space in European cities was primarily shaped during the period following the Industrial Revolution when publicly funded parks were provided for general use. These areas were conceived as places to facilitate both physical and social wellbeing among the working class by being a resource for fresh air, a place of beauty, a space for relatively passive recreation such as picnicking, carriage riding or walking, and an area for social interaction (Conway 2000). These parks were characterised by formal ornamental horticulture (e.g. ornamental flowerbeds, rose gardens; Kendle & Forbes 1997). Since the Victorian era, many European urban parks have predominantly been in the form of recreational grounds consisting of mown grass areas for a range of more active recreation (Kendle & Forbes 1997), although chronic underinvestment in the UK's system of 27 000 urban parks resulted in the loss and degradation of much infrastructure during recent decades (CABE Space 2005b).

In the USA, urban green spaces were a response to the development of large cities, prompted by concerns about access to nature and clean air (Pincetl & Gearin 2005). In the twentieth century, public park design emphasised function over experiences of nature, based on the idea that social reform depended on the availability of areas for structured play (Draper 1996), and only recently were more naturalistic open spaces developed alongside urban parks, most notably in suburban areas (Rome 2001). Most recently, green space has been seen as a tool for social regeneration, particularly in cities where provision is very low (Wolch et al. 2002). For example, parks cover only about 5% of Los Angeles, USA, falling well short of national standards (Wolch et al. 2002).

Access to urban green spaces

Visiting green spaces is a popular activity, but access varies markedly among the human population. Many cities in the developing world are characterised by low levels of green space provision (Jenks & Burgess 2000). For example, open space provision totals about six square metres per urban inhabitant of Kuala Lumpur (Dali 2004), and only one square metre per inhabitant of Bangkok, Thailand (Fraser 2002). In 2001, urban green space accounted for about one-third of the surface area of Guangzhou, a city with a population of 3.22 million people in Guangdong, China; this is one of the highest proportions in China (Jim & Chen 2006b). There is a strong drive to create a comprehensive urban green space network in South Africa (South African Cities Network 2006), yet there are no clear standards for green space provision, nor the policy framework to implement them (Sutton 2008).

Levels of green space provision among the white population of Los Angeles exceed that of the African-American population by a factor of 20 (Wolch et al. 2002), and recent efforts by the Californian government are thus focused jointly on raising the quantity of access to natural green spaces, but also on ensuring equity across society (Garcia & White 2007). A negative relationship between green space availability and human socioeconomic deprivation has been found in New Delhi, India (Sliuzas & Kuffer 2008), although the opposite result was found in a study of Sheffield, with public green space provision being biased toward less affluent sectors of society, at least in terms of its quantity (Barbosa et al. 2007). In part these differences might reflect the different histories of green space provision among countries. However these patterns arise, correspondence between the distribution of human social diversity and green space provision is one component of a wider relationship between human socioeconomics and patterns of biodiversity across urban landscapes (Jim & Liu 2001; Pickett et al. 2001; Kinzig et al. 2005).

Although simply accessing green space itself is a form of interaction with nature, the quality of that interaction is critical. This will depend on both proximity of the green space to potential users, and also its recreational value. The role of proximity in determining green space visitation seems to vary. More than 90% of respondents to a survey in Guangzhou indicated that proximity was important or very important in their decision to visit an urban green space (Jim & Chen 2006a), and studies in both the UK and USA have found that people generally desire a green space within 3–5 minutes' walk of their home or workplace (Godbey et al. 1992; Comedia & Demos 1995). Despite this, 36% of respondents to a UK study reported that their nearest urban green space is not the one they visit most frequently (GreenSpace 2007), indicating that decisions are not based on distance criteria alone. Reflecting this, the quantitative targets for green space provision that dominated UK urban planning policy from 1925 to 1976 were replaced by a requirement to provide a hierarchy of parks based on accessibility to a range of different amenities provided across the green space network (Turner 1992, 2006). A series of government-commissioned reports in the UK has resulted in a recent emphasis there on the quality and not just the quantity of urban green spaces (Urban Task Force 1999; CABE Space 2004, 2005a, 2005b, 2007; GreenSpace 2007).

Outcomes of interactions with nature in urban green spaces

The human need for green spaces close to where we live has arguably been instrumental in moderating the impacts of urbanisation on nature, resulting in the retention of large parks, well-wooded remnants and encapsulated patches of undeveloped land within many city boundaries. Although small patches of green spaces typically make up a high proportion of overall vegetation cover across a city (Fuller *et al.* 2010), larger patches promote the integrity of ecological processes, and the pattern of green space provision will directly affect the ability of species to persist in urban environments. For example, larger green spaces in Birmingham, UK, contained a higher diversity of plants, and had a higher probability of occupancy by individual plant species (Bastin & Thomas 1999). Inadequate retention of green spaces as an area becomes urbanised results in local extinction of species that are habitat specialists, require large habitat patches, utilise the interiors rather than the edges of patches or are associated with complex vegetation structures (Pickett *et al.* 2001; McKinney 2002; Chace & Walsh 2006).

Although green spaces can clearly moderate some of the negative effects of urbanisation on biological assemblages, the way in which they are used and managed is critical to their biodiversity value, and this is where human activity and biodiversity must be simultaneously considered. An extreme example might be that a repeatedly mown and trampled football field will not support as many plant species as an unmown long grassy margin; the value for recreation trumps biodiversity needs in this instance. However, there are many more subtle cases. For example, the species richness of woodpeckers, forest birds and hole-nesting birds in urban parks in Örebro, Sweden, was positively correlated with the number of trees, and an increase in the proportion of survey plots with a shrub layer was related to an increase in bird species richness (Sandström *et al.* 2006). Species richness of birds in urban green spaces in Greater London was positively related to the amount of unmown rough grassy habitat (Chamberlain *et al.* 2007). Each of these features of green spaces (tree cover, the presence of a shrub layer and mowing regimes) is subject to management decisions that must simultaneously consider human visitor needs and biological value (CABE Space 2005a).

Safety of person and property is one of the main issues in urban green space management, and there is some evidence that a high density of vegetation in urban parks is associated with negative perceptions of safety (Schroeder & Anderson 1984; Lyytimäki et al. 2008). Vegetation influences perceptions of safety by affecting the ability to see long distances, or the ease of locomotion (e.g. Nasar et al. 1993, Nasar & Jones 1997). This has led to the clearance of vegetation and opening up of lines of sight being a dominant paradigm in the management of many urban green spaces (Nasar & Fisher 1993; Kuo & Sullivan 2001), no doubt with negative impacts on the ability of a green space to support biodiversity.

However, this paradigm is contradicted by research on landscape preference, in that vegetation structure and structural designs that obscure the view, thereby providing 'mystery' and interest, are part of the reason that some landscapes are well liked (Kaplan & Kaplan 1989; Herzog & Miller 1998). There is a need for a more refined understanding of the perceptions of safety in urban green space, and several strands of research have begun to explore the relationships among safety, preference, visitor use and naturalistic designs that promote biodiversity and ecosystem service provision in greater depth. For example, Bjerke et al. (2006) found a greater preference for more densely vegetated park landscapes among middle-aged, educated individuals as well as those with an interest in wildlife and a more pro-environmental attitude. Jim and Chen (2006a) found a similar preference for more naturalistic design in China, while Özgüner and Kendle (2006) concluded that both naturalistic and more manicured designs are equally appreciated. In contrast, signs of neglect may also influence perceptions of safety. Respondents to a survey in Sheffield mentioned signs of neglect just as much as vegetation-related factors in contributing to places feeling unsafe (Irvine et al. 2010). These related to the types of people and their activities (e.g. antisocial behaviour), the impact of poor maintenance (e.g. cracked hardscape under children's play area, dilapidated buildings), and also the potential impact of litter (e.g. broken glass). Given the interconnectedness of these various elements, future research should continue to study these issues in combination, rather than in isolation, and investigation of cultural variation is required.

The heavy use of urban green spaces creates a paradox in the sense that amenity use can reduce biodiversity value, but without the amenity value to humans many of these green areas would cease to exist. Disturbance is intimately tied up with the concept of open access to green spaces, and it is not clear how the two can be separated. Disturbance effects include the simple presence of large numbers of people, startling noises, trampling, disturbance by pets, dumping of garden waste, litter, campsites and firewood gathering (Matlack

1993; Ditchkoff *et al.* 2006). Such disturbances extend well into vegetation patches in urban green spaces, and markedly exceed natural edge effects in penetration and severity (Matlack 1993). Disturbance data can be used to formulate park management guidelines particularly where sensitive species are known to be present in a green space. For example, flushing distances of birds have been used to suggest design parameters for urban parks in Madrid, Spain (Fernández-Juricic *et al.* 2001).

Naturalistic or more ecologically appropriate management of urban green spaces necessitates a change in expectations of visitors and a willingness to accept some 'messiness' in the landscape, to let go of a certain degree of control, and to accept the surprises and vagaries of natural processes. For example, the inclusion of culturally relevant 'cues to care' into more naturalistic design can result in higher preference for and acceptance of such landscapes (Nassauer 1997), and Sheffield park users were willing to accept management for biodiversity where it was compatible with human uses (Irvine *et al.* 2010). Green space management that includes both human and biological factors is already enshrined in law in several developed nations, although it is rare in the developing world (Pierce *et al.* 2005). Plans for urban green space management that fail to include citizens' perceptions will not be widely accepted or understood (Balram & Dragicevic 2005).

Public green space and private gardens might, to some extent, provide alternatives for contact with nature in urban settings. Indeed, in Sheffield there is a negative correlation between the extent of public green space and private garden space across the city (Barbosa *et al.* 2007). However, the degree to which these radically different kinds of green space can substitute for one another is unclear as the two very frequently play different roles (Kellett 1982). For example, public green space can promote community integration while social interactions in gardens are focused around a private social network (Bernardini & Irvine 2007). Research that explicitly compares measures of perceptions, activities, biodiversity contact and human wellbeing in public versus private green spaces will be required to assess the consequences of changes in the proportion of urban green space that is provided via the two different routes.

Maintaining a private garden

Despite their generally small size in comparison with publicly available urban green spaces, private domestic gardens (backyards) are probably the most easily accessed and close at hand elements of nature for many urban dwellers. Frequent contact with plants and animals is possible, and the garden forms the arena for many childhood experiences of nature (Bernardini & Irvine 2007; Louv 2008). Individual gardens are a dense mosaic of many microhabitat types, with UK gardens averaging six, but reaching up to 12 land cover types despite median garden size being only 100–200 m^2 (Loram *et al.* 2007, 2008b). Although urban gardens are a habitat type with no real natural analogue (Gaston *et al.* 2007a), they can be surprisingly biodiverse (Owen 1991). A sample of 267 private gardens in the UK together contained more than 1000 plant species, many non-native or planted (Loram *et al.* 2008a). As such, gardens can provide people with exposure to a concentrated and intensified spectrum of natural processes through seasonal and life cycles, and thus contribute substantively to quality of life (Martin & Stabler 2004).

In the USA, at least 78% of households participate in some form of gardening activity (Clayton 2007), and 87% of UK households have access to a garden (Davies *et al.* 2009). Forty per cent of African urban dwellers engage in urban agriculture in private gardens (Mougeot 1994), and this proportion has continued to increase since the 1990s, in line with decreasing food security and ongoing urbanisation of the African population (Foeken *et al.* 2002). In 1998, about 30% of the population of 239 000 people in Nakuru, Kenya was engaged in some form of urban agriculture, producing 8000 tonnes of crops per annum, enough to satisfy about 30% of the city's food requirements (Foeken *et al.* 2002). There is substantial variation within cities, with 27% of households in inner city locations in five UK cities reporting no access to any type of outside space, while the corresponding figures for outer suburbs and areas situated between the suburbs and the inner city were 3.1% and 4.5% respectively (Gaston *et al.* 2007b).

According to Francis (1988), gardens are a venue to exert and maintain control in contrast to the world beyond the garden fence, although Catanzaro and Ekanem (2004) found no evidence that control of nature is important to gardeners. Whether or not control is an important driving factor, vegetation managed by citizens in their private gardens forms a significant component of all urban plant assemblages. Gardens cover about one-quarter of a typical UK city, and form up to a half of all urban green space coverage (Loram *et al.* 2007), as well as affecting patterns of global environmental change (Niinemets & Peñuelas 2009). Enhancing the quality of garden habitat emerged as more likely to improve ecological connectivity than the specific establishment of green corridors across urban landscapes in British Columbia, Canada (Rudd *et al.* 2002). Plant species composition, richness, evenness and density are continually influenced by human intervention in private gardens (Faeth *et al.* 2005), and floras are typically dominated by non-native species (Loram *et al.* 2008a). Heavy subsidies of nutrients and control of competition by gardeners leads to the persistence of species at lower densities than could occur in unmanaged populations (Smith *et al.* 2006b; Loram *et al.* 2008a). Human control over urban plant assemblages therefore appears to overwhelm geographic, historical and climatic variation among cities, with species richness, diversity and composition of garden plant assemblages being highly conserved across five disparate UK cities (Loram *et al.* 2008a).

In contrast to the findings from Europe, urban gardeners in Boa Vista, Brazil, concentrate on only a few species of exotic trees, primarily because of their superior fruit production (Semedo & Barbosa 2007). Likewise, urban gardens in Limete, Democratic Republic of Congo, show low plant species richness, with only 18 tree species in 201 plots (Makumbelo et al. 2002). A recent assertion that urban gardens in China support few species of birds, bees and butterflies was accompanied with a call for research into how such places might be improved for biodiversity conservation (Wang et al. 2007).

Cultural differences in how gardens are used relate in part to the historical provision of gardens as housing stock was built in successive planning cycles, and in part to the pervasive economic necessity associated with using vegetable and livestock products from small garden plots. In developing countries, urban gardening is often carried out for subsistence purposes, e.g. producing food from urban gardens in Mozambique (Sheldon 2003), the Philippines (Miura et al. 2003) and Peru (Works 1990). Garden maintenance in Africa for subsistence or as a small-scale commercial enterprise occurs wherever small plots of land are available, around the margins of the home, or in vacant plots or those owned by a third party. As cities grow, former peri-urban areas become densely settled, plots are subdivided and agriculture continues in the land available in the vicinity of the home (Foeken et al. 2002).

An urban garden in Cape Town of only 30 m^2 is capable of providing half the vegetable subsistence requirements for a household (Eberhard 1989, cited in Slater 2001), although potential productivity is higher than currently realised, and there is much interest in improving yields and horticulture techniques (Nugent 2000). Extensions of subsistence gardening include sale of produce at local markets, as occurs in the closely settled peri-urban regions around Bobo-Dioulasso, Burkina Faso (Freidberg 2001), and many African cities depend on 'garden belts' around their perimeter for a high proportion of their food requirements (Guyer 1987; Linares 1996). Small-scale urban agriculture competes with intense land-use pressures in rapidly growing cities across the developing world. Although urban agriculture is potentially economically viable in South Africa, intense competition for land in Cape Town, ironically in part from nearby nature reserves, means that plans for its continued expansion are uncertain there (Reuther & Dewar 2005).

While there are tremendous cultural differences in the way gardens are managed, and in the motivation for cultivating particular plants, their ecological ramifications are scarcely understood. Uses of private gardens varies markedly in California, such that they are seen by white-collar suburbanites as places of escape or retreat from the stresses of the day, but by poor, urban African-Americans as a source of food, a tie to cultural practices passed down from parents and grandparents, as well as a safe haven in an otherwise less than safe neighbourhood (McNally 1987, 1990). Economic factors can drive biodiversity in private gardens in rather direct ways. For example, in urban Barcelona higher-income households can afford more water-consuming plants, and tend to have gardens dominated by Atlantic species, while lower-income households tend to use species adapted to the drier local Mediterranean climate (Domene *et al.* 2005). Thus a major dichotomy in garden plant assemblages is driven by the affordability of irrigation.

Managing nature in private gardens

Gardening provides multiple opportunities for interacting with nature. Decisions about garden management are made at the household level and are strongly influenced by the personal characteristics of the householders (Figure 7.2). Fieldwork in Sheffield revealed a wide range of garden management regimes from minimal active planting or vegetation control, through assiduous planting and management of vegetation for a range of purposes, to completely paving over the space and continually treating it with herbicide to ensure no vegetation grows within the garden (Irvine *et al.* 2010).

Perhaps most obviously, and whatever the management motive, gardeners control the range of plant species and the abundance of individual species present in their garden. Gardeners typically favour non-native plants, either alone or in combination with natives (Zagorski *et al.* 2004; Head & Muir 2006), and this is reflected in the high proportion of non-native species available at horticultural suppliers (Thompson *et al.* 2003), planted in private gardens (Loram *et al.* 2008a) and escaping therefrom to form established populations in the wild (Dehnen-Schmutz *et al.* 2007). While it seems logical that native plants will provide food for coevolved native invertebrates, assertions that non-native garden plants can have a significant negative impact on the natural environment (e.g. Steinberg 2006) require more detailed investigation, and there is much debate about the relevance of exotic status alone in predicting biological quality of garden plants (Gaston *et al.* 2007a). In one of the few direct tests of this issue, the richness and abundance of invertebrates in Sheffield gardens were not dependent on the proportion of plants that were of native origin (Smith *et al.* 2006a, 2006c). In addition, the presence of structurally complex vegetation was more important than its species composition in predicting occupancy of urban sites by small mammals and terrestrial reptiles in Brisbane, Australia (Garden *et al.* 2007).

Garden management decisions depend strongly on garden size. Across five cities in the UK, many garden features were more likely to be present in larger gardens (cultivated borders, lawns, paths, uncultivated land, compost heaps/bins, ponds, vegetable patches, greenhouses, unmown grass, tall trees, and linear features within the garden such as hedges and walls; Loram *et al.* 2008b). As well as affecting the biodiversity value of gardens, such features

Figure 7.2 Divergent approaches to garden management are evident in the contrast between neighbouring gardens in suburban Sheffield, UK. Such variation in how gardens are managed makes for a highly heterogeneous resource for wildlife when summed across an urban landscape. Photograph used with permission of R. A. Fuller.

contribute to ecosystem service provision by gardens. For example, paving of front gardens in the UK has led to significant concerns about elevated stormwater runoff and has led to additional management costs for local authorities (Perry & Nawaz 2008). About 12% of respondents to a Sheffield survey indicated that they parked their car within their garden space (Table 7.1). The likelihood of large trees being in a garden reduces significantly in smaller gardens (Smith *et al.* 2005), so provision of carbon sequestration depends on how garden management decisions interact with garden size. Urban areas experience heat island effects, in which temperatures are elevated in comparison with surrounding landscapes, particularly at night and in colder weather (Pickett *et al.* 2001). Decisions by individual householders can moderate these effects insofar as cover by trees and impervious surfaces determines the magnitude of the effect (Chen & Wong 2006). Urban parcel sizes are already small in comparison with those in peri-urban and rural areas (Luck & Wu 2002), and ongoing reductions in garden sizes through increasing residential density will no doubt influence future land cover change.

Table 7.1 *Activities carried out at least weekly in Sheffield, UK, gardens in summer and winter.*

Activity	Summer	Winter
Relax	76	4
Gardening	70	10
Observe nature	52	40
Think about things	51	10
Socialise	42	2
Play sports	23	2
Park car	13	12

Note: Figures are percentages of 671 respondents to a postal questionnaire sent to 2016 households with garden access in Sheffield.

Concern has recently been raised about the consequences of garden management decisions for areas outside the garden itself. For example, the use of peat fertilisers in horticulture is widespread despite the fact that large-scale peat extraction causes loss of biodiversity and ecosystem services (Chapman *et al.* 2003). The likelihood of plant species escaping horticultural use and becoming established in the wild in the UK is positively related to their frequency of occurrence in the nineteenth-century horticultural market, indicating that social trends in gardening can influence the rate of establishment of non-native species (Dehnen-Schmutz *et al.* 2007).

The degree to which decisions about the management of different gardens are independent is uncertain, but critical if we are to understand how garden management decisions influence biodiversity and determine the ecological quality of urban landscapes. Gardens form an interesting gradient between the public arena of the street and the private interior of the home. The garden itself is in many ways a semi-public, semi-private space; many urban gardens are overlooked by neighbours in the back or people passing by in the front, yet are frequently enclosed by a hedge or fence. Because of the semi-public nature of gardens, perceptions of social norms are one source of dependency that could lead to homogenisation of garden management within neighbourhoods, but starker differences among neighbourhoods. We are not aware of any work on this issue, although intriguing work in Tasmania has shown that front and back gardens show consistently different floristic composition and vegetation structure (Daniels & Kirkpatrick 2006). While there was no difference in plant species richness between front and back gardens, the former were dominated by trees and shrubs, and lacked features indicating production and function. Relegation of production and function to backyards is also evident in the USA (Dorney *et al.* 1984; Richards *et al.* 1984), and may relate to security concerns (Daniels & Kirkpatrick 2006). Interestingly, there was no evidence that Tasmanian front gardens were more showy (of complex ornamental design rather than productive or utilitarian) than back gardens; rather it appeared that the degree of showiness reflected householder attitudes to, and interest in, gardening, and was thus manifest in all their garden space. However, the proportion of gardens that were showy in the front and not showy in the back was negatively related to suburb age.

Rented and shared accommodation are major exceptions to the rule that domestic gardens are privately owned. In Tasmania, the proportion of gardens in a suburb with non-showy front and back components increased with the proportion of households that were rented (Daniels & Kirkpatrick 2006). This could either represent a difference in the characteristics of the people occupying owner versus rented accommodation, or a difference in motivation when the property is not personally owned.

Motivations for garden management

While urban agriculture is motivated primarily by the need for food, it can also provide employment, food security, income (direct or fungible), and impetus for urban sustainability and development in general (Smit et al. 1996). About one-third of crops are self-consumed in Nakuru, Kenya (Foeken et al. 2002), and garden crops can add 20% to household income in Bangkapi, Bangkok (Fraser 2002). As well as utilitarian benefits, there are substantial additional advantages such as empowerment of women via control over household food consumption, the formation of social networks and community development (Slater 2001). For example, small-scale urban gardening in Brazil promotes social interaction between urban and rural communities as well as providing subsistence (Winklerprins 2003). Entrepreneurial urban agriculture is a growing phenomenon in the USA, and provides a focus for community activity as well as enhancing local food security (Kaufman & Bailkey 2000; Saldivar-Tanaka & Krasny 2004).

Where it is not directly focused on growing food for subsistence, gardening is perceived as both work and recreation. Fifty-six per cent of householders in Perth, Australia, indicated gardening to be both work and recreation, but over 70% regarded it as a valuable way to spend time (ARCWIS 2002). We found comparable results in Sheffield, and the distribution of responses was surprisingly similar across the two cities (Figure 7.3). Studying customers recruited from a garden centre in Ohio, USA, Clayton (2007) identified a range of motivations for gardening. In order of importance, these were spending time outdoors, observing nature or natural processes, relaxation, controlling appearance of the garden, working with the hands, novelty, producing food or herbs, demonstrating effort and demonstrating expertise. Spending time outdoors, observing nature and relaxation were significantly more important

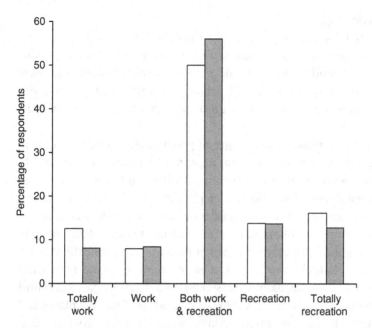

Figure 7.3 Attitudes to gardening are remarkably similar in studies across two cities. Responses were to the same stem question, 'Do you consider gardening to be work or a recreation activity?' Open bars are responses by 615 householders in Perth, Australia (data from ARCWIS 2002), and shaded bars are responses by 663 householders in Sheffield, UK.

than all of the others. This closely echoes our finding from Sheffield that observing nature was the only important use of gardens in winter, when many other garden activities ceased (Table 7.1). Reasons for specific garden management decisions in Clayton's (2007) Ohio study included, in order of importance, achieving colourful appearance, safety, reducing or eliminating weeds, maintaining a healthy ecosystem, enhancing property value, ease of maintenance, demonstrating care, minimising resource use, low cost, using native plants and meeting neighbourhood standards. Perhaps surprisingly, individuals were neutral as to whether the practices used in their yard affected the local environment. Practical concerns as drivers for gardening practices were negatively correlated with use of the garden for interacting with nature, suggesting that the two forms of motivation for garden management are somewhat opposed (Clayton 2007).

There is marked variation across society in the use of 'wildlife-friendly' garden features and the propensity to engage with nature, but rather few investigations of the motivations for particular management decisions. From a policy perspective, both the conservation of urban biodiversity and the enhancement of public health depend on a better understanding of this issue.

Benefits of gardening

Gardening is generally perceived as being important to overall wellbeing. Responses to a similar question, though ranked on different scales, about the importance of spending time in the garden to overall wellbeing were remarkably similar for householders in Perth and Sheffield, with the majority of respondents indicating that it was quite important or very important (Figure 7.4).

Gardening produces measurable changes to psychological processes, enhancing self-esteem, self-efficacy and personal identity (Bernardini & Irvine 2007). Texan gardeners reported higher satisfaction with life and better self-reported health than non-gardeners, and had higher physical activity levels (Waliczek et al. 2005). Among gardeners the act of gardening itself brings satisfaction and a pleasant break from household chores or routine (Kaplan 1973; ARCWIS 2002), and the availability of places to garden increases neighbourhood satisfaction (Frey 1981). Gardens are full of cues to connect people with their personal history. For example, daily or seasonal cycles, visits by particular bird species or flowering of a favourite plant bring reminders of people, places and events (Francis 1988; Bernardini & Irvine 2007). Twenty-eight of forty interviewees in Leicester, UK, expressed positive place identity associated with their garden (Bernardini & Irvine 2007). Other benefits include the joy of anticipation and sense of accomplishment that come through the processes of garden planning, planting and harvesting, sustained engagement with the natural environment and reduction in stress levels (Kaplan 1973; Kaplan & Kaplan 1989; Catanzaro & Ekanem 2004). Kaplan and Kaplan (1989, p. 170) suggest that 'an important source of satisfaction derived from gardening is that it holds one's attention in a multitude of ways, even when the garden lies dormant'.

A range of health measures in Scotland, UK, are related to household tenure, such that rented housing in the public sector is associated with particularly high levels of health problems. However, these effects are partially offset by the availability of gardens, with 56.4% of respondents without access to a garden describing their health as fair/poor, compared with 37.3% among those with access to a garden (Macintyre et al. 2003). Having access to and visiting one's garden frequently were both associated with lower self-reported sensitivity to stress among people across nine Swedish cities (Stigsdotter & Grahn 2004).

Benefits of gardens need not involve direct interaction. For example, having a 'green' view increased cognitive functioning of children living in single-family homes in the USA (Wells 2000). Similarly, having access to a view over gardens is positively associated with residents' sense of community (Kaplan 1985), and satisfaction with the neighbourhood, as well as an individual's ability to function effectively (e.g. plan), among individuals living in multi-storey, multi-family buildings (Kaplan 2001).

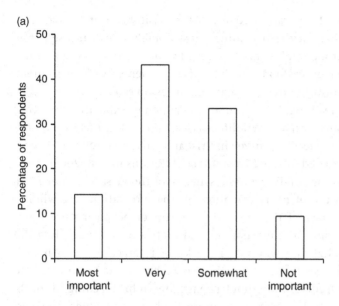

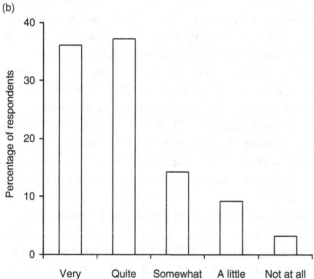

Figure 7.4 The importance of spending time in the garden to overall wellbeing in studies across two cities for (a) 616 householders in Perth, Australia (data from ARCWIS 2002), and (b) 667 householders in Sheffield, UK.

Feeding urban wildlife

In one of the most direct people–nature interactions, many people provide food for urban wildlife. Food is provided for birds at an estimated 12.6 million of the 26 million homes in the UK (Davies *et al.* 2009), 43–50% of the households in the USA and 25–57% of the households in Australia (Jones & Reynolds 2008). Such activity is not limited to private gardens; 36% of visitors to urban parks in Sheffield indicated that they feed wildlife, most notably ducks and squirrels (Irvine *et al.* 2010). Although birds are the most frequent recipients, mammals are also given supplementary food, with 11% of respondents to a British survey of urban residents indicating they fed mammals (Gaston *et al.* 2007b), and 35% of Americans who feed wildlife around their home explicitly targeting mammals (United States Fish and Wildlife Service 2001). Three of 134 respondents to a garden wildlife feeding survey in Brisbane, Australia, reported specifically feeding blue-tongued skinks *Tiliqua scincoides* (Rollinson *et al.* 2003).

Observing and feeding wildlife predicts neighbourhood satisfaction (Frey 1981), but we are aware of no information on the motivations for wildlife feeding. Because it is one of the most direct forms of people–nature interactions, an appreciation of the perceptions and motivations surrounding the activity could throw light on why people seek out interactions with nature close to where they live. Rather more is known about the socioeconomic correlates of wildlife feeding. Householders engaging in bird feeding in southeast Michigan tended to be older, were more likely to be women and had achieved higher educational qualifications than those not participating (Lepczyk *et al.* 2004), although bird feeding activity was not related to the number of people in the household, their occupation or the size of the dwelling.

What are the costs and benefits of wildlife feeding? On the one hand, increasing public participation in wildlife feeding activity is an explicit policy target in the UK, based on the idea that enhancing this interaction with nature will simultaneously benefit both human wellbeing and biodiversity (DEFRA 2002). On the other hand, wildlife feeding is highly controversial in some countries. For example, it is strongly discouraged or illegal in Australian protected areas, particularly where dangerous wildlife are present (Mallick & Driessen 2003; Petrie *et al.* 2003). It has been suggested that the strongly seasonal climate of northern latitudes makes supplementary feeding a logical activity to 'help' wildlife through the cold winter months when food is scarce, but that this paradigm does not translate well to less seasonal environments such as those in Australia, and as a consequence bird feeding is not beneficial (Parsons 2007). However, there is no good evidence that wildlife feeding has fundamentally different effects on Australian faunas compared with those of Europe or North America. In fact, the evidence base surrounding the impacts of supplementary wildlife feeding in general, both positive and negative, is so weak that it is hard to draw any definitive conclusions at this stage. Studies that can clearly disentangle feeding effects from urbanisation, habitat and individual quality effects are urgently required (Jones & Reynolds 2008).

Despite the controversy surrounding wildlife feeding in Australia, many visits to national parks are motivated by wildlife feeding, and participation rates among the Australian public remain high (Woodall 1995; Rollinson *et al.*

2003; O'Leary & Jones 2006). In response to this, best practice guidelines have been published, alongside arguments that habitat enhancement rather than feeding is the best way to encourage wildlife to visit a domestic garden (Parsons 2007; Plant 2008). The evidence base to support the details within such best practice guidelines is acknowledged as limited, and work on this issue is urgently needed (Jones 2008; Jones & Reynolds 2008). For example, although the spread of mycoplasmal conjunctivitis in house finches *Carpodacus mexicanus* in the USA has been linked to bird feeders (Fischer *et al.* 1997), and tubular feeders in particular (Hartup *et al.* 1998), none of 25 samples taken from 25 bird feeding stations in Sheffield contained *Salmonella* spp. (Cannon 2005).

Such widespread resource provision is likely to have significant impacts on urban wildlife populations (Cannon 1999; Shochat 2004; Faeth *et al.* 2005). A study in Sheffield found a positive relationship between levels of bird feeding and the abundance of species known regularly to take supplementary food from gardens (Fuller *et al.* 2008). There was no association between bird feeding and the abundance of birds that did not regularly take supplementary food, and there was also no effect on avian species richness. Experimental work to determine the extent to which direct feeding of wildlife changes assemblage structure at a landscape scale would be valuable. Such work should incorporate differences in the frequency of feeding, given that 46% of households feeding birds across five UK cities did so less than once per week (Gaston *et al.* 2007b).

Feeding wildlife in private gardens also shapes non-urban ecosystems. The seeds of around 30 species of plant are regularly used in the feeding of birds in the UK, but seed mixes often contain small numbers of many other species. More than 400 plant species are believed to have been imported in this way with bird seed into Britain from various countries around the world, and subsequently have been found growing in the wild (Hanson & Mason 1985). Moreover, given that they are grown precisely because of their attractiveness to birds, cash crops aimed at the bird seed market, such as sunflower *Helianthus* species, are at risk of *in situ* damage by wild birds (Blackwell *et al.* 2003). Avicides are used to control such damage, but this is a highly controversial practice (Fenwick 2001), and ironic given that part of the crop is destined for bird feeders in domestic gardens.

Wildlife feeding could potentially be harnessed explicitly on a large scale to influence the conservation status of a particular species occurring within urban areas. Urban environments support nationally important populations of some species (Bland *et al.* 2004; Cannon *et al.* 2005; Chamberlain *et al.* 2005), yet the degree to which this people–nature interaction could be used for conservation remains unknown, and seems an exciting avenue for further work. Such interactions might also contribute to targets relating to urban liveability, yet we know almost nothing about the benefits to people of engaging in wildlife feeding.

People–nature interactions: common threads

People–nature interactions shape urban environments

It is fast becoming apparent that the environmental outcomes of people–nature interactions in urban areas are significant. Most obviously, many vegetation communities in urban areas are deliberately planted, or regularly managed, be they in public green spaces or in private gardens. Thus the composition of urban plant assemblages, and the distribution of vegetation types and habitats across the urban landscape, are continually shaped by human management decisions. Such management is often motivated by the need for public accessibility, safety and amenity, and the presence of large numbers of people leads to disturbance effects of various kinds. The dynamics of decisions about garden management have a large potential impact, especially if we consider how they sum to generate landscape level effects (Niinemets & Peñuelas 2009). Gardens are places for self-expression, and thus they are places where the human psyche is linked intimately with the distribution and abundance of species in the environment.

Thus to a first approximation, it is human decisions that structure most of Bell's (1976) 'fabricated or disrupted' urban nature. Fingerprints of societal changes over time are reflected in components of urban nature such as changing preferences for planting certain species, the age, density and maturity of street trees, and evolving paradigms of urban park management (Richards 1983; Zipperer & Zipperer 1992). Despite this, there remains relatively little direct integration of human processes into the study of urban ecology in the 10 years since Hostetler's (1999) call for action.

Interacting with nature benefits people

It is clear that people derive a range of benefits from interacting with nature. Benefits to mental and physical health and wellbeing are relatively well documented (for reviews see Chapter 9; Irvine & Warber 2002b; Brown & Grant 2005), but the implications for quality of life more broadly have scarcely been studied. Strategies to improve quality of life among urban dwellers urgently require this knowledge. Other aspects of human quality of life that also depend to varying extents on access to urban nature include economic sustainability, social cohesion, the success of urban regeneration programmes, crime reduction, sense of community and environmental awareness (Department of Transport, Local Government and the Regions 2006; Maller et al. 2006).

Given the many documented benefits of local interactions with nature, low levels of ambient biodiversity close to where people live can be interpreted as an axis of deprivation. In Sheffield there is a negative relationship between bird species richness and the level of deprivation among neighbourhoods, as measured by the Index of Multiple Deprivation (Irvine et al. 2010). Less privileged sectors of society have lower levels of ambient biodiversity around the places where they live, suggesting that lack of access to biodiversity can compound

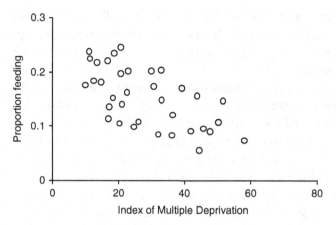

Figure 7.5 Socioeconomic deprivation predicts engagement with nature. The proportion of households at which birds are fed declines among 35 neighbourhood types in Sheffield, UK, with increasing deprivation. Reproduced with permission from Fuller *et al.* (2008).

more traditional measures of social and economic deprivation. This contrasts, however, with the finding that access to local green space was better in more deprived neighbourhoods in the same city (Barbosa *et al.* 2007). Socioeconomic deprivation is also linked to the likelihood of interacting with nature. For example, people in more deprived neighbourhoods of Sheffield were also less likely to feed garden birds, perhaps reflecting either constraints to the activity among those neighbourhoods, or a response to few bird species being present in those areas (Figure 7.5; Fuller *et al.* 2008). Thus, there is some evidence that interactions with nature might decline in less affluent sectors of human society, potentially compounding other components of deprivation.

Clearly, there is also potential for negative outcomes of some people–nature interactions, such as those with household and garden pests, encounters with dangerous species, or negative perceptions of safety in densely vegetated areas. Management of such conflicts between humans and wildlife in urban areas has been amply reviewed elsewhere (Conover 2001).

People and urban nature are interdependent

The increasingly well-documented associations between human and biological organisation suggest that there is a close interplay between the configurations of ecosystems and human communities in urban environments (Hostetler 1999; Hope *et al.* 2003). Evidence of such mutual feedback mechanisms in people–nature interactions suggests that individual components of the interactions cannot properly be studied independently. For example, escape of non-native species in urban environments can create problems that are expensive to deal with, but resolving the issue depends on a good understanding of the decision-making process surrounding species selection (Ehrenfeld 2008). Expending time and effort to attract wildlife into a domestic garden is perhaps more likely to continue if the activity is rewarded by the presence of certain target species (Rollinson et al. 2003).

Perhaps one of the most important questions in this regard is how perceptions resulting from experiences gained in the urban environment translate into wider attitudes to nature and biodiversity conservation (Lohr & Pearson-Mims 2005; Cheesman & Key 2007; also see Clergeau et al. 2001). Are people more likely to conserve nature when they have positive and direct experiences of biodiversity on a regular basis? Childhood interactions with nature are cited as important formative experiences by those active in conservation later in life (Tanner 1980; Chawla 1999), and so the 'extinction of experience' resulting from low levels of contact between today's children and biodiversity has been cited as one plausible reason for the lack of a coherent global strategy for biological conservation (Pyle 1978; Miller 2005). Perhaps what is needed is a re-learning of how to interact with the natural environment. Not only is it important to understand factual information about the landscape, such as the names of birds visiting one's back garden, but also to have procedural or skill-based knowledge (e.g. how to set up a compost pile) that may facilitate more active engagement. Research on this issue is timely, because there has been a recent upsurge of interest in how local living relates to global conservation, most notably through contributions to climate change. If such links can successfully be made, local experiences of biodiversity could well prove to be important in determining the future of the global response to the biodiversity crisis (Ehrlich 2002).

People–nature interactions are diverse

There are many more interactions between people and nature than we have been able to cover specifically here, including knowing that nature is nearby, watching wildlife on television, hanging pictures of nature on the wall, viewing nature through a window, visiting urban zoos and botanic gardens, campaigning for environmental change, joining local conservation management groups and engaging in citizen science. Many activities are hard to delineate given that humans are almost perpetually in contact with some form of nature in one way or another. A clear typology of people–nature interactions is lacking, but would be useful as research in this field continues to develop.

People–nature interactions vary with socioeconomic factors and the cultural background of participants. The literature is strongly biased toward Western cultures in terms of overall quantity, but there is also cultural variation in emphasis (Teel et al. 2007). Studies in Western societies tend to focus on human quality of life and wellbeing, and maintaining the biological quality of the urban landscape, while those in non-Western societies frequently emphasise subsistence, economic gain and developmental potential of interactions between people and nature. Perhaps most notably, there is a large literature on urban agriculture in developing countries (see, for example, the 526 references in Obudho & Foeken 1999). Because of its inherently interdisciplinary nature, work on people–nature interactions is scattered through the natural science and social science literatures (for example, the citations in this chapter arise from journals in 25 major subject areas). Our impression is that there is far more work by social scientists considering the importance of nature for humans than by biologists considering the importance of human decisions in structuring biological communities.

Concluding remarks

Experiences of nature have declined in quantity and quality over the past few decades, and concerns are mounting that this process will accelerate as the world's population becomes increasingly urbanised. Anthropogenic simplification of ecosystems through harvesting and agriculture, the setting aside of protected areas for biodiversity conservation, and urbanisation of the world's population have all contributed to a reduction in contact between people and biodiversity, and a growing separation between people and nature that shows no sign of slowing. This separation appears to have two major consequences, (i) a reduction in the contribution of experiences of nature to human quality of life, and (ii) a lack of broad-based support for biodiversity conservation (Miller 2005).

Biodiversity conservation is a response to anthropogenic impacts on ecosystems, and as such depends on a good understanding of the motivations and drivers of human behaviours that lead to such impacts. Given that most of the world's population lives in urban areas, this is where we must look to study human relationships with biodiversity, because the outcomes of interactions with nature appear to be important in shaping our values and attitudes toward nature conservation. Implementing solutions to the biodiversity crisis will depend on interdisciplinary research efforts as well as systems of implementation that can actively trade off ecological value and benefits to human wellbeing (Polasky *et al.* 2008). Much remains to be done, and we encourage increased collaboration among researchers whose disciplines intersect on these issues.

Acknowledgements

R.F. is supported by the Applied Environmental Decision Analysis research hub, funded through the Commonwealth Environment Research Facilities programme, Australia, and additionally by a grant from the University of Queensland. We thank O. Barbosa, Z. Davies, P. Devine-Wright, C. Fuller, K. Gaston, S. Herbert, S. Kark, J. Rhodes, J. Tratalos, S. Warber and P. Warren for helpful discussions on people–nature interactions, and useful comments were provided by two reviewers.

References

Alberti, M. and Marzluff, J. M. (2004). Ecological resilience in urban ecosystems: linking urban patterns to human and ecological functions. *Urban Ecosystems*, **7**, 241–65.

Alberti, M., Marzluff, J. M., Shulenberger, E. *et al.* (2003). Integrating humans into ecology: opportunities and challenges for studying urban ecosystems. *BioScience*, **53**, 1169–79.

Angold, P. G., Sadler, J. P., Hill, M. O. *et al.* (2006). Biodiversity in urban habitat patches. *Science of the Total Environment*, **360**, 196–204.

ARCWIS (2002). *Perth Domestic Water-use Study: Household Appliance Ownership and Community Attitudinal Analysis 1999–2000*. Perth, Australia: Australian Research Centre for Water in Society, CSIRO.

Balram, S. and Dragicevic, S. (2005). Attitudes toward urban green spaces: integrating questionnaire survey and collaborative GIS techniques to improve attitude measurements. *Landscape and Urban Planning*, **71**, 147–62.

Barbosa, O., Tratalos, J. A., Armsworth, P. R. *et al.* (2007). Who benefits from access to green space? A case study from Sheffield, UK. *Landscape and Urban Planning*, **83**, 187–95.

Bastin, L. and Thomas, C. D. (1999). The distribution of plant species in urban vegetation fragments. *Landscape Ecology*, **14**, 493–507.

Bell, D. (1976). *The Coming of Post-industrial Society*. New York: Basic Books.

Bell, S., Morris, N., Findlay, C. *et al.* (2004). *Nature for People: The Importance of Green Spaces to East Midlands Communities*. Peterborough, UK: English Nature.

Bengston, D. N., Fletcher, J. O. and Nelson, K. C. (2004). Public policies for managing urban growth and protecting open space: policy instruments and lessons learned in the United States. *Landscape and Urban Planning*, **69**, 271–86.

Bernardini, C. and Irvine, K. N. (2007). The 'nature' of urban sustainability: private or public greenspaces? *Sustainable Development and Planning III*, **102**, 661–74.

Bird, W. (2004). *Natural Fit: Can Green Space and Biodiversity Increase Levels of Physical Activity?* Sandy, UK: Royal Society for the Protection of Birds.

Bird, W. (2007). *Natural Thinking: Investigating the Links Between the Natural Environment, Biodiversity and Mental Health*. Sandy, UK: Royal Society for the Protection of Birds.

Bixler, R. D. and Floyd, M. F. (1997). Nature is scary, disgusting, and uncomfortable. *Environment and Behavior*, **29**, 443–67.

Bjerke, T., Østdahl, T., Thrane, C. and Strumse, E. (2006). Vegetation density of urban parks and perceived appropriateness for recreation. *Urban Forestry and Urban Greening*, **5**, 35–44.

Blackwell, B. F., Huszar, E., Linz, G. M. and Dolbeer, R. A. (2003). Lethal control of red-winged blackbirds to manage damage to sunflower: an economic evaluation. *Journal of Wildlife Management*, **67**, 818–28.

Bland, R. L., Tully, J. and Greenwood, J. J. D. (2004). Birds breeding in British gardens: an underestimated population? *Bird Study*, **51**, 96–106.

Bradshaw, G. A. and Bekoff, M. (2000). Integrating humans and nature: reconciling the boundaries of science and society. *Trends in Ecology and Evolution*, **15**, 309–10.

Braun, B. (2005). Environmental issues: writing a more-than-human urban geography. *Progress in Human Geography*, **29**, 635–50.

British Trust for Ornithology (2006). We spend £200 million a year on wild bird food. Thetford, UK: BTO Press release No. 2006/12/76, December 2006.

Brown, C. and Grant, M. (2005). Biodiversity and human health: what role for nature in healthy urban planning? *Built Environment*, **31**, 326–38.

Bryant, M. M. (2006). Urban landscape conservation and the role of ecological greenways at local and metropolitan scales. *Landscape and Urban Planning*, **76**, 23–44.

Bucur, V. (2006). *Urban Forest Acoustics*. New York: Springer.

CABE Space (2004). *The Value of Public Space: How High Quality Parks and Public Spaces Create Economic, Social and Environmental Value*. London: Commission for Architecture and the Built Environment.

CABE Space (2005a). *Decent Parks? Decent Behaviour?* London: Commission for Architecture and the Built Environment.

CABE Space (2005b). *Start with the Park: Creating Sustainable Urban Green Spaces in Areas of Housing Growth and Renewal*. London: Commission for Architecture and the Built Environment.

CABE Space (2007). *Living with Risk: Promoting Better Public Space Design*. London: Commission for Architecture and the Built Environment.

Cannon, A. (1999). The significance of private gardens for bird conservation. *Bird Conservation International*, **9**, 287–98.

Cannon, A. R. (2005). Wild birds in urban gardens: opportunity or constraint? Unpublished Ph.D. thesis, University of Sheffield.

Cannon, A. R., Chamberlain, D. E., Toms, M. P., Hatchwell, B. J. and Gaston, K. J. (2005). Trends in the use of private gardens by wild birds in Great Britain 1995-2002. *Journal of Applied Ecology*, **42**, 659–71.

Catanzaro, C. and Ekanem, E. (2004). Home gardeners value stress reduction and interaction with nature. *Acta Horticulturae*, **639**, 269–75.

Chace, J. F. and Walsh, J. J. (2006). Urban effects on native avifauna: a review. *Landscape and Urban Planning*, **74**, 46–69.

Chamberlain, D. E., Gosler, A. G. and Glue, D. E. (2007). Effects of the winter beechmast crop on bird occurrence in British gardens. *Bird Study*, **54**, 120–6.

Chamberlain, D. E., Vickery, J. A., Glue, D. E. *et al.* (2005). Annual and seasonal trends in the use of garden feeders by birds in winter. *Ibis*, **147**, 563–75.

Chapman, S., Buttler, A., Francez, A.-J. *et al.* (2003). Exploitation of northern peatlands and biodiversity maintenance:

a conflict between economy and ecology. *Frontiers in Ecology and the Environment*, **1**, 525–32.

Chawla, L. (1999). Life paths into effective environmental action. *Journal of Environmental Education*, **31**, 15–26.

Cheesman, O. D. and Key, R. S. (2007). The extinction of experience: a threat to insect conservation? In A. J. A. Stewart, T. R. New and O. T. Lewis, eds., *Insect Conservation Biology*. London: Royal Entomological Society, pp. 322–48.

Chen, Y. and Wong, N. H. (2006). Thermal benefits of city parks. *Energy and Buildings*, **38**, 105–20.

Chiesura, A. (2004). The role of urban parks for the sustainable city. *Landscape and Urban Planning*, **68**, 129–38.

Cilliers, S. S. and Bredenkamp, G. J. (2000). Vegetation of road verges on an urbanisation gradient in Potchefstroom, South Africa. *Landscape and Urban Planning*, **46**, 217–39.

Clayton, S. (2007). Domesticated nature: motivations for gardening and perceptions of environmental impact. *Journal of Environmental Psychology*, **27**, 215–24.

Clergeau, P., Mennechez, G., Sauvage, A. and Lemoine, A. (2001). Human perception and appreciation of birds: a motivation for wildlife conservation in urban environments of France. In J. M. Marzluff, R. Bowman and R. Donnelly, eds., *Avian Ecology and Conservation in an Urbanizing World*. Norwell, MA: Kluwer Academic Publishers, pp. 69–88.

Clergeau, P., Sauvage, A., Lemoine, A. *et al.* (1997). Quels oiseaux dans la ville? *Les Annales de la Reserche Urbaine*, **74**, 119–30.

Collins, J. P., Kinzig, A., Grimm, N. B. *et al.* (2000). A new urban ecology. *American Scientist*, **88**, 416–25.

Comedia and Demos (1995). *Park Life: Urban Parks and Social Renewal*. London: Comedia and Demos.

Conover, M. R. (2001). *Resolving Human Wildlife Conflicts: The Science of Wildlife Damage Management*. Boca Raton, FL: CRC Press.

Conway, H. (2000). Parks and people: the social functions. In J. Woodstra and K. Fieldhouse, eds., *The Regulation of Public Parks*. London: E. & F. N. Spon, pp. 9–20.

Cornelis, J. and Hermy, M. (2004). Biodiversity relationships in urban and suburban parks in Flanders. *Landscape and Urban Planning*, **69**, 385–401.

Dali, M. M. (2004). Urban open spaces uses as a function of lifestyle and space characteristics: the Malaysian context. *Proceedings of Open Space People Space: An International Conference on Inclusive Environments*. Openspace Research Centre, Edinburgh College of Art, Edinburgh, Scotland. 27–29 October 2004. Paper 31.

Daniels, G. D. and Kirkpatrick, J. B. (2006). Comparing the characteristics of front and back domestic gardens in Hobart, Tasmania, Australia. *Landscape and Urban Planning*, **78**, 344–52.

Davies, Z. G., Fuller, R. A., Loram, A. *et al.* (2009). A national scale inventory of resource provision for biodiversity within domestic gardens. *Biological Conservation*, **142**, 761–71.

DEFRA (2002). *Working with the Grain of Nature: A Biodiversity Strategy for England*. London: Department for Environment, Food and Rural Affairs.

Dehnen-Schmutz, K., Touza, J., Perrings, C. and Williamson, M. (2007). The horticultural trade and ornamental plant invasions in Britain. *Conservation Biology*, **21**, 224–31.

Department of Transport, Local Government and the Regions (2006). *Green Spaces, Better Places: Final Report of the Urban Green Spaces Taskforce*. London: Department of Transport, Local Government and the Regions.

Ditchkoff, S. S., Saalfield, S. T. and Gibson, C. J. (2006). Animal behavior in urban ecosystems: modifications due to human-induced stress. *Urban Ecosystems*, **9**, 5–12.

Domene, E., Saurí, D. and Pares, M. (2005). Urbanization and sustainable resource use: the case of garden watering in the metropolitan region of Barcelona. *Urban Geography*, **26**, 520–35.

Dorney, J. R., Guntenspergen, G. R., Keough, J. R. and Stearns, F. (1984). Composition and structure of an urban woody plant community. *Urban Ecology*, **8**, 69–90.

Draper, J. (1996). The art and science of park planning in the United States: Chicago's small parks, 1902–1905. In M. C. Sies and C. Silver, eds., *Planning the Twentieth-century American City*. Baltimore, MD: Johns Hopkins University Press, pp. 98–119.

Dunnett, N., Swanwick, C. and Woolley, H. (2002). *Improving Urban Parks, Play Areas and Green Spaces*. London: Department for Transport, Local Government and the Regions.

Dye, J. L. and Johnson, T. (2007). *A Child's Day: 2003 (Selected Indicators of Child Well-being)*. Washington, DC: US Census Bureau.

Ehrenfeld, J. G. (2008). Exotic invasive species in urban wetlands: environmental correlates and implications for wetland management. *Journal of Applied Ecology*, **45**, 1160–9.

Ehrlich, P. R. (2002). Human natures, nature conservation, and environmental ethics. *BioScience*, **52**, 31–43.

Faeth, S. H., Warren, P. S., Stochat, E. and Marussich, W. A. (2005). Trophic dynamics in urban communities. *BioScience*, **55**, 399–407.

Fenwick, G. H. (2001). *Opposition to Proposed DRC 1339 South Dakota Blackbird Poisoning*. Open letter to United States Department of Agriculture on behalf of American Bird Conservancy.

Fernández-Juricic, E., Jimenez, M. D. and Lucas, E. (2001). Bird tolerance to human disturbance in urban parks of Madrid, Spain: management implications. In J. M. Marzluff, R. Bowman and R. Donnelly, eds., *Avian Ecology and Conservation in an Urbanizing World*. Norwell, MA: Kluwer Academic, pp. 259–73.

Fischer, J. R., Stallknecht, D. E., Luttrell, M. P., Dhondt, A. A. and Converse, K. A. (1997). Mycoplasmal conjunctivitis in wild songbirds: the spread of a new contagious disease in a mobile host population. *Emerging Infectious Diseases*, 3, 69–72.

Foeken, D. W. J., Owuor, S. O. and Klaver, W. (2002). *Crop Cultivation in Nakuru Town, Kenya: Practice and Potential*. Leiden, the Netherlands: African Studies Centre.

Francis, M. (1988). The garden in the mind and in the heart. In H. Van Hoorgdalem, N. L. Prack, Th. J. M. Van Der Voordt and H. B. R. Van Wegen, eds., *Looking Back to the Future, IAPS 10/1988 Proceedings*, Vol. 2. Delft: Delft University Press, pp. 495–500.

Fraser, E. D. G. (2002). Urban ecology in Bangkok, Thailand: community participation, urban agriculture and forestry. *Environments*, 30, 37–50.

Freidberg, S. E. (2001). Gardening on the edge: the social conditions of unsustainability on an African urban periphery. *Annals of the Association of American Geographers*, 91, 349–69.

Frey, J. E. (1981). Preferences, satisfaction, and the physical environments of urban neighborhoods. Unpublished Ph.D. thesis, University of Michigan, Ann Arbor.

Fuller, R. A. and Gaston, K. J. (2009). The scaling of green space coverage in European cities. *Biology Letters*, 5, 352–5.

Fuller, R. A., Irvine, K. N., Devine-Wright, P., Warren, P. H. and Gaston, K. J. (2007). Psychological benefits of greenspace increase with biodiversity. *Biology Letters*, 3, 390–4.

Fuller, R. A., Tratalos, J. and Gaston, K. J. (2009). How many birds are there in a city of half a million people? *Diversity and Distributions*, 15, 328–37.

Fuller, R. A., Tratalos, J., Warren, P. H. et al. (2010). Environment and biodiversity. In M. Jenks and C. Jones, eds., *Dimensions of the Sustainable City*. Dordrecht, the Netherlands: Springer, pp. 75–103.

Fuller, R. A., Warren, P. H., Armsworth, P. R., Barbosa, O. and Gaston, K. J. (2008). Garden bird feeding predicts the structure of urban avian assemblages. *Diversity and Distributions*, 14, 131–7.

Garcia, R. and White, A. (2007). *Healthy Parks, Schools and Counties: Mapping Green Access and Equity for California*. Los Angeles, CA: The City Project.

Garden, J. G., McAlpine, C. A., Possingham, H. P. and Jones, D. N. (2007). Habitat structure is more important than vegetation composition for local-level management of native terrestrial reptile and small mammal species living in urban remnants: a case study from Brisbane, Australia. *Austral Ecology*, 32, 669–85.

Gaston, K. J., Cush, P., Ferguson, S. et al. (2007a). Improving the contribution of urban gardens for wildlife: some guiding propositions. *British Wildlife*, 18, 171–7.

Gaston, K. J., Fuller, R. A., Loram, A. et al. (2007b). Urban domestic gardens (XI): variation in urban wildlife gardening in the United Kingdom. *Biodiversity and Conservation*, 16, 3227–38.

Godbey, G., Grafe, A. and James, W. (1992). *The Benefits of Local Recreation and Park Services: A Nationwide Study of the Perceptions of the American Public*. Pennsylvania State University, PA: College of Health and Human Development.

GreenSpace (2007). *The Park Life Report*. Reading, UK: GreenSpace.

Greenwood, J. J. D. (2007). Citizens, science and bird conservation. *Journal of Ornithology*, 148, S77–S124.

Gregory, R. D. and Baillie, S. R. (1998). Large-scale habitat use of some declining British birds. *Journal of Applied Ecology*, 35, 785–99.

Guyer, J. I. (1987). *Feeding Africa's Cities: Studies in Regional Social History*. Bloomington: Indiana University Press.

Hale, M. (ed.) (1993). *Ecology in Education*. Cambridge, UK: Cambridge University Press.

Hanson, C. G. and Mason, J. L. (1985). Bird seed aliens in Britain. *Watsonia*, 15, 237–52.

Hartup, B. K., Mohammed, H. O., Kollias, G. V. and Dhondt, A. A. (1998). Risk factors associated with mycoplasmal conjunctivitis in house finches. *Journal of Wildlife Diseases*, **34**, 281–8.

Head, L. and Muir, P. (2006). Suburban life and the boundaries of nature: resilience and rupture in Australian backyard gardens. *Transactions of the Institute of British Geographers*, **31**, 505–24.

Helden, A. J. and Leather, S. R. (2004). Biodiversity on urban roundabouts – Hemiptera, management and the species–area relationship. *Basic and Applied Ecology*, **5**, 367–77.

Herzog, T. R. and Miller, E. J. (1998). The role of mystery in perceived danger and environmental preference. *Environment and Behavior*, **30**, 429–49.

Hope, D., Gries, C., Zhu, W. et al. (2003). Socioeconomics drive urban plant diversity. *Proceedings of the National Academy of Sciences of the USA*, **100**, 8788–92.

Hostetler, M. (1999). Scale, birds, and human decisions: a potential for integrative research in urban ecosystems. *Landscape and Urban Planning*, **45**, 15–19.

Irvine, K. N. and Warber, S. L. (2002a). The healing power of nature. In P. B. Kaufman, C. W. Coon and J. N. Govil, eds., *Creating a Sustainable Future: Living in Harmony with the Earth*. New Delhi: Sci Tech Publishing LLC, pp. 313–23.

Irvine, K. N. and Warber, S. L. (2002b). Greening healthcare: practicing as if the natural environment really mattered. *Alternative Therapies*, **8**, 76–83.

Irvine, K. N., Devine-Wright, P., Payne, S. R. et al. (2009). Green space, soundscape and urban sustainability: an interdisciplinary, empirical study. *Local Environment*, **14**, 155–72.

Irvine, K. N., Fuller, R. A., Devine-Wright, P. et al. (2010). Ecological and psychological value of urban green space. In M. Jenks and C. Jones, eds., *Dimensions of the Sustainable City*.

Dordrecht, the Netherlands: Springer, pp. 215–37.

Jenks, M. and Burgess, R. (eds.) (2000). *Compact Cities: Sustainable Urban Forms for Developing Countries*. London: E. & F. N. Spon Press.

Jim, C. Y. and Chen, W. Y. (2006a). Recreation–amenity use and contingent valuation of urban greenspaces in Guangzhou, China. *Landscape and Urban Planning*, **75**, 81–96.

Jim, C. Y. and Chen, W. Y. (2006b). Perception and attitude of residents toward urban green spaces in Guangzhou (China). *Environmental Management*, **38**, 338–49.

Jim, C. Y. and Liu, H. T. (2001). Patterns and dynamics of urban forests in relation to land use and development history in Guangzhou City, China. *Geographical Journal*, **167**, 358–75.

Jones, D. N. (2008). Feed the birds? *Wingspan*, **18**, 16–19.

Jones, D. N. and Reynolds, S. J. (2008). Feeding birds in our towns and cities: a global research opportunity. *Journal of Avian Biology*, **39**, 265–71.

Kaplan, R. (1973). Some psychological benefits of gardening. *Environment and Behavior*, **5**, 145–62.

Kaplan, R. (1985). Nature at the doorstep: residential satisfaction and the nearby environment. *Journal of Architectural and Planning Research*, **2**, 115–27.

Kaplan, R. (2001). The nature of the view from home: psychological benefits. *Environment and Behavior*, **33**, 507–42.

Kaplan, R. and Kaplan, S. (1989). *The Experience of Nature: A Psychological Perspective*. New York: Cambridge University Press.

Kaplan, R., Bardwell, L. V., Ford, H. A. and Kaplan, S. (1996). The corporate back-40: employee benefits of wildlife enhancement efforts on corporate land. *Human Dimensions of Wildlife*, **1**, 1–13.

Kaufman, J. and Bailkey, M. (2000). *Farming Inside Cities: Entrepreneurial Urban Agriculture in the*

United States. Cambridge, MA: Lincoln Institute of Land Policy.

Kellett, J. E. (1982). The private garden in England and Wales. *Landscape and Urban Planning,* **9,** 105–23.

Kendle, T. and Forbes, S. (1997). *Urban Nature Conservation: Landscape Management in the Urban Countryside.* London: E. & F. N. Spon.

Kinzig, A. P., Warren, P., Martin, C., Hope, D. and Katti, M. (2005). The effects of human socioeconomic status and cultural characteristics on urban patterns of biodiversity. *Ecology and Society,* **10,** 23.

Kirkpatrick, J. B. (2004). Vegetation change in an urban grassy woodland 1974–2000. *Australian Journal of Botany,* **52,** 597–608.

Kuo, F. E. and Sullivan, W. C. (2001). Environment and crime in the inner city. does vegetation reduce crime? *Environment and Behavior,* **33,** 343–67.

La Rouche, G. P. (2003). *Birding in the United States: A Demographic and Economic Analysis.* Washington, DC: United States Fish and Wildlife Service.

Lepczyk, C. A., Mertig, A. G. and Liu, J. (2004). Assessing landowner activities related to birds across rural-to-urban landscapes. *Environmental Management,* **33,** 110–25.

Linares, O. F. (1996). Cultivating biological and cultural diversity: urban farming in Casamance, Senegal. *Africa,* **66,** 104–21.

Lohr, V. I. (2007). Benefits of nature: what we are learning about why people respond to nature. *Journal of Physiological Anthropology,* **26,** 83–5.

Lohr, V. I. and Pearson-Mims, C. H. (2005). Children's active and passive interactions with plants influence their attitudes and actions toward trees and gardening as adults. *HortTechnology,* **15,** 472–6.

Loram, A., Thompson, K., Warren, P. H. and Gaston, K. J. (2008a). Urban domestic gardens (XII): the richness and composition of the flora in five UK cities. *Journal of Vegetation Science,* **19,** 321–30.

Loram, A., Tratalos, J., Warren, P. H. and Gaston, K. J. (2007). Urban domestic gardens (X): the extent & structure of the resource in five major cities. *Landscape Ecology,* **22,** 601–15.

Loram, A., Warren, P. H. and Gaston, K. J. (2008b). Urban domestic gardens (XIV): the characteristics of gardens in five cities. *Environmental Management,* **42,** 361–76.

Louv, R. (2008). *Last Child in the Woods: Saving Our Children from Nature-deficit Disorder.* Chapel Hill, NC: Algonquin Books.

Luck, M. and Wu, J. (2002). A gradient analysis of urban landscape pattern: a case study from the Phoenix metropolitan region, Arizona, USA. *Landscape Ecology,* **17,** 327–39.

Lyytimäki, J., Petersen, L. K., Normander, B. and Bezák, P. (2008). Nature as a nuisance? Ecosystem services and disservices to urban lifestyle. *Journal of Integrative Environmental Sciences,* **5,** 161–72.

Macintyre, S., Ellaway, A., Hiscock, R., *et al.* (2003). What features of the home and the area might help to explain observed relationships between housing tenure and health? Evidence from the west of Scotland. *Health and Place,* **9,** 207–18.

Makumbelo, E., Lukoki, L., Paulus, J. S. and Luyindula, N. (2002). Inventory of vegetable species cultivated in plot gardens in urban environment. Instance of Limete Commune- Kinshasa- Democratic Republic of Congo. *Tropicultura,* **20,** 89–95.

Maller, C., Townsend, M., Pryor, A., Brown, P. and St Leger, L. (2006). Healthy nature healthy people: 'contact with nature' as an upstream health promotion intervention for populations. *Health Promotion International,* **21,** 45–54.

Mallick, S. A. and Driessen, M. M. (2003). Feeding of wildlife: how effective are the 'Keep Wildlife Wild' signs in Tasmania's national parks? *Ecological Management and Restoration,* **4,** 199–204.

Manfredo, M. J., Teel, T. L. and Bright, A. D. (2003). Why are public values toward wildlife changing? *Human Dimensions of Wildlife,* **8,** 287–306.

Martin, C. A. and Stabler, L. B. (2004). Urban horticultural ecology: interactions between plants, people and the physical environment. *Acta Horticulturae*, **639**, 97–102.

Mason, C. F. (2000). Thrushes now largely restricted to the built environment in eastern England. *Diversity and Distributions*, **6**, 189–94.

Mason, J., Moorman, C., Hess, G. and Sinclair, K. (2007). Designing suburban greenways to provide habitat for forest-breeding birds. *Landscape and Urban Planning*, **80**, 153–64.

Matlack, G. R. (1993). Sociological edge effects: spatial distribution of human impact in suburban forest fragments. *Environmental Management*, **17**, 829–35.

McKinney, M. L. (2002). Urbanization, biodiversity, and conservation. *BioScience*, **52**, 883–90.

McKinney, M. L. (2005). Urbanization as a major cause of biotic homogenization. *Biological Conservation*, **127**, 247–60.

McNally, M. (1987). Participatory research and natural resource planning. *Journal of Architectural and Planning Research*, **4**, 322–8.

McNally, M. (1990). Valued places. In M. Francis and R. T. Hester, eds., *The Meaning of Gardens: Ideas, Place and Action*. Cambridge, MA: MIT Press, pp. 172–6.

Mertig, A. G., Dunlap, R. E. and Morrison, D. E. (2002). The environmental movement in the United States. In R. E. Dunlap and W. Michaelson, eds., *Handbook of Environmental Sociology*. Westport, CT: Greenwood Press, pp. 448–81.

Miller, J. R. (2005). Biodiversity conservation and the extinction of experience. *Trends in Ecology and Evolution*, **20**, 430–4.

Miller, J. R. and Hobbs, R. J. (2002). Conservation where people live and work. *Conservation Biology*, **16**, 330–7.

Miura, S., Kunii, O. and Wakai, S. (2003). Home gardening in urban poor communities of the Philippines. *International Journal of Food Sciences and Nutrition*, **54**, 77–88.

Mougeot, L. J. A. (1994). African city farming from a world perspective. In A. G. Egziabher, D. Lee-Smith, D. G. Maxwell *et al.*, eds., *Cities Feeding People: An Examination of Urban Agriculture in East Africa*. Ottawa, Canada: International Development Research Centre, pp. 1–24.

Nasar, J. L. and Fisher, B. (1993). 'Hot spots' of fear and crime: a multi-method investigation. *Journal of Environmental Psychology*, **13**, 187–206.

Nasar, J. L. and Jones, K. M. (1997). Landscapes of fear and stress. *Environment and Behavior*, **29**, 291–323.

Nasar, J. L., Fisher, B. and Grannis, M. (1993). Proximate physical cues to fear of crime. *Landscape and Urban Planning*, **26**, 161–78.

Nassauer, J. I. (ed.) (1997). *Placing Nature: Culture in Landscape Ecology*. Washington, DC: Island Press.

Niinemets, Ü. and Peñuelas, J. (2009). Gardening and urban landscaping: significant players in global change. *Trends in Plant Science*, **13**, 60–5.

Nugent, R. (2000). The impact of urban agriculture on the household and local economies. In N. Bakker, M. Dubbeling, S. Gündel, U. Sabel-Koschella and H. de Zeeuw, eds., *Growing Cities, Growing Food: Urban Agriculture on the Policy Agenda*. Feldafing, Germany: Deutsche Stiftung für internationale Entwicklung, pp. 67–97.

Obudho, R. A. and Foeken, D. W. J. (1999). *Urban Agriculture in Africa: A Bibliographical Survey*. Nairobi: Centre for Urban Research.

O'Leary, R. and Jones, D. N. (2006). The use of supplementary foods by Australian magpies *Gymnorhina tibicen*: implications for wildlife feeding in suburban environments. *Austral Ecology*, **31**, 208–16.

Owen, J. (1991). *The Ecology of a Garden: The First Fifteen Years*. Cambridge, UK: Cambridge University Press.

Özgüner, H. and Kendle, A. D. (2006). Public attitudes towards naturalistic versus designed landscapes in the city of Sheffield

(UK). *Landscape and Urban Planning*, **74**, 139–57.

Parsons, H. (2007). *Best Practice Guidelines for Enhancing Urban Bird Habitat: Scientific Report*. Sydney: Birds in Backyards Program.

Perry, T. and Nawaz, R. (2008). An investigation into the extent and impacts of hard surfacing of domestic gardens in an area of Leeds, United Kingdom. *Landscape and Urban Planning*, **86**, 1–13.

Petrie, M., Walsh, D. and Hotchkis, D. (2003). Encountering wildlife without feeding. *Land for Wildlife, Note 20*. Brisbane: Queensland Parks and Wildlife Service.

Pickett, S. T. A., Cadenasso, M. L., Grove, J. M. et al. (2001). Urban ecological systems: linking terrestrial, ecological, physical, and socioeconomic components of metropolitan areas. *Annual Review of Ecology and Systematics*, **32**, 127–57.

Pierce, S. M., Cowling, R. M., Knight, A. T. et al. (2005). Systematic conservation planning products for land-use planning: interpretation for implementation. *Biological Conservation*, **125**, 441–58.

Pimm, S. L. and Raven, P. (2000). Biodiversity: extinction by numbers. *Nature*, **403**, 843–5.

Pincetl, S. and Gearin, E. (2005). The reinvention of public green space. *Urban Geography*, **26**, 365–84.

Plant, M. (2008). Good practice when feeding wild birds. *Wingspan*, **18**, 20–3.

Polasky, S., Nelson, E., Camm, J. et al. (2008). Where to put things? Spatial land management to sustain biodiversity and economic returns. *Biological Conservation*, **141**, 1505–24.

Pyle, R. M. (1978). The extinction of experience. *Horticulture*, **56**, 64–7.

Pyle, R. M. (2003). Nature matrix: reconnecting people and nature. *Oryx*, **37**, 206–14.

Reuther, S. and Dewar, N. (2005). Competition for the use of public open space in low-income urban areas: the economic potential of urban gardening in

Khayelitsha, Cape Town. *Development Southern Africa*, **23**, 97–122.

Richards, N. A. (1983). Diversity and stability in a street tree population. *Urban Ecology*, **7**, 159–71.

Richards, N. A., Mallette, J. R., Simpson, R. J. and Macie, E. A. (1984). Residential greenspace and vegetation in a mature city: Syracuse, New York. *Urban Ecology*, **8**, 99–125.

Rodaway, P. (1994). *Sensuous Geographies: Body, Sense and Place*. London: Routledge.

Rollinson, D. J., O'Leary, R. and Jones, D. N. (2003). The practice of wildlife feeding in suburban Brisbane. *Corella*, **27**, 52–8.

Rome, A. (2001). *The Bulldozer and the Countryside: Suburban Sprawl and the Rise of American Environmentalism*. Cambridge, MA: Cambridge University Press.

Rudd, H., Vala, J. and Schaefer, V. (2002). Importance of backyard habitat in a comprehensive biodiversity conservation strategy: a connectivity analysis of urban green spaces. *Restoration Ecology*, **10**, 368–75.

Saldivar-Tanaka, L. and Krasny, M. E. (2004). Culturing community development, neighborhood open space, and civic agriculture: the case of Latino community gardens in New York City. *Agriculture and Human Values*, **21**, 399–412.

Sali, M. J., Kuehn, D. M. and Zhang, L. (2008). Motivations for male and female birdwatchers in New York state. *Human Dimensions of Wildlife*, **13**, 187–200.

Sandström, U. G., Angelstam, P. and Mikusiński, G. (2006). Ecological diversity of birds in relation to the structure of urban green space. *Landscape and Urban Planning*, **77**, 39–53.

Saunders, C. D., Brook, A. T. and Myers, O. E. Jr (2006). Using psychology to save biodiversity and human well-being. *Conservation Biology*, **20**, 702–5.

Schroeder, H. W. and Anderson, L. M. (1984). Perception of personal safety in urban recreation sites. *Journal of Leisure Research*, **16**, 178–94.

170 R. A. FULLER AND K. N. IRVINE

Schultz, P. W. (2000). Empathizing with nature: the effects of perspective taking on concern for environmental issues. *Journal of Social Issues*, **56**, 391–406.

Schultz, P. W. and Zelezny, L. (1998). Values and proenvironmental behavior: a five-country survey. *Journal of Cross-cultural Psychology*, **29**, 540–58.

Semedo, R. J. da C. G. and Barbosa, R. I. (2007). Fruit trees in urban home gardens of Boa Vista, Roraima, Brazilian Amazonia. *Acta Amazonica*, **37**, 497–504.

Shapiro, A. M. (2002). The Californian urban butterfly fauna is dependent on alien plants. *Diversity and Distributions*, **8**, 31–40.

Sheldon, K. (2003). Markets and gardens: placing women in the history of urban Mozambique. *Canadian Journal of African Studies*, **37**, 358–95.

Shochat, E. (2004). Credit or debit? Resource input changes population dynamics of city-slicker birds. *Oikos*, **106**, 622–6.

Silk, D. (1986). The potential of urban agriculture: growing vegetables – and hope. *United Nations University Work in Progress*, **10**, 6.

Slater, R. J. (2001). Urban agriculture, gender and empowerment: an alternative view. *Development Southern Africa*, **18**, 635–50.

Sliuzas, R. and Kuffer, M. (2008). Analysing the spatial heterogeneity of poverty using remote sensing: typology of poverty areas using selected RS based indicators. In C. Jurgens, ed., *Remote Sensing: New Challenges of High Resolution*. Bochum, Germany: Geographisches Institut, Ruhr-Universität Bochum, pp. 158–67.

Smit, J., Ratta, A. and Nasr, J. (1996). *Urban Agriculture: Food, Jobs and Sustainable Cities*. New York: United Nations Development Programme.

Smith, R. M., Gaston, K. J., Warren, P. H. and Thompson, K. (2005). Urban domestic gardens (V): relationships between landcover composition, housing and landscape. *Landscape Ecology*, **20**, 235–53.

Smith, R. M., Gaston, K. J., Warren, P. H. and Thompson, K. (2006a). Urban domestic gardens (VIII): environmental correlates of invertebrate abundance. *Biodiversity and Conservation*, **15**, 2515–45.

Smith, R. M., Thompson, K., Hodgson, J. G., Warren, P. H. and Gaston, K. J. (2006b). Urban domestic gardens (IX): composition and richness of the vascular plant flora, and implications for native biodiversity. *Biological Conservation*, **129**, 312–22.

Smith, R. M., Warren, P. H., Thompson, K. and Gaston, K. J. (2006c). Urban domestic gardens (VI): environmental correlates of invertebrate species richness. *Biodiversity and Conservation* **15**, 2415–38.

South African Cities Network (2006). *State of the Cities Report*. Cape Town, South Africa.

Speirs, L. J. (2003). Sustainable planning: the value of green space. *Sustainable Planning and Development*, **6**, 337–46.

Steinberg, T. (2006). *American Green*. New York: Norton.

Stigsdotter, U. A. and Grahn, P. (2004). A garden at your doorstep may reduce stress: private gardens as restorative environments in the city. *Proceedings of Open Space People Space: An International Conference on Inclusive Environments*. Openspace Research Centre, Edinburgh College of Art, Edinburgh, Scotland. 27–29 October 2004. Paper 15.

Stilgoe, J. R. (2001). Gone barefoot lately? *American Journal of Preventive Medicine*, **20**, 243–4.

Sutton, C. M. (2008). Urban open space: a case study of Msunduzi municipality, South Africa. Unpublished M.Sc. thesis, Queen's University, Canada.

Tanner, T. (1980). Significant life experiences. *Journal of Environmental Education*, **11**, 20–4.

Teel, T. L., Manfredo, M. J. and Stinchfield, H. M. (2007). The need and theoretical basis for exploring wildlife value orientations cross-culturally. *Human Dimensions of Wildlife*, **12**, 297–305.

Thompson, K., Austin, K. C., Smith, R. M. *et al.* (2003). Urban domestic gardens (I): putting small-scale plant diversity in context. *Journal of Vegetation Science*, **14**, 71–8.

Turner, T. (1992). Open space planning in London: from standards per 1000 to green strategy. *Town Planning Review*, **63**, 365–86.

Turner, T. (2006). Greenway planning in Britain: recent work and future plans. *Landscape and Urban Planning*, **76**, 240–51.

Turner, W. R., Nakamura, T. and Dinetti, M. (2004). Global urbanization and the separation of humans from nature. *BioScience*, **54**, 585–90.

Urban Task Force (1999). *Towards an Urban Renaissance: Final Report of the Urban Task Force*. London: E. & F. N. Spon.

US Department of the Interior, Fish and Wildlife Service (2006). *National Survey of Fishing, Hunting, and Wildlife-associated Recreation*. Washington, DC: United States Government Printing Office.

US Fish and Wildlife Service (2001). *National Survey of Fishing, Hunting and Wildlife Associated Recreation*. Washington, DC: United States Government Printing Office.

Vähä-Piikkiö, I., Kurtto, A. and Hahkala, V. (2004). Species number, historical elements and protection of threatened species in the flora of Helsinki, Finland. *Landscape and Urban Planning*, **68**, 357–70.

Vitousek, P. M., Ehrlich, P. R., Ehrlich, A. H. and Matson, P. A. (1986). Human appropriation of the products of photosynthesis. *BioScience*, **36**, 368–73.

Vitousek, P. M., Mooney, H. A., Lubchenco, J. and Melillo, J. M. (1997). Human domination of earth's ecosystems. *Science*, **277**, 494–9.

Waliczek, T. M., Zajicek, J. M. and Lineberger, R. D. (2005). The influence of gardening activities on consumer perceptions of life satisfaction. *HortScience*, **40**, 1360–5.

Wang, X., Li, D., Sheng, L. *et al.* (2007). Significance of birds, bees and butterflies in urban gardens and their attraction and protection. *Scientia Silvae Sinicae*, **43**, 134–43.

Wells, N. M. (2000). At home with nature: effects of 'greenness' on children's cognitive functioning. *Environment and Behavior*, **32**, 775–95.

Williams, K. J. H. and Cary, J. (2002). Landscape preferences, ecological quality, and biodiversity protection. *Environment and Behavior*, **34**, 257–74.

Wilson, E. O. (1984). *Biophilia*. Cambridge, MA: Harvard University Press.

Winklerprins, A. M. G. A. (2003). House-lot gardens in Santarem, Para, Brazil: linking rural with urban. *Urban Ecosystems*, **6**, 43–65.

Wolch, J., Wilson, J. P. and Fehrenbach, J. (2002). *Parks and Park Funding in Los Angeles: An Equity Mapping Analysis*. Los Angeles, CA:University of California Sustainable Cities Program and GIS Research Laboratory.

Woodall, P. R. (1995). Results of the QOS garden bird survey, 1979–1980, with particular reference to South-east Queensland. *Sunbird*, **25**, 1–17.

Works, M. A. (1990). Dooryard gardens in Moyobamba, Peru. *Focus*, **40**, 12–17.

Yang, W. and Kang, J. (2005). Soundscape and sound preferences in urban squares: a case study in Sheffield. *Journal of Urban Design*, **10**, 61–80.

Zagorski, T., Kirkpatrick, J. B. and Stratford, E. (2004). Gardens and the bush: gardeners' attitudes, garden types and invasives. *Australian Geographical Studies*, **42**, 207–20.

Zinn, H. C. and Shen, X. S. (2007). Wildlife value orientations in China. *Human Dimensions of Wildlife*, **12**, 331–8.

Zipperer, W. C. and Zipperer, C. E. (1992). Vegetation responses to changes in design and management of an urban park. *Landscape and Urban Planning*, **22**, 1–10.

CHAPTER EIGHT

Urban ecology and human social organisation

PAIGE S. WARREN, SHARON L. HARLAN,
CHRISTOPHER BOONE, SUSANNAH B. LERMAN,
EYAL SHOCHAT AND ANN P. KINZIG

Introduction

Consider a tree, growing in a forest. Its fate, in terms of growth rate, longevity and reproduction, is regulated by a suite of biotic and abiotic factors, such as soil moisture, competition with other plants and interactions with natural enemies. Its branches, leaves and roots contribute to nutrient and water cycles. The tree provides habitat for animals – insects, birds, mammals. Understanding these interactions falls squarely in the domain of classical ecology. Now, imagine that the tree and its surrounding forest is part of a city, the trees interspersed with houses, streets, lawns and gardens. Here, the fate of the tree becomes strongly intertwined with the lives and decisions of humans. Its longevity, reproductive rate and contributions to the larger ecosystem are mediated by a complex suite of anthropogenic processes. In this setting, the tree's life cycle may differ from that of all of its ancestors: its 'birth' perhaps in a commercial nursery, its dispersal through an economic system of marketing and transportation, its growth determined by local application of fertiliser by a resident or a lawn care firm, its ultimate removal (death/decay) perhaps mediated by city policies on tree hazards, its reproduction quelled entirely. Yet, the tree and its urban forest still contribute to ecosystem processes. It provides habitat for animals, and its leaves, branches and roots still influence the flow of water through the urban ecosystem. Humans, therefore, both consciously and unconsciously sculpt biodiversity and ecosystem functioning in cities (and elsewhere).

We use this story of a tree in an urban forest, not to focus attention on trees, but to illustrate how human social organisation drives the characteristics of a city's green infrastructure (e.g. vegetation, animals, soil and water). Urban ecology has emerged in recent years as an interdisciplinary or trans-disciplinary

Urban Ecology, ed. Kevin J. Gaston. Published by Cambridge University Press.
© British Ecological Society 2010.

endeavour that aims to integrate understanding of human social organisation with traditional ecological approaches to the study of ecosystems.

Over the past century and more, social scientists in virtually all disciplines, including sociology, geography, anthropology, economics, environmental psychology, landscape architecture, urban planning and design, have generated rich and sometimes conflicting bodies of knowledge and theory about human life and the environment in cities. Ecologists have come to study the city far later in the game, and frequently find challenges far different from those in their traditional arenas of study. How should an ecologist proceed in such an environment, where understanding ecological processes necessitates crossing the boundary into social sciences? Here we provide a necessarily brief, but ecologist-friendly, survey of key theories from the social sciences and a review of findings from recent integrated empirical research. As a point of entry, we focus on the spatial pattern and temporal dynamics of urban ecosystems, these being fundamental components of disciplinary-based ecology.

The social organisation of cities
Historical context

Most scholars agree that cities, defined as dense settlements incorporating great physical and human social diversity, originated in Mesopotamia five to six thousand years ago. A necessary precursor to cities was the invention of agriculture, which generated a surplus of food that could be traded for other types of food, goods or services. From the earliest city-states to modern metropolitan areas, the agglomeration economies of urban settlements allow for increasing divisions of labour and productivity, which are key drivers of urbanisation (Batty 2008). Until the Industrial Revolution, relatively few people could afford to live in cities, dependent on others to grow most of their food. Significant advances in the productivity of agriculture and manufacturing, made possible through steam power and fossil fuels, pushed surplus labour from farms and drew workers to cities in larger numbers. By 1850, England became the first urban nation in the world, with more than half its population living in towns and cities (Bédarida 1979). By 1920, the United States passed this urban threshold, and in 2008, half of humanity was housed in urban areas (UNFPA 2007).

Distinct hierarchies and social inequalities are as much characteristics of modern-day Lagos as ancient Babylon. Yet despite very visible and persistent poverty, urban areas continue to offer opportunity for rural migrants or shantytown dwellers to find work, education, health care and entertainment. Efforts to reduce rural to urban migration have largely failed, and have often worsened economic conditions for the very poor. Inadequate planning and mismanagement of resources have amplified poor living standards and environmental impacts that can accompany urban migration, especially in the developing world. Those on the lowest rungs of the socioeconomic ladder

---

Box 8.1 Divisions in cities

Marcuse and van Kempen (2002) illustrate the complexity of urban forms, drawing on the following definitions of function, status and culture. *Pre-industrial cities* were divided by **economic function**, according to the kind of market activity taking place in particular locations. Public commercial spaces were identifiable and demarcated from residences. Neighbourhoods were often organised around residents engaged in similar kinds of craft occupations. More complex divisions of land by types of industry, residences, open spaces, highways and so forth are characteristic of *industrial cities*, which formalise restrictions on the functional use of space with legal restrictions (zoning). (See also the section on 'Temporal dynamics' for further discussion of historical changes in division by land-use.)

Hierarchical status divisions in cities embody the power of higher-status residents spatially to separate themselves from subordinate status groups and less desirable functional activities. **Socioeconomic status** (a set of indicators measured by household income, occupation and education) taps an elusive underlying dimension of social status. **Cultural divisions** are not themselves hierarchical, though they complicate urban spatial patterns because they may or may not overlap with status hierarchies and economic function. Cultural divisions are most readily identified by language, ethnicity and visible symbols of associated beliefs and lifestyles.

Most urban ecological studies to date have sought to relate ecological outcomes, such as patterns of species diversity, to socioeconomic indicators (see 'Social stratification and ecological outcomes'). Fewer studies have addressed cultural divisions (see 'Other social dynamics').

---

suffer most from environmental consequences of urbanisation, including polluted air, water and soil (UNFPA 2007). It is increasingly recognised that quality of life in urban areas depends in part on an adequate supply of ecosystem services (Martine 2008). However, inclusion of the benefits of ecosystem services into urban planning is rare, and addressing social and environmental inequities and vulnerabilities remains a challenging and pressing issue.

Spatial organisation

Evidence of partitioned urban space exists in ancient cities and extends with remarkable consistency through time and across virtually every historical urban tradition (Marcuse & Van Kempen 2002). Urban archaeological records identify 'zones' that correspond to administrative and economic functions and demarcate residential areas with different population characteristics (Keith 2003; Cowgill 2007; Smith 2008). In a sweeping analysis of world cities, past and present, Marcuse and Van Kempen (2002) concluded that three social divisions in cities – function, status and culture (see Box 8.1) – constitute the foundation for spatial boundaries. While these boundaries, which demarcate power differences in segments of society, have been blurred at times in the past and 'may be merging again in today's economies' (Marcuse & Van Kempen 2002, p. 14), significant segregation remains in many modern cities (Johnston *et al.* 2007). These authors and others (e.g. Briggs 2004) acknowledge that cities, both past and present, differ widely in the contours, permeability and characteristics of their sociospatial boundaries. Social complexities introduced by industrialisation and global capitalism have increased the variety and degree of sociospatial differentiation within cities (Johnston *et al.* 2007). For ecologists interested in the environmental consequences of spatial partitioning in cities, these are relevant points of agreement across historical cases and different branches of urban theory:

- Spatial patterns are socially constructed by human actors;
- Spatial patterns often reflect social hierarchies;
- Particular spatial configurations of social activities fragment environments within a city and increasingly beyond city boundaries, fragmenting environments in peripheral areas.

Social theories differ, however, in their explicit recognition of the institutional forces through which fragmented urban space is intentionally constructed and controlled (Joseph 2008).

American sociological theory on cities and spatial differentiation originated with the human ecologists associated with the University of Chicago in the early twentieth century (Park 1915; Burgess 1928). They were the first to detail a theoretical model of urban growth and development based on concepts borrowed from ecologists, including competition for space, and the succession and segregation of new population groups with city expansion and increasingly diverse economic activities (Palen 2005). Among the contributions of the Chicago School was conceptually linking:

urban spaces to distinctive social groups, creating a spatiality to the urban form and to cultural difference that was previously undeveloped. Urban space came to be seen as divided and organized by social boundaries that were connected to class, race, ethnicity, and degree of assimilation . . . in a way that recognized the extent of concentration and segregation, and linked cultural assimilation to spatial movement from the inner city to the outer rings.

(Joseph 2008, p. 9)

Largely inspired by this tradition, subsequent sociologists have created a richly contextualised body of literature that explores both the positive aspects of residential segregation in the form of ethnic enclaves and the negative aspects of concentrated poverty and racial discrimination in slums and ghettos (e.g. Whyte 1943; Wirth 1956; Duneier 1999; Johnston *et al.* 2007). These studies

---

Box 8.2 Defining terms

A **central city** is the core municipality of a large metropolitan region.
Growth around a central city was originally conceived as a spreading set of
concentric rings corresponding to functional and status divisions. This model
is no longer subscribed to by most urban geographers. **Suburbs**, largely
residential communities, historically arose on the margins of central cities through
abandonment of central cities by wealthy elites. **Suburban sprawl** or
suburbanisation refers to the largely post-World War II phenomenon of the
expansion of metropolitan regions through low-density development.
Post-industrial cities, such as Los Angeles, are more often **polycentric**, with
reduced dependencies of suburbs on a single central city such as London,
Paris or Chicago. These metropolitan regions have multiple nodes of
higher-density development dispersed throughout a mixed landscape of
low-density and remnant rural lands. Many quantitative definitions exist for terms
like urban and suburban, but authors generally agree on these qualitative
definitions (McIntyre *et al.* 2000).

---

of urban social environments serve as valuable precursors and counterparts to
studies of various environmental inequalities among city neighbourhoods.

The specific urban core–periphery spatial model proposed by the human
ecologists was modified and ultimately rejected by later social theorists who
contrasted pre-industrial cities with industrial cities (e.g. Sjoberg 1960) and
those who studied the emergence of Los Angeles, USA, and other world megacities with distinctly different sociospatial patterns (e.g. Soja 2000). Thus, the
urban to rural gradient is an overly simplistic assumption in the social sciences
because urban spatial patterns vary according to the contours of multiple and
overlapping functional and cultural divisions with socioeconomic hierarchies
(Box 8.1). In polycentric cities, for instance (Box 8.2), fragmentation appears to
characterise urban space better than linear distances from a central city.
Second, the human ecologists' assumption that spatial segregation is the result
of 'natural' selection as opposed to economic and political power exercised by
elites (Chaskin 1997) was also abandoned by later theorists who focused on how
the political economy of urban development is intentionally structured to
benefit elites and disadvantage society's marginalised groups (Palen 2005).
Logan and Molotch (1987) argued in a widely acclaimed book that land in the
USA, as well as in the developing nations, is highly commodified, privatised and
controlled by development corporations with the acquiescence of local governments. In response to profit-making opportunism, municipal governments
respond to the 'urban growth machine', leaving local residential communities
without consideration or resources to obtain a good quality of urban life.

Political economists differ on whether they attribute urban social and environmental degradation primarily to investment capital decisions by local governments (Harvey 1973; Logan & Molotch 1987) or to withdrawal of industrial capital from the urban economy (Wilson 1997), but they speak with one voice about the deliberate construction of urban sociospatial inequalities. Access to existing environmental amenities and proximity to disamenities (such as pollution) is stratified locally between privileged neighbourhoods and neighbourhoods that are politically disenfranchised by legally enforced restrictions on land, socially burdened population groups and lack of economic resources (Boone 2008).

The production of sociospatially segregated cities is a multi-scale process, influenced by underlying choices about urban growth policies, society's provisions for the education and employment of urban residents, and the political economic realities that manipulate consumers' desires and tastes. Household decisions about where to live are experientially conditioned by all these factors even *before* they are also limited by household income, knowledge and lifestyle preferences (Bourdieu 1984; Gottdiener & Hutchinson 2001). Racism and legacies of colonialist ideologies, reflected in real estate markets and loan practices, often play on the majority's prejudice and fear to exacerbate the isolation of minority communities (Krysan & Farley 2002; Quillian 2002).

The extent and salience of contemporary urban residential segregation by social class and race/ethnicity has been amply documented in the USA (Massey & Denton 1993; Lee et al. 2008), in other developed nations (Musterd 2005; Johnston et al. 2007) and in some cities in developing countries (King 1976; Monteiro 2008). Social scientists have focused attention on the ramifications of sociospatial structuring of cities for vulnerabilities to problems such as air pollution and toxic waste sites (Bolin et al. 2005; Grineski et al. 2007). Attention to sociospatial patterns has followed more slowly in the ecological literature. In later sections, we examine spatial associations between social and ecological patterns, such as vegetation structure and composition of biotic communities.

Temporal dynamics

Cities grow, decline and change in response to long- and short-term cycles and rhythms. Tall grass on lawns of abandoned homes and swimming pools filled with algae are reminders of the cyclical nature of capitalist economies. Karl Marx remarked on the cyclical patterns of economies 140 years ago and others have since paid careful attention to how the 'crisis of overaccumulation', the result of surplus capital unable to find profitable markets, affects processes and patterns of urbanisation (Castells 1977; Smith 1996). For Harvey (1982, 2003), the crisis of overaccumulation in cities culminates in large-scale infrastructure development as a means, if only temporary, of averting an economic downturn. New infrastructure projects provide room for spatial expansion of capital as well as a temporal deferral of economic decline, what he terms a spatiotemporal fix. Kuznets (1967) recognised the same infrastructure investment as a 15- to 25-year long wave economic cycle. Other economists have predicted economic cycles for technology (Kondratieff's 40–60 years), fixed investment (Juglar's 5–7 years) and inventory (Kitchin's 3–5 years) (Hall 1988). What is important for our understanding of cities is that these waves ripple through the built environment, expanding suburban limits and renewing older neighbourhoods during up years, and leading to abandoned homes on the periphery and disinvestment in the down years.

Similar to all landscapes, cities record generations of decisions and actions in the built forms that remain. Older cities like London, UK, or Beijing, China, show remnants of urbanisation that are hundreds or even thousands of years old (Vance 1990). The morphology of cities reflects the dominant economic function and prevailing technologies, especially for transportation, in eras of building and growth. Indeed, most cities are about an hour across (Marchetti's Constant), corresponding to the average travel time people will tolerate on a daily basis, regardless of the transportation mode (Newman & Jennings 2008). Configuration of a city like Phoenix, USA, characterised by low-density suburbs and segregated land-use, is largely a product of post-World War II growth in an era of widespread automobile use, inexpensive energy and rising incomes. Pre-industrial cities, by contrast, were constructed when movement of goods and labour happened on foot or by beast of burden, resulting in high densities, narrow roads and mixed land-use. While these pre-industrial morphologies may now be ringed by low-density suburbs, they continue to have important impacts on metropolitan function. Old historical cities like metropolitan Boston, USA, or Paris, France, poorly accommodate the automobile, but are well suited to mass transit, walking and mixed land-uses, elements of the compact city that are now celebrated and promoted in smart growth policies and the planning and design philosophy of New Urbanism (Boone & Modarres 2006). Similar to ecosystems, past urban practices continue to have legacy effects on present-day city structure and function.

Urban areas are dynamic by their nature, but some components are persistent. Road networks, for instance, can persist and define urban morphology for hundreds or even thousands of years. Private residences, on the other hand, may last only two or three generations before being replaced with new structures. Social organisation in cities can also display the same persistence and dynamism. In the USA, white and later black middle-class abandonment of central cities occurred swiftly after World War II, drawn to the suburbs by federally guaranteed mortgages and subsidised automobile transportation, in addition to the tactics of estate agents (realtors) looking to capitalise on residents' fear, racial prejudices and discrimination (Jackson 1985). Earlier research on neighbourhood succession, drawing on ecological theory, neglected some of the institutional and political factors, such as redlining (the twentieth-century practice of delineating areas, usually black inner city neigh-bourhoods, where banks would not invest) that accelerated the process of neighbourhood change (Hillier 2003a, 2003b). While thresholds of poverty and home ownership rates appear to be triggers of neighbourhood transform-ation, much research remains to be done on the institutional and structural drivers of change (Galster *et al.* 2000). Another promising avenue of research from the environmental justice literature is the link between environmental disamenities and neighbourhood dynamics over the long term (Boone 2008).

Urban growth on the periphery continues to accelerate throughout the world, but central cities (see Box 8.2) are far from static. Because of the econo-mies of agglomeration they provide, central business districts attract high-end commercial, office and service sectors. Many older neighbourhoods in central cities have enticed investors to purchase and renovate ageing houses. Gentrification of inner city neighbourhoods is one example that has garnered a good deal of attention, in part because the rejuvenation of historic houses can drive up rents and drive out poorer residents. One theory on gentrification argues that the return to the central city is a function of the rent-gap, or the difference between actual and realised value of inner city property (Smith 1987). Another argues that changing demographics, especially the rise of a new cre-ative class, is the primary force behind infill housing, gentrification and demand for inner city living (Ley 2003; Florida 2005). Net effects, however, are small compared with the rapid expansion of suburbs. By 1970 the suburbs of the USA contained more people than central cities or rural areas, and by 2000 more people lived in the suburbs than urban and rural areas combined. Over the course of 80 years, the USA moved from being an urban to a suburban nation.

Global context

Urban theorising in the social sciences since 1990 has demonstrated a broader global perspective on cities, space and the physical environments as these concepts relate to the world system of cities in the hierarchy of nations (Palen 2005). Political ecologists have assessed the impacts of global capitalism on the economies of developing nations that are undergoing rapid urbanisation and industrial transformation (O'Connor 1994). Political ecology 'seeks to under-stand the human processes leading to the destruction and creation of material environments' (Kirkby *et al.* 2001), and in the developing world, as in industrial-ised nations, this amounts to the struggle between powerful agents over the control of land (Bryant 1998).

Encroachment of human settlement beyond city boundaries is not a new problem, but landscape fragmentation on the urban fringe is occurring now on an unprecedented global scale (Alberti *et al.* 2003; Zhao *et al.* 2006). Suburban sprawl in industrialised nations, which turned natural and agricultural land-scapes into millions of houses, is one of the two great land-use changes of the late twentieth century (Rudel 2009). The other is tropical deforestation by conversion

Figure 8.1 The growth of shantytowns and squatter settlements, such as this one in Mirpur, Bangladesh, exemplify some of the starkest inequities in environmental conditions found in modern cities, particularly in the developing world. New urban migrants in search of work struggle to gain accommodation in shantytowns and squatter settlements. Photograph used with permission of A. Y. Hoque.

to agriculture, which according to Rudel's (2009) analysis results from the same political economic process of local elites forming 'growth coalitions' for profit. Uncontrolled urban sprawl around cities in industrialising nations, however, is the third-order spatial transition that further degrades landscapes and increases the vulnerability of marginal populations and fragile ecosystems.

According to Massey (1996), the world has entered a new era of urban extremes, in which rapid urbanisation combined with rising levels of income inequality create 'hypersegregation', or an 'ecology of inequality', in world cities. The economic, social, human health and, we argue, ecological consequences of sociospatial stratification are profound. The consequences of these changes for the physical environment can hardly be overstated. One billion people, or almost one-third of the world's urban population, live in slums, defined as abject poverty where there is no fresh water, sanitation or public infrastructure (UN-HABITAT 2003). This is projected to double in 30 years to one-third of all humanity, with the vast majority of these people living in Asia, Africa and South America (Whelan 2004). Many of the poor are housed in inner city slums but even greater changes are taking place on the periphery of cities as new urban migrants in search of work struggle to gain accommodation in shantytowns and squatter settlements (Figure 8.1). 'Squatter settlements are mainly uncontrolled low-income residential areas with an ambiguous legal status regarding land occupation . . .' (UNCHS (Habitat) 1981, p. 15). The conundrum faced by environmental scientists and conservationists is that *both* planned growth (urban growth coalitions) and unplanned growth (settlements by desperate, marginalised people) seem to lead to increasing sociospatial divisions within cities that can create social and environmental problems.

Linking social organisation and ecological outcomes

We now take a 'one-way' look at how the socioeconomic and temporal processes outlined above lead to various ecological outcomes, from patterns of nutrient loading to plant communities to animal communities inhabiting urban spaces. We review spatial and temporal processes jointly because it is clear that they are closely intertwined in the formation of the patterns detected in recent studies. Work that links human social organisation to ecological outcomes has generally fallen into three classes of study: (i) impacts of zoning policies as exhibited by differences among land-use categories, (ii) correlates of social stratification, such as associations between ecological variables and income level, and (iii) environmentally oriented human actions at the household level, such as gardening, fertilising lawns or feeding birds. These three roughly track a gradient in spatial scale of the effects with the household or property parcel scale as the smallest zone of effect (Figure 8.2).

The patterns emerging from recent literature suggest a profound influence of humans on the structure and functioning of urban ecosystems, and they point to a new catalogue of environmental inequalities – of access to greenness and biodiversity – playing out alongside inequities in exposure to toxic chemical releases and other classic examples that led to the founding of the environmental justice movements of the late twentieth century. But the survey we present here also points to the many broad gaps in our knowledge.

Differentiation by land use

Historical economic and technological changes have differentiated the growth forms of cities. For heuristic purposes, a simple classification of urban morphologies yields two major types: pre-industrial cities with high densities, narrow roads and mixed land-use, sometimes surrounded by modern expanses of lower-density suburbs; and post-World War II cities dominated by low-density suburbs and segregated land-use (see 'Temporal dynamics' above). Ecologists have generally addressed these distinct morphologies by applying a land-use classification, an urbanisation gradient approach or a combination of these approaches (Rebele 1994; Pickett *et al.* 2001; Cadenasso *et al.* 2007). A broad range of ecological outcomes are strongly correlated with land-use type, and can usually be characterised as varying with increasing levels of urbanisation,

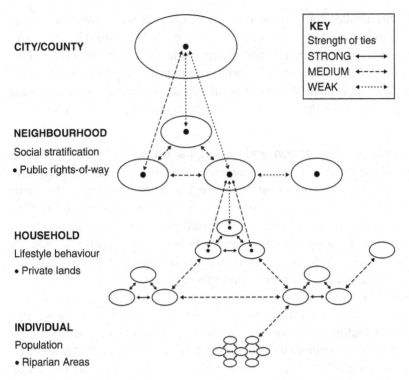

CITY/COUNTY

NEIGHBOURHOOD

Social stratification

• Public rights-of-way

HOUSEHOLD

Lifestyle behaviour

• Private lands

INDIVIDUAL

Population

• Riparian Areas

KEY
Strength of ties
STRONG ⟷
MEDIUM ⟵- - -⟶
WEAK ⟵·······⟶

Figure 8.2 An illustration of the connections between different levels of social organisation in cities. Modified from Grove *et al.* (2006) with permission.

as measured by increasing density of human populations, built structure and impervious surface cover (Blair 1996; Pickett *et al.* 2001; McKinney 2002).

Urbanisation drives local extinctions of many native plants while introducing other, non-native species (Kaye *et al.* 2006). Exotic species contribute to the high species richness of urban communities (Chapter 6; Hope *et al.* 2003), which may exceed diversity in wildlands. Land-use type also affects plant communities and diversity. In the Georgia Piedmont, USA, species richness of woody plants decreased with impervious surface and landscape diversity. Alien species richness was the highest in urban sites, whereas native stem densities were lower in urban and agricultural lands, compared with unmanaged forests (Burton & Samuelson 2008). Urban and suburban areas can also affect plant species composition in adjacent forest patches, by serving as sources for the spread of non-native species into forests. In Ottawa, Canada, forest fragments surrounded by urban habitat had a 50% greater proportion of introduced plant species than fragments surrounded by agricultural and forest habitats (Duguay *et al.* 2007).

Whereas urban plant communities are normally more diverse than wildland communities because of the introduction of many species, the opposite pattern has been observed for most animal taxa. Several studies on arthropods suggest that urbanisation causes a reduction in diversity, although within the urban core, land-use type may also influence diversity. In Phoenix, Arizona, Shochat *et al.* (2004) found spider diversity in xeric residential yards (those with drought-tolerant plants) to be as high as in the adjacent Sonoran desert, whereas mesic residential yards (those with water-dependent plants) had spider diversity as low as in agricultural fields. In Birmingham, UK, the richness and diversity of carabid beetles were lower in urban and suburban habitats than in rural zones (Sadler *et al.* 2006). As with many other taxa (McKinney 2002; DeStefano & DeGraaf 2003; Fuller *et al.* 2007a), the decrease in arthropod diversity in urban settings is normally associated with an increase in the total density. Faeth *et al.* (2005) suggested, based on field experiments from Phoenix, that this increase is the result of bottom-up control (the increase in plant densities and overall productivity), despite the strong top-down control (higher bird population densities in the urban habitat compared with the Sonoran desert).

Bird diversity also tends to decrease with urbanisation (Chace & Walsh 2006), although, along a land-use gradient, bird diversity may peak at moderate levels of urbanisation. In Seattle, Washington, Marzluff (2005) found that bird diversity peaked at intermediate levels of urbanisation, where the proportion of forest cover in the landscape was still relatively high. Early successional birds (species that are found in a variety of habitats around the area) contributed to the high species richness in this part of the landscape. As the proportion of the built environment in the landscape increased, loss of many of these species and specialist forest species exceeded immigration rates of synanthropic species, and bird diversity declined. Blair and Johnson (2008) also showed similar patterns from Ohio, Minnesota and California, suggesting that suburban land-uses, with their intermediate levels of development, serve as points of extirpation for woodland birds as well as entry points for invasive species into urban systems.

Patterns of association between land-use and nutrient cycles are more variable, depending on ecological context. In Baltimore, Maryland, a pre-industrial city in a predominantly forested region, nitrate concentration is lower in dense urban areas than in either suburban or agricultural areas (Groffman *et al.* 2004), owing to differential inputs. The source of urban nitrogen is mostly atmospheric, whereas suburban and agricultural areas include deposition but also fertiliser (Pickett *et al.* 2008). In arid regions, such as Phoenix, Arizona (another post-World War II city), concentrations of soil nutrients are less strongly associated with building density, instead following differences in landscaping designs. Concentrations of carbon, nitrogen and phosphorus in the soil are higher in mesic residential and agricultural lands than in xeric residential and desert habitats, although legacies of land-use influence these patterns (Kaye *et al.* 2006). These differences are driven not only by biophysical variables, but also by sociocultural variables underlying differences in cover of turfgrass, tree and impervious surface. Carbon accumulates in human-dominated soils, although most of it has short residence times in mesic yard and agricultural soils (Kaye *et al.* 2006).

In summary, many ecological outcomes can be tied to urbanisation gradients and differences in land-use, and these urban morphologies in turn derive from historical legacies of changing modes of transportation technology and changing economies. Plant and animal communities shift in relatively predictable ways with increasing levels of urbanisation – increasing densities of exotic species and synanthropic native species, declines in native species diversity, with a possible peak in total biodiversity at intermediate levels of urbanisation (reviewed in Chapter 5; McKinney 2002; Shochat *et al.* 2006; Pickett *et al.* 2008). Water quality and aquatic food web structure degrades with increasing levels of urbanisation (Paul & Meyer 2001; Groffman *et al.* 2003). Other ecological outcomes, such as soil nutrient concentrations, show fewer clear generalisations across cities and land-use types (Kaye *et al.* 2006). Variations on land-use themes, such as the differences between mesic and xeric landscaping designs in cities of the southwestern United States, illustrate the limitations of solely employing land-use classifications or gradient approaches.

Social stratification and ecological outcomes

As we search for ecological signatures of the sociospatial divisions in cities, the first step has been to document correlations between patterns of social stratification and various ecological factors. Income is a commonly used indicator of social stratification, along with ethnicity and education level. Members of the Chicago School suggested that environmental quality would follow wealth, leaving impoverished zones of cities also in poor environmental condition (see 'Spatial organisation' above). However, it took some time before ecologists attended to this prediction.

A focus on vegetation emerged relatively early on in the urban ecological literature, with a few early studies documenting correlations between income level and properties of the vegetation in city neighbourhoods (Whitney & Adams 1980; Talarchek 1990). In the 1990s, a rapid proliferation of large-scale ecologically driven studies in cities generated large statistical samples of vegetation structure and other ecological properties across metropolitan regions. The vast majority of these studies have focused analysis on relationships of vegetation to human population density or to characteristics of built structures (e.g. along urbanisation gradients). This greater attention to urbanisation gradients has also been true for studies of animal communities, soil properties, nutrient processes and water quality (Pickett *et al.* 2001; Shochat *et al.* 2006).

Inference of underlying social processes from ecological studies taking an urbanisation gradient approach is limited by the dynamic relationships over time between social stratification and built structure (see 'Temporal dynamics' above). Although it might be expected that gradients in built structure are associated with poverty and wealth, processes such as gentrification introduce complexity to this relationship (Ley 2003). Also, as Talarchek (1990, p. 66) points out, 'sequential occupancy of various income and social groups could, in time, result in no relationship between socioeconomic status and vegetation.' Furthermore, urban or suburban neighbourhoods of similar population density and forms of built structures can harbour distinct biotic communities, with their characteristics driven by a combination of historical legacies and contemporary management by residents (Whitney & Adams 1980; DeGraaf & Wentworth 1986).

Direct tests of association between social stratification and local ecological conditions increasingly show evidence for the predicted pattern of environmental quality following wealth, though with some complexities emerging. Documented associations with income level and related socioeconomic indicators include: vegetation cover (Talarchek 1990; Iverson & Cook 2000; Smith *et al.* 2005; Grove *et al.* 2006; Jenerette *et al.* 2007; Tratalos *et al.* 2007), plant diversity (Whitney & Adams 1980; Hope *et al.* 2003; Martin *et al.* 2004), bird diversity (Hope *et al.* 2004; Kinzig *et al.* 2005; Melles 2005), densities of some bird species (Loss *et al.* 2009), supplementary feeding of birds (Fuller *et al.* 2008; Lepczyk *et al.* in press; but see Gaston *et al.* 2007), small mammal diversity (Nilon & Huckstep 1998), nitrogen application to lawns (Law *et al.* 2004), aquatic food web complexity (Overmyer *et al.* 2005), ambient noise levels (Forkenbrock & Schweitzer 1999; Warren *et al.* 2006) and temperature regime (Jenerette *et al.* 2007). The levels of confidence in these associations vary. For example, associations with nitrogen application and aquatic food web complexity are represented by just one study each, with low statistical replication (Law *et al.* 2004; Overmyer *et al.* 2005).

The amount of vegetation shows the strongest general patterns of association with social stratification. Across a wide variety of studies in cities around the world, wealthier neighbourhoods appear to harbour more greenery: greater tree cover (Talarchek 1990; Iverson & Cook 2000; Smith *et al.* 2005; Grove *et al.* 2006; Kirkpatrick *et al.* 2007), more areas with trees and other vegetation (Grove & Burch 1997) and greater vegetation density (Hope *et al.* 2003; Martin *et al.* 2004; Jenerette *et al.* 2007). There are profound social and ecological implications of this association. Greater vegetation cover, and tree cover in particular, increases the delivery of many key ecosystem services, from regulating water flows and climate, to providing cleaner air (with some exceptions), to harbouring greater biodiversity (e.g. Jenerette *et al.* 2007; Tratalos *et al.* 2007; Wang *et al.* 2008). Trees and other vegetation may provide a variety of other human benefits, including fostering increased physical activity (Frumkin

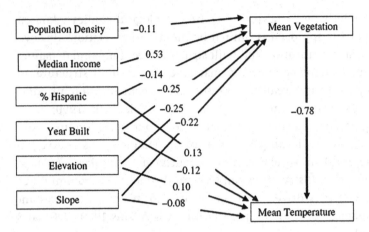

Figure 8.3 Social stratification and housing age are strong predictors of vegetation cover in Phoenix, Arizona, leading to effects on local climates and ecosystem services. Reproduced with permission from Jenerette *et al.* (2007).

2003), lowering urban crime (Kuo *et al.* 1998) and increasing property values (Morancho 2003; but see Troy & Grove 2008). In Phoenix, Arizona, higher-income areas having higher vegetation cover are cooler (Figure 8.3; Jenerette *et al.* 2007), leading to inequities in risk of heat exhaustion in this hot desert city (Harlan *et al.* 2006). Jenerette *et al.* (2007, p. 362) point out that 'along with residing in warmer locations, lower-income inhabitants may be less able to afford air-conditioning and may have fewer opportunities to avoid high temperatures'.

A few studies have found that plant community structure may also be associated with income level. In Phoenix, Arizona, two studies find that plant diversity is strongly associated with income level at the census block group level in parks and residential yards, and across a random selection of urban sites (Hope *et al.* 2003, 2004; Martin *et al.* 2004). Higher-income neighbourhoods are also more likely to have cacti or other desert-adapted plants (Martin *et al.* 2004). Particular forms of garden or landscaping are also associated with income and other social stratification variables in two Australian studies (Kirkpatrick *et al.* 2007; Luck *et al.* 2009). Gardens with trees are more common in high-income suburbs, as are 'simple native' gardens and 'shrub-bush-tree' gardens (Kirkpatrick *et al.* 2007). All of these garden types are likely to support native bird species in that region (Daniels & Kirkpatrick 2006). Low-income areas, and those with high proportions of unemployed and renting residents, are more likely to have 'non-gardens' and 'no-input exotic gardens' (Kirkpatrick *et al.* 2007). In the classic work of Whitney and Adams (1980), two types of tree communities are associated with higher-income areas in Akron, Ohio: 'old oak' and 'mixed suburban', and these two types contain the sites with highest tree species diversity. The strong association of animals with particular plant assemblages would suggest that these differences are likely to have effects on animal community structure and other ecological processes that are as profound as the effects of total vegetation cover.

A small but increasing number of studies finds associations between social stratification and animal communities: mammal species diversity in St Louis, Missouri (Nilon & Huckstep 1998) and bird species diversity in Phoenix (Hope et al. 2004; Kinzig et al. 2005), Vancouver, British Columbia (Melles 2005) and Chicago, Illinois (Loss et al. 2009). In Phoenix and Vancouver, total bird species richness is higher in parks as well as in residential neighbourhoods in census units with higher income and education level and with lower proportions of non-white residents, all classic indicators of social stratification (Kinzig et al. 2005; Melles 2005). Chicago, by contrast, seems to have higher native species richness in lower-income areas, with higher exotic species richness in higher-income areas (Loss et al. 2009). The differences in the findings may be due to differences in study design. But they may also reflect the complex interactions of social stratification with dynamic change over time in neighbourhood occupancy. Phoenix is a post-World War II city, dominated by suburbs with relatively invariant lot sizes (Gammage 1999). Most of the variation in bird habitat in residential areas of Phoenix is likely to be due to vegetation characteristics in yards, characteristics known to covary with income level in Phoenix and elsewhere (Figure 8.4; Kinzig et al. 2005). Chicago, by contrast, is a pre-industrial city with steeper gradients in built structure from one portion of the city to another. Moreover, the study sites used by Loss et al. (2009) included a variety of land-uses, such as residential areas, parklands and vacant lands, while the Phoenix and Vancouver studies focused on residential areas. Since urbanisation gradients are clearly a dominant factor in structuring bird communities (Chapter 5; McKinney 2002), further studies may find a more accurate signature of social stratification on bird communities by carefully controlling for other factors like land-use and housing density (e.g. Mennis 2006).

Many of the spatial patterns described above are difficult to disentangle from temporal ones. Time and space are intertwined in the development of socio-spatial divisions. Processes like disinvestment and suburban flight can lead to the formation of lower-income neighbourhoods in formerly moderate- to high-income ones (Pickett et al. 2008). Vegetation structure appears to vary with housing age in nonlinear ways (Grove et al. 2006), with tree cover increasing over the early decades, but reaching an inflection point in Baltimore, Maryland, at around 40 years old (Figure 8.5). For neighbourhoods of a fixed density, vegetation can mature and become more complex, supporting more species, over time. This process of maturation can lead to time-lags in the relationships between social processes and vegetation patterns (Troy et al. 2007; Boone et al. 2009; Luck et al. 2009). Effects of time may also be dependent on ecological region. New housing on old fields has a different impact on bird communities than new housing on former forest land (DeGraaf & Wentworth 1986; Loss et al. 2009).

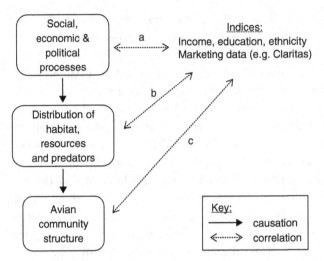

Figure 8.4 Diagram of hypothesised processes responsible for observed correlations between income level and patterns of bird community structure. Social, economic and political processes drive sociospatial divisions in cities (see also Box 8.1), leading to variation in vegetation structure, resources and predators. These in turn drive variation in bird community structure. Socioeconomic indicators such as income level serve as indicators of sociospatial divisions (a); a growing body of literature finds that vegetation structure is correlated with socioeconomic indicators (b); and recent studies have found correlations between bird species diversity and income level at the neighbourhood (census unit) scale (c) (Kinzig *et al.* 2005; Melles 2005; Loss *et al.* 2009), suggesting that sociospatial divisions have implications for biotic structure at higher trophic levels as well.

In arid environments, time may lead to a steady attrition of plants from periodic drought, with wealthier residents capable of combating this attrition with investment in water and re-planting (Martin *et al.* 2004). Water conservation policies in arid cities may also lead to shifting emphases in plantings toward native or drought-resistant species which generally support greater diversity of native animals (McIntyre & Hostetler 2001; Daniels & Kirkpatrick 2006). Wealthier residents may be better able to accommodate these shifts; as noted above, cacti (both native and drought resistant) are more common in wealthier neighbourhoods in Phoenix (Martin *et al.* 2004). Historic shifts in the patterns of housing and growth lead to apparent effects of age on plant and animal communities, not due to differences in the maturation of vegetation. For example, in cities like Chicago and Vancouver, age of neighbourhood is associated with housing density (older neighbourhoods at higher density), itself a potential factor driving patterns of diversity (Melles 2005; Loss *et al.* 2009). Thus, as cities age, processes of vegetation maturation overlay sociospatial patterns of investment and disinvestment, perhaps leading to the non-linear relationships between age and vegetation cover found in Baltimore.

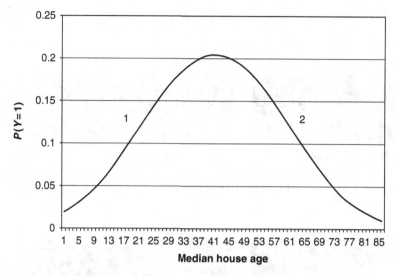

Figure 8.5 Nonlinear changes in vegetation through time in Baltimore, Maryland, USA, a pre-industrial city. Graph shows the probability that tree cover is equal to 100% of private land as a function of median age of block group, based on quasi-likelihood logit regression. Tree cover increases steadily for the first few decades, reaching an inflection point in neighbourhoods of about 40 years old. Modified with permission from Grove *et al.* (2006).

Other social dynamics

Alongside the socioeconomic divisions at the level of the city are finer-scaled variations associated with the activities that people engage in around their homes. For example, several authors have noted that front yards have become a site for the 'expression of individuality and ideology in a consumer society' (Kirkpatrick *et al.* 2007, p. 314; see also Bhatti & Church 2004). Grove *et al.* (2006, p. 580) argue that 'a household's land management decisions are influenced by its desire to uphold the prestige of its community and outwardly express its membership in a given lifestyle group'. Grove *et al.* go further to argue that the social pressures exerted to adhere to group norms can have a stronger effect on some environmental conditions than social stratification. They coin the phrase 'ecology of prestige' to describe this alternative process (Grove *et al.* 2006). In a series of studies, they show that social stratification is a strong indicator of 'plantable space' or the amount of pervious cover in parcels (Troy *et al.* 2007). By contrast, they find that the actual vegetation cover (trees versus grass versus bare ground) is associated with finer social classifications according to lifestyle (Grove *et al.* 2006; Troy *et al.* 2007).

Larsen and Harlan (2006) also refer to the 'symbolic presentation of self' represented by front yards and gardens. They find differences in preferred

1. Front yard – desert.

3. Front yard – oasis.

2. Front yard – lawn.

4. Front yard – courtyard.

Figure 8.6 Exemplars of landscape types used in a survey of landscape preferences and behaviour in Phoenix, Arizona, USA. Both macro-level processes (legacies from developers' plantings), and micro-level processes (residents' preferences) influenced actual landscape types in residents' front yards. Preferences for frontyard types varied significantly with income level, with lower-income homeowners tending to prefer lawn, middle-income homeowners preferring desert, and higher-income homeowners evenly divided between desert and oasis. Reproduced with permission from Larsen & Harlan (2006).

frontyard and backyard types associated with income level (Figure 8.6). They caution, however, that particularly in newer portions of cities, developers' choices function as a macro-level force constraining the degree to which, for example, frontyard landscapes reflect the choices of individual residents (see also Kirby *et al.* 2006). Larsen and Harlan (2006) found evidence of macro-level forces shaping actual frontyard landscapes through developers' planting legacies.

Market analysis has been advanced by some authors as a method of quantifying lifestyle and group identity and mapping these additional sociospatial divisions in cities (Weiss 2000; Kinzig *et al.* 2005; Grove *et al.* 2006; Fuller *et al.* 2008). Marketing datasets incorporate records of purchasing decisions, reasoning that people's purchases reflect their group identities and ideologies (Grove *et al.* 2006). These datasets have been criticised, however, for failing to account for informal economic networks, thereby underestimating purchasing power in lower-income communities (Pawasarat & Quinn 2001).

Distinctions between the 'ecology of inequality' and the 'ecology of prestige' are their emphases on processes operating at different spatial scales and different levels of social organisation (Figure 8.2). The ecology of prestige emphasises the actions and activities engaged in by urban residents at local scales that shape local environments, typically the household and neighbourhood (Figure 8.2; Grove *et al.* 2006). In contrast, theories of social stratification emphasise the political and economic forces operating at larger scales to constrain both real and perceived options available to urban residents, thereby yielding strong inequities in local environmental conditions (Massey 1996; see also 'Spatial organisation' above). Larsen and Harlan (2006) refer to these two sets of processes as 'micro-level' and 'macro-level'.

Ethnic and culturally mediated differences in preferences and values form a related but distinct component of the sociospatial organisation of cities. As outlined above, racial and ethnic divisions tend to be reinforced through processes of social stratification. While this often leads to poorer economic conditions for minority groups, such divisions also contribute to colourful heterogeneities in both built and unbuilt portions of cities (Whyte 1943; Wirth 1956; Duneier 1999; Johnston *et al.* 2007). This human diversity is imprinted on the architecture and landscape architecture of ethnic enclaves. The field of environmental psychology also explores culturally mediated preferences for different environments, such as the amount and configuration of trees versus other vegetation (Kaplan & Talbot 1988). Much work remains to be done in this area to understand whether and how these differences in environmental preferences lead to varying ecological outcomes, although there are suggestions that this is the case (Fraser & Kenney 2000).

Ecological outcomes from household decisions

Many small-scale decisions about land management occur at the household or property parcel level through gardening and maintaining lawns, keeping swimming pools, feeding birds and maintaining outdoor pets. These actions can provide resources such as habitat and food for birds, bugs, mammals, amphibians and other wildlife (Gaston *et al.* 2007), but they can also have negative impacts on other ecosystem functions such as water quality (e.g. runoff of pesticides and fertiliser), and animal populations (e.g. pesticides and introduced predators, such as cats; Law *et al.* 2004; Lepczyk *et al.* 2004; Overmyer *et al.* 2005). Because residential landscapes represent a large percentage of urban land cover (Cannon 1999; Martin *et al.* 2003; Chamberlain *et al.* 2004), private landowners collectively have a profound impact on urban ecosystems.

Landscaping is a multi-billion-dollar a year activity. In the UK, an estimated 60% of all households participate in gardening and spend £2.62 billion a year on gardening products (Horticulture Trades Association Garden Industry Monitor, unpublished data, cited in Gaston *et al.* 2005). The National Gardening

Association (2007) estimated that 71% of Americans participate in gardening activities, and that they spent $34 billion on lawn care and landscaping products in 2007. The manner in which a particular garden is managed and maintained influences urban biodiversity. Two studies in Australia found associations between native plantings and the presence of native bird species (French *et al.* 2005; Daniels & Kirkpatrick 2006). Burghardt *et al.* (2009) examined urban food webs in southern Pennsylvania, comparing butterfly larvae and bird distribution in gardens landscaped with native plants to those in gardens landscaped with exotic species. They also found that native landscaping supported higher abundances of native birds and butterflies.

In cultivating lawns and gardens, humans generally remove weeds and pests, as evidenced by the amount of pesticides and herbicides applied to residential gardens. Unfortunately, these activities have negative implications for biodiversity both directly (loss of invertebrate fauna) and through altered trophic dynamics (Blackburn & Arthur 2001; Lepczyk *et al.* 2004). For example, pesticide applications greatly decrease insect diversity and density, thus eliminating a potential food source for insectivorous birds. Furthermore, as evidenced by Carson's (1962) exposure of DDT and the many toxicology studies since (e.g. Bishop *et al.* 1998a, 1998b; Brickle *et al.* 2000), pesticides have devastating impacts on the reproductive health, growth and survival of birds and other taxa.

Bird feeding is a popular activity with more than 52 million people feeding wild birds in the USA (US Fish and Wildlife Service 2002) and about 12 million UK households (48%) feeding garden birds (Davies *et al.* 2009). Like gardening, feeding birds is a multi-billion-dollar endeavour. An estimated $30 billion per annum is spent on seed and feeders according to a US Fish and Wildlife survey in 2001 (US Fish and Wildlife Service 2002). Numerous studies have shown that feeding birds benefits the people participating in the activity. Bird feeding fosters a connection between people and the natural world (Fuller *et al.* 2007b). This exposure leads to an increased understanding of ecological issues and thus allows people to incorporate sound ecological initiatives into public policy (Rosenzweig 2003; Turner *et al.* 2004; Miller 2005). In a study across Sheffield, UK, Fuller *et al.* (2008) found that densities of typical bird feeder species were strongly correlated with the densities of bird feeders, although species richness did not have a significant relationship. Lepczyk *et al.* (2004) surveyed private landowners in southeast Michigan to determine their influence on bird populations. Two out of three respondents engaged in some level of bird feeding, and fed birds on average nine months out of the year. One elegant study of supplemental feeding replicated the behaviour of how individual landowners feed birds (Jansson *et al.* 1981). The authors concluded that winter survival probabilities of willow tits *Parus montanus* and crested tits *Lophophanes cristatus* increased because of the supplemental foods. Backyard bird feeding most likely has contributed to the northern expansion of both the northern cardinal *Cardinalis cardinalis* and American goldfinch *Caruelis tristis* in North America (Morneau *et al.* 1999).

A few studies have shown negative impacts of bird feeding (reviewed by Jones & Reynolds 2008; Robb *et al.* 2008). The concentrated food source leads to greater contact, with potential for increased transmission of disease. In eastern North America, house finches *Carpodacus mexicanus* have encountered an increase in mycoplasmal conjunctivitis (Dhondt *et al.* 2005). Bird feeders also provide resources for avian predators (e.g. corvids), pest and exotic species (Daniels & Kirkpatrick 2006; Marzluff & Neatherlin 2006). The higher densities of these predators might lead to increased nest predation, although this has yet to be tested. Effects of bird feeders on predation risk may vary according to life stage. For adult birds, the aggregation of birds around feeders reduces the risk per individual through dilution effects and increased levels of vigilance (Dunn & Tessaglia 1994). Woods *et al.* (2003) actually found a negative relationship between bird feeders and risk of cat predation.

The regular food source provided by bird feeders has been implicated in the cessation of migration in some populations of some bird species (Jokimäki *et al.* 1996). This leads to higher densities of sedentary species which possibly increases competition with migratory species (Jansson *et al.* 1981), potentially contributing to reduced species diversity (Clergeau *et al.* 1998). Concerns voiced by some bird enthusiasts regarding the dependence of birds on feeders have not been supported. Studies in the UK (blue tits *Cyanistes caeruleus*), USA (Florida scrub jay *Aphelocoma coerulescens*) and Australia (Australian magpie *Cracticus tibicen*) demonstrated that the majority of food provisioning to nestlings consisted of natural food items despite the prevalence of human-subsidised food items (Cowie & Hinsley 1988; Fleischer *et al.* 2003; O'Leary & Jones 2006).

Poor nutritional content of food provided by humans might have detrimental effects on health and behaviour, though long-term damage is unknown. When Australian magpies were provided with processed meats, this led to elevated levels of plasma cholesterol (Ishigame *et al.* 2006). When Florida scrub jays were given high-fat protein food supplements, they foraged 12% less than birds in wildlands without supplementation (Schoech *et al.* 2004). Net effects of supplemental feeding on avian populations and communities remains an important area for further research.

Although bird feeding is by far the most popular wildlife gardening activity (Gaston *et al.* 2007), a number of other features also provide resources for wildlife. Lepczyk *et al.* (2004) found that 50% of those surveyed provided nest boxes in their yards. The nest boxes could provide significant assistance for cavity nesters in urban and suburban landscapes.

Collectively, garden features that enhance urban wildlife presence and survival could have a profound effect on urban biodiversity. Furthermore, landowners participating in at least one wildlife gardening activity are more prone to participate in additional activities (Lepczyk *et al.* 2004; Gaston *et al.* 2007), thus increasing the benefits for a variety of taxa. Participation in these activities may provide a variety of benefits, directly and indirectly to the people engaging in them (Miller 2005; Fuller *et al.* 2007b).

Synthesis and directions for future research

Our review of the social organisation of cities finds a rich literature in the social sciences addressing potential processes underlying spatial structuring of ecological outcomes in cities. They point to three social divisions in cities – function, status and culture (Box 8.1). Signatures for all three social divisions can be found in ecological outcomes. Much more attention has been devoted, however, to ecological impacts of functional divisions (e.g. land-use effects) than to status or cultural divisions. Interactions between these divisions and temporal dynamics of technology and economy underlie variations in urban forms. This differentiation in form among cities, along with ecoregional effects and effects of city size on social organisation, may account for some of the differences found among cities in patterns of biodiversity and ecosystem functioning. Ecologists have not yet attempted to derive generalisations about the ecological consequences of differences in urban morphology – e.g. pre-industrial versus post-World War II cities. One finding of concern is that socioeconomic inequities extend to ecological inequities, with lower-income portions of cities typically having less vegetation and in some cases lower biodiversity than wealthier portions. There are significant potential consequences of these inequities for the delivery of ecosystem services, from climate regulation to clean air and to aesthetic and recreational services.

Beneath these general findings, however, lie broad gaps in knowledge. Ecologists have yet fully to capitalise on the potential for integration with the social sciences. Even basic description of the patterns of association between biotic structure and socioeconomic divisions still has far to go before truly generalisable findings can be drawn. Studies should move beyond the well-trodden ground of bird and plant community structure to address ecological processes like trophic dynamics, extinction–colonisation processes and ecosystem functions such as nutrient cycling. Finally, greater attention needs to be paid to the developing countries, where the most rapid urban growth is occurring and where disparities in income, health and environmental conditions on the margins of cities are far more stark than those in industrialised countries.

Acknowledgements

We thank G. Cowgill, J. Novic, M. Smith, B. Stanley, B. Stark and A. York for enlightening discussions and many references from our interdisciplinary research project, 'Urban Organization through the Ages'. We thank M. Nation for her assistance with gathering material for this chapter; any omissions, however, are ours and not hers. This material is based upon work supported by the National Science Foundation (NSF) under grant number DEB-0423704, Central Arizona–Phoenix Long-Term Ecological Research (CAP LTER) and the Baltimore Ecosystem Study (BES LTER), grant number DEB-0423476. Any opinions, findings and conclusions or recommendation expressed in this material are those of the authors and do not necessarily reflect the views of the National Science Foundation. We thank K. Gaston and two anonymous reviewers for many helpful comments.

References

Alberti, M., Marzluff, J. M., Shulenberger, E. et al. (2003). Integrating humans into ecology: opportunities and challenges for studying urban ecosystems. BioScience, 53, 1169–79.

Batty, M. (2008). The size, scale, and shape of cities, Science, 319, 769–71.

Bédarida, F. (1979). A Social History of England, 1851–1975. London and New York: Methuen.

Bhatti, M. and Church, A. (2004). Home, the culture of nature and meanings of gardens in late modernity. Housing Studies, 19, 37–51.

Bishop, C. A., Boermans, H. J., Ng, P., Campbell, G. D. and Struger, J. (1998a). Health of tree swallows (Tachycineta bicolor) nesting in pesticide-sprayed apple orchards in Ontario, Canada. I. Immunological parameters. Journal of Toxicology and Environmental Health – Part A – Current Issues, 55, 531–59.

Bishop, C. A., Van Der Kraak, G. J., Ng, P., Smits, J. E. G. and Hontela, A. (1998b). Health of tree swallows (Tachycineta bicolor) nesting in pesticide-sprayed apple orchards in Ontario, Canada. II. Sex and thyroid hormone concentrations and testes development. Journal of Toxicology and Environmental Health – Part A, 55, 561–81.

Blackburn, J. and Arthur, W. (2001). Comparative abundance of centipedes on organic and conventional farms, and its possible relation to declines in farmland bird populations. Basic and Applied Ecology, 2, 373–81.

Blair, R. B. (1996). Land use and avian species diversity along an urban gradient. Ecological Applications, 6, 506–19.

Blair, R. B. and Johnson, E. M. (2008). Suburban habitats and their role for birds in the urban-rural habitat network: points of local invasion and extinction? Landscape Ecology, 23, 1157–69.

Bolin, B., Grineski, S. and Collins, T. (2005). The geography of despair: environmental racism and the making of South Phoenix, Arizona, USA. Research in Human Ecology, 12, 156–68.

Boone, C. G. (2008). Environmental justice as process and new avenues for research. Environmental Justice, 3, 149–54.

Boone, C. G., Cadenasso, M. L., Grove, J. M., Schwartz, K. and Buckley, G. L. (2009). Landscape, vegetation characteristics, and group identity in an urban and suburban watershed: why the 60s matter. Urban Ecosystems. DOI: 10.1007/s11252-009-0118-7.

Boone, C. G. and Modarres, A. (2006). City and Environment. Philadelphia: Temple University Press.

Bourdieu, P. (1984). Distinction: A Social Critique of the Judgment of Taste, trans. Richard Nice. Cambridge, MA: Harvard University Press.

Brickle, N. W., Harper, D. G. C., Aebischer, N. J. and Cockayne, S. H. (2000). Effects of agricultural intensification on the breeding success of corn buntings Miliaria calandra. Journal of Applied Ecology, 37, 742–55.

Briggs, X. de S. (2004). Civilization in color: the multicultural city in three millennia. City and Community, 3, 311–42.

Bryant, R. L. (1998). Power, knowledge and political ecology in the third world: a review. Progress in Physical Geography, 22, 79–94.

Burgess, E. W. (1928). Residential segregation in American cities. *Annals of the American Academy of Political and Social Sciences*, **140**, 105–15.

Burghardt, K. T., Tallamy, D. W. and Shriver, W. G. (2009). Impact of native plants on bird and butterfly biodiversity in suburban landscapes. *Conservation Biology*, **23**, 219–24.

Burton, M. L. and Samuelson, L. J. (2008). Influence of urbanization on riparian forest diversity and structure in the Georgia Piedmont, US. *Plant Ecology*, **195**, 99–115.

Cadenasso, M. L., Pickett, S. T. A. and Schwarz, K. (2007). Spatial heterogeneity in urban ecosystems: reconceptualizing land cover and a framework for classification. *Frontiers in Ecology and the Environment*, **5**, 80–8.

Cannon, A. (1999). The significance of private gardens for bird conservation. *Bird Conservation International*, **9**, 287–97.

Carson, R. (1962). *Silent Spring*. New York: Houghton Mifflin.

Castells, M. (1977). *The Urban Question*. London: Edward Arnold.

Chace, J. F. and Walsh, J. J. (2006). Urban effects on native avifauna: a review. *Landscape and Urban Planning*, **74**, 46–69.

Chamberlain, D. E., Cannon, A. R. and Toms, M. P. (2004). Associations of garden birds with gradients in garden habitat and local habitat. *Ecography*, **27**, 589–600.

Chaskin, R. J. (1997). Perspectives on neighborhood and community: a review of the literature. *Social Service Review*, **71**, 521–47.

Clergeau, P., Savard, J.-P. L., Mennechez, G. and Falardeau, G. (1998). Bird abundance and diversity along an urban–rural gradient: a comparative study between two cities on different continents. *The Condor*, **100**, 413–25.

Cowgill, G. (2007). The urban organization of Teotihuacan, Mexico. In E. C. Stone, ed., *Settlement and Society: Essays Dedicated to Robert McCormick Adams*. Los Angeles: Costen Institute of Archaeology, pp. 261–95.

Cowie, R. J. and Hinsley, S. A. (1988). The provision of food and the use of bird feeders in suburban gardens. *Bird Study*, **35**, 163–8.

Craul, P. J. (1992). *Urban Soil in Landscape Design*. New York: Wiley.

Daniels, G. D. and Kirkpatrick, J. B. (2006). Does variation in garden characteristics influence the conservation of birds in suburbia? *Biological Conservation*, **133**, 326–35.

Davies, Z. G., Fuller, R. A., Loram, A. *et al.* (2009). A national scale inventory of resource provision for biodiversity within domestic gardens. *Biological Conservation*, **142**, 761–71.

DeGraaf, R. M. and Wentworth, J. M. (1986). Avian guild structure and habitat associations in suburban bird communities. *Urban Ecology*, **9**, 399–412.

DeStefano, S. and DeGraaf, R. M. (2003). Exploring the ecology of suburban wildlife. *Frontiers in Ecology and the Environment*, **1**, 95–101.

Dhondt, A. A., Altizer, S., Cooch, E. G. *et al.* (2005). Dynamics of a novel pathogen in an avian host: mycoplasmal conjunctivitis in house finches. *Acta Tropica*, **94**, 77–93.

Duguay, S., Eigenbrod, F. and Fahrig, L. (2007). Effects of surrounding urbanization on non-native flora in small forest patches. *Landscape Ecology*, **22**, 589–99.

Duneier, M. (1999). *Sidewalk*. New York: Farrar, Straus and Giroux.

Dunn, E. H. and Tessaglia, D. L. (1994). Predation of birds at feeders in winter. *Journal of Field Ornithology*, **65**, 8–16.

Faeth, S. H., Warren, P. S., Shochat, E. and Marussich, W. A. (2005). Trophic dynamics in urban communities. *BioScience*, **55**, 399–407.

Fleischer, A. L., Bowman, R. and Woolfenden, G. E. (2003). Variation in foraging behavior, diet, and time of breeding of Florida scrub-jays in suburban and wildland habitats. *The Condor*, **105**, 515–27.

Florida, R. L. (2005). *Cities and the Creative Class*. New York: Routledge.

Forkenbrock, D. J. and Schweitzer, L. A. (1999). Environmental justice in transportation planning. *Journal of the American Planning Association*, **65**, 96–111.

Fraser, E. D. G. and Kenney, W. A. (2000). Cultural background and landscape history as factors affecting perceptions of the urban forest. *Journal of Arboriculture*, **26**, 106–12.

French, K., Major, R. and Hely, K. (2005). Use of native and exotic garden plants by suburban nectarivorous birds. *Biological Conservation*, **121**, 545–59.

Frumkin, H. (2003). Healthy places: exploring the evidence. *American Journal of Public Health*, **93**, 1451–6.

Fuller, R. A., Evans, K. L., Davies, R. G. et al. (2007a). Bird densities are associated with household densities. *Global Change Biology*, **13**, 1685–95.

Fuller, R. A., Irvine, K. N., Devine-Wright, P., Warren, P. H. and Gaston, K. J. (2007b). Psychological benefits of greenspace increase with biodiversity. *Biology Letters*, **3**, 390–4.

Fuller, R. A., Warren, P. H., Armsworth, P. R., Barbosa, O. and Gaston, K. J. (2008). Garden bird feeding predicts the structure of urban avian assemblages. *Diversity and Distributions*, **14**, 131–7.

Galster, G. C., Quercia, R. G. and Cortes, A. (2000). Identifying neighborhood thresholds: an empirical exploration. *Housing Policy Debate*, **11**, 701–32.

Gammage, G. Jr (1999). *Phoenix in Perspective*. Phoenix, AZ: Herberger Center for Design.

Gaston, K. J., Fuller, R. A., Loram, A. et al. (2007). Urban domestic gardens (XI): variation in urban wildlife gardening in the UK. *Biodiversity and Conservation*, **16**, 3227–38.

Gaston, K. J., Smith, R. M., Thompson, K. and Warren, P. H. (2005). Urban domestic gardens (II): experimental tests of methods for increasing biodiversity. *Biodiversity and Conservation*, **14**, 395–413.

Gottdiener, M. and Hutchison, R. (2001). *The New Urban Sociology*. Boulder, CO: Westview Press.

Grineski, S., Bolin, B. and Boone, C. (2007). Criteria air pollution and marginalized populations: environmental inequity in metropolitan Phoenix, Arizona. *Social Science Quarterly*, **88**, 535–54.

Groffman, P. M., Bain, D. J., Band, L. E. et al. (2003). Down by the riverside: urban riparian ecology. *Frontiers in Ecology and the Environment*, **1**, 315–21.

Groffman, P. M., Law, N. L., Belt, K. T., Band, L. E. and Fisher, G. T. (2004). Nitrogen fluxes and retention in urban watershed ecosystems. *Ecosystems*, **7**, 393–403.

Grove, J. M. and Burch, W. R. (1997). A social ecology approach to urban ecosystem and landscape analyses. *Urban Ecosystems*, **4**, 259–75.

Grove, J. M., Troy, A. R., O'Neil-Dunne, J. P. M. et al. (2006). Characterization of households and its implications for the vegetation of urban ecosystems. *Ecosystems*, **9**, 578–97.

Hall, P. (1988). The intellectual history of long waves. In M. Young and T. Schuller, eds., *The Rhythms of Society*. London and New York: Routledge, pp. 37–52.

Harlan, S. L., Brazel, A. J., Prashad, L., Stefanov, W. L. and Larsen, L. (2006). Neighborhood microclimates and vulnerability to heat stress. *Social Science and Medicine*, **63**, 2847–63.

Harvey, D. (1973). *Social Justice and the City*. Baltimore: Johns Hopkins Press.

Harvey, D. (1982). *The Limits to Capital*. Chicago: University of Chicago Press.

Harvey, D. (2003). *The New Imperialism*. New York: Oxford University Press.

Hillier, A. (2003a). Spatial analysis of historical redlining: a methodological exploration. *Journal of Housing Research*, **1**, 137–68.

Hillier, A. (2003b). Who received loans? Home owners' loan corporation lending and discrimination in Philadelphia in the 1930s. *Journal of Planning History*, **1**, 3–24.

Hope, D., Gries, C., Warren, P. et al. (2004). How do humans restructure the biodiversity of the Sonoran desert? In *Connecting Mountain Islands and Desert Seas: Biodiversity and Management of the Madrean Archipelago II*, Vol. RMRS-P-26. Tucson, AZ: USDA Forest Service Proceedings (Fort Collins, CO), pp. 189–94.

Hope, D., Gries, C., Zhu, W. X. et al. (2003). Socioeconomics drive urban plant diversity. *Proceedings of the National Academy of Sciences of the USA*, **100**, 8788–92.

Ishigame, G., Baxter, G. S. and Lisle, A. T. (2006). Effects of artificial foods on the blood chemistry of the Australian magpie. *Austral Ecology*, **31**, 199–207.

Iverson, L. R. and Cook, E. A. (2000). Urban forest cover of the Chicago region and its relation to household density and income. *Urban Ecosystems*, **4**, 105.

Jackson, K. T. (1985). *Crabgrass Frontier: The Suburbanization of the United States.* New York: Oxford University Press.

Jansson, C., Ekman, J. and Vonbromssen, A. (1981). Winter mortality and food-supply in tits *Parus*-spp. *Oikos*, **37**, 313–22.

Jenerette, G. D., Harlan, S. L., Brazel, A. *et al.* (2007). Regional relationships between surface temperature, vegetation, and human settlement in a rapidly urbanizing ecosystem. *Landscape Ecology*, **22**, 353–65.

Johnston, R., Poulsen, M. and Forrest, J. (2007). The geography of ethnic residential segregation: a comparative study of five countries. *Annals of the Association of American Geographers*, **97**, 713–38.

Jokimäki, J., Suhonen, J., Inki, K. and Jokinen, S. (1996). Biogeographical comparison of winter bird assemblages in urban environments in Finland. *Journal of Biogeography*, **23**, 379–86.

Jones, D. N. and Reynolds, S. J. (2008). Feeding birds in our towns and cities: a global research opportunity. *Journal of Avian Biology*, **39**, 265–71.

Joseph, L. (2008). Finding space beyond variables: an analytical review of urban space and social inequalities. *Spaces for Difference: An Interdisciplinary Journal*, **1**, 29–50.

Kaplan, R. and Talbot, J. F. (1988). Ethnicity and preference for natural settings: a review and recent findings. *Landscape and Urban Planning*, **15**, 107–17.

Kaye, J. P., Groffman, P. M., Grimm, N. B., Baker, L. A. and Pouyat, R. V. (2006). A distinct urban biogeochemistry? *Trends in Ecology and Evolution*, **21**, 192–9.

Keith, K. (2003). The spatial patterns of everyday life in old Babylonian neighborhoods. In

M. L. Smith, ed., *The Social Construction of Ancient Cities.* Washington, DC: Smithsonian Institution Press, pp. 56–80.

King, A. D. (1976). *Colonial Urban Development: Culture, Social Power and Environment.* Boston: Routledge and Kegan Paul.

Kinzig, A. P., Warren, P., Martin, C., Hope, D. and Katti, M. (2005). The effects of human socioeconomic status and cultural characteristics on urban patterns of biodiversity. *Ecology and Society*, **10**(1), 23 (online).

Kirby, A., Harlan, S. L., Larsen, L. *et al.* (2006). Examining the significance of housing enclaves in the metropolitan United States of America. *Housing, Theory and Society*, **23**, 19–33.

Kirkby, J., O'Keefe, P. and Howorth, C. (2001). Introduction: rethinking environment and development in Africa and Asia. *Land Degradation and Development*, **12**, 195–203.

Kirkpatrick, J. B., Daniels, G. D. and Zagorski, T. (2007). Explaining variation in front gardens between suburbs of Hobart, Tasmania, Australia. *Landscape and Urban Planning*, **79**, 314–22.

Krysan, M. and Farley, R. (2002). The residential preferences of blacks: do they explain persistent segregation? *Social Forces*, **80**, 937–80.

Kuo, F. E., Bacaicoa, M. and Sullivan, W. C. (1998). Transforming inner-city landscapes – trees, sense of safety, and preference. *Environment and Behavior*, **30**, 28–59.

Kuznets, S. S. (1967) [1930]. *Secular Movements in Production and Prices: Their Nature and Their Bearing upon Cyclical Fluctuations.* Reprints of economic classics. New York: A. M. Kelley.

Larsen, L. and Harlan, S. L. (2006). Desert dreamscapes: residential landscape preference and behavior. *Landscape and Urban Planning*, **78**, 85–100.

Law, N. L., Band, L. E. and Grove, J. M. (2004). Nitrogen input from residential lawn care practices in suburban watersheds in Baltimore County, MD. *Journal of Environmental Management*, **47**, 737–55.

Lee, S. W., Ellis, C. D., Kweon, B. S. and Hong, S. K. (2008). Relationship between landscape structure and neighborhood satisfaction in urbanized areas. *Landscape and Urban Planning*, **85**, 60–70.

Lepczyk, C. A., Mertig, A. G. and Liu, J. G. (2004). Assessing landowner activities related to birds across rural-to-urban landscapes. *Environmental Management*, **33**, 110–25.

Lepczyk, C. A., Warren, P. S., Machabée, L., Kinzig, A. P. and Mertig, A. Who feeds the birds? A comparison between Phoenix, Arizona and Southeastern Michigan. *Studies in Avian Biology*, in press.

Ley, D. (2003). Artists, aestheticisation and the field of gentrification. *Urban Studies*, **40**, 2527–44.

Logan, J. and Molotch, H. (1987). *Urban Fortunes: The Urban Economy of Place*. Berkeley: University of California Press.

Loss, S. R., Ruiz, M. O. and Brawn, J. D. (2009). Associations between avian diversity and social, economic, and environmental characteristics of an urban landscape. *Biological Conservation*, **142**, 2578–85.

Luck, G. W., Smallbone, L. T. and O'Brien, R. (2009). Socio-economics and vegetation change in urban ecosystems: patterns in space and time. *Ecosystems* (online). **DOI** 10.1007/s10021-009-9244-6.

Marcuse, P. and van Kempen, R. (2002). *Of States and Cities: The Partitioning of Urban Space*. London and New York: Oxford University Press.

Martin, C. A., Peterson, K. A. and Stabler, L. B. (2003). Residential landscaping in Phoenix, Arizona, U.S.: practices and preferences relative to covenants, codes, and restrictions. *Journal of Arboriculture*, **29**, 9–17.

Martin, C. A., Warren, P. S. and Kinzig, A. P. (2004). Neighborhood socioeconomic status is a useful predictor of perennial landscape vegetation in residential neighborhoods and embedded small parks of Phoenix, AZ. *Landscape and Urban Planning*, **69**, 355–68.

Martine, G. (2008). *The New Global Frontier: Urbanization, Poverty and Environment in the 21st Century*. London and Sterling, VA: Earthscan.

Marzluff, J. M. (2005). Island biogeography for an urbanizing world: how extinction and colonization may determine biological diversity in human-dominated landscapes. *Urban Ecosystems*, **8**, 157–77.

Marzluff, J. M. and Neatherlin, E. (2006). Corvid response to human settlements and campgrounds: causes, consequences, and challenges for conservation. *Biological Conservation*, **130**, 301–14.

Massey, D. S. (1996). The age of extremes: concentrated affluence and poverty in the twenty-first century. *Demography*, **33**, 395–412.

Massey, D. S. and Denton, N. A. (1993). *American Apartheid: Segregation and the Making of the Underclass*. Cambridge, MA: Harvard University Press.

McIntyre, N. E. and Hostetler, M. E. (2001). Effects of urban land use on pollinator (Hymenoptera: Apoidea) communities in a desert metropolis. *Basic and Applied Ecology*, **2**, 209–18.

McIntyre, N. E., Knowles-Yanez, K. and Hope, D. (2000). Urban ecology as an interdisciplinary field: differences in the use of 'urban' between the social and natural sciences. *Urban Ecosystems*, **4**, 5–24.

McKinney, M. L. (2002). Urbanization, biodiversity, and conservation. *BioScience*, **52**, 883–90.

Melles, S. (2005). Urban bird diversity as an indicator of social diversity and economic inequality in Vancouver, British Columbia. *Urban Habitats*, **3**, 25–48.

Mennis, J. (2006). Socioeconomic-vegetation relationships in urban, residential land: the case of Denver, Colorado. *Photogrammetric Engineering and Remote Sensing*, **72**, 911–21.

Miller, J. R. (2005). Biodiversity conservation and the extinction of experience. *Trends in Ecology and Evolution*, **20**, 430–4.

Morancho, A. B. (2003). A hedonic valuation of urban green areas. *Landscape and Urban Planning*, **66**, 35–41.

Morneau, F., Decarie, R., Pelletier, R. *et al.* (1999).
Changes in breeding bird richness and
abundance in Montreal parks over a period
of 15 years. *Landscape and Urban Planning*,
44, 111–21.

Musterd, S. (2005). Social and ethnic segregation
in Europe: levels, causes, and effects. *Journal
of Urban Affairs*, **27**, 331–48.

National Gardening Association (2007). *The
Residential Lawn and Landscape Services and the
Value of Landscaping*, Vol. 2009. National
Gardening Association.

Newman, P. and Jennings, I. (2008). *Cities as
Sustainable Ecosystems: Principles and Practices*.
Washington, DC: Island Press.

Nilon, C. H. and Huckstep, S. (1998). Impacts of
site disturbance on the small mammal
fauna of urban woodlands. In J. Brueste,
H. Feldmann and O. Uhlmann, eds., *Urban
Ecology*. Berlin: Springer-Verlag, pp. 623–7.

O'Connor, J. (1994). Is sustainable capitalism
possible? In M. O'Connor, ed., *Is Capitalism
Sustainable? Political Economy and the Politics of
Ecology*. New York: The Guildford Press.

O'Leary, R. and Jones, D. N. (2006). The use of
supplementary foods by Australian magpies
Gymnorhina tibicen: implications for wildlife
feeding in suburban environments. *Austral
Ecology*, **31**, 208–16.

Overmyer, J. P., Noblet, R. and Armbrust, K. L.
(2005). Impacts of lawn-care pesticides
on aquatic ecosystems in relation to
property value. *Environmental Pollution*,
137, 263–72.

Palen, J. J. (2005). *The Urban World*. New York:
McGraw-Hill.

Park, R. E. (1915). The city: suggestions for the
investigation of human behavior in the city
environment. *American Journal of Sociology*,
20, 577–612.

Paul, M. J. and Meyer, J. L. (2001). Streams in the
urban landscape. *Annual Review of Ecology and
Systematics* **32**, 333–65.

Pawasarat, J. and Quinn, L. M. (2001). Exposing
urban legends: the real purchasing power
of central city neighborhoods. In *Discussion
Papers*. Washington, DC: Brookings

Institution, Center on Urban and
Metropolitan Policy, p. 22.

Pickett, S. T. A., Cadenasso, M. L., Grove, J. M.
et al. (2001). Urban ecological systems:
linking terrestrial ecological, physical, and
socioeconomic components of
metropolitan areas. *Annual Review of Ecology
and Systematics*, **32**, 127–57.

Pickett, S. T. A., Cadenasso, M. L., Grove, J. M.
et al. (2008). Beyond urban legends: an
emerging framework of urban ecology, as
illustrated by the Baltimore Ecosystem
Study. *BioScience*, **58**, 139–50.

Quillian, L. (2002). Why is black–white residential
segregation so persistent? Evidence on three
theories from migration data. *Social Science
Research*, **31**, 197–229.

Rebele, F. (1994). Urban ecology and special
features of urban ecosystems. *Global Ecology
and Biogeography Letters*, **4**, 173–87.

Robb, G. N., McDonald, R. A., Chamberlain, D. E.
and Bearhop, S. (2008). Food for thought:
supplementary feeding as a driver of
ecological change in avian populations.
Frontiers in Ecology and the Environment,
6, 476–84.

Rosenzweig, M. L. (2003). *Win–Win Ecology: How the
Earth's Species Can Survive in the Midst of Human
Enterprise*. Oxford: Oxford University Press.

Rudel, T. (2009). States and settlement
expansion: suburban sprawl and tropical
reforestation in a comparative perspective.
American Journal of Sociology, **115**, 129–35.

Sadler, J. P., Small, E. C., Fiszpan, H., Telfer, M. G.
and Niemelä, J. (2006). Investigating
environmental variation and landscape
characteristics of an urban–rural gradient
using woodland carabid assemblages.
Journal of Biogeography, **33**, 1126–38.

Schoech, S. J., Bowman, R. and Reynolds, S. J.
(2004). Food supplementation and possible
mechanisms underlying early breeding in
the Florida Scrub-Jay (*Aphelocoma coerulescens*).
Hormones and Behavior, **46**, 565–73.

Shochat, E., Stefanov, W. L., Whitehouse, M. E. A.
and Faeth, S. H. (2004). Urbanization and
spider diversity: influences of human

modification of habitat structure and productivity. *Ecological Applications*, **14**, 268–80.

Shochat, E., Warren, P. S., Faeth, S. E., McIntyre, N. E. and Hope, D. (2006). Urban mechanistic ecology: from pattern to emerging processes. *Trends in Ecology and Evolution*, **21**, 186–91.

Sjoberg, G. (1960). *The Preindustrial City: Past and Present*. New York: Free Press.

Smith, M. E. (2008). *Aztec City-state Capitals*. Gainsville, FL: University Press of Florida.

Smith, N. (1987). Gentrification and the tent gap. *Annals of the Association of American Geographers*, **77**, 462–5.

Smith, N. (1996). *The New Urban Frontier: Gentrification and the Revanchist City*. London and New York: Routledge.

Smith, R. M., Gaston, K. J., Warren, P. H. and Thompson, K. (2005). Urban domestic gardens (V): relationships between landcover composition, housing and landscape. *Landscape Ecology*, **20**, 235–53.

Soja, E. (2000). *Postmetropolis: Critical Studies of Cities and Regions*. Oxford: Blackwell Publishers.

Talarchek, G. M. (1990). The urban forest of New Orleans – an exploratory analysis of relationships. *Urban Geography*, **11**, 65–86.

Tratalos, J., Fuller, R. A., Warren, P. H., Davies, R. G. and Gaston, K. J. (2007). Urban form, biodiversity potential and ecosystem services. *Landscape and Urban Planning*, **83**, 308–17.

Troy, A. and Grove, J. M. (2008). Property values, parks, and crime: a hedonic analysis in Baltimore, MD. *Landscape and Urban Planning*, **87**, 233–45.

Troy, A. R., Grove, J. M., O'Neil-Dunne, J. P. M. et al. (2007). Predicting opportunities for greening and patterns of vegetation on private urban lands. *Environmental Management*, **40**, 394–412.

Turner, W. R., Nakamura, T. and Dinetti, M. (2004). Global urbanization and the separation of humans from nature. *BioScience*, **54**, 585–90.

UNCHS (Habitat) (1981). *Upgrading of Urban Slums and Squatter Areas*. UN Commission on Human Settlements.

UNFPA (2007). *State of the World Population 2007: Unleashing the Potential of Urban Growth*. New York: United Nations Population Fund.

US Fish and Wildlife Service (2002). *2001 National Survey of Fishing, Hunting, and Wildlife-associated Recreation*. Report of the US Department of the Interior, Fish and Wildlife Service, and US Department of Commerce, Bureau of the Census.

UN-HABITAT (2003). *The Challenge of Slums – Global Report on Human Settlements*. London: Earthscan.

Vance, J. E. (1990). *The Continuing City: Urban Morphology in Western Civilization*. Baltimore: Johns Hopkins University Press.

Wang, J., Endreny, T. A. and Nowak, D. J. (2008). Mechanistic simulation of tree effects in an urban water balance model. *Journal of the American Water Resources Association*, **44**, 75–85.

Warren, P. S., Katti, M., Ermann, M. and Brazel, A. (2006). Urban bioacoustics: it's not just noise. *Animal Behaviour*, **71**, 491–502.

Weiss, M. J. (2000). *The Clustered World: How We Live, What We Buy, and What It All Means About Who We Are*. Boston: Little, Brown, and Co.

Whelan, S. (2004). One third of the world's urban population lives in a slum. International Committee of the Fourth International (ICFI), available at http://www.wsws.org/articles/2004/feb2004/slum-f17.shtml

Whitney, G. G. and Adams, S. D. (1980). Man as a maker of new plant-communities. *Journal of Applied Ecology*, **17**, 431–48.

Whyte, W. F. (1943). *Street Corner Society: The Social Structure of an Italian Slum*. Chicago: University of Chicago Press.

Wilson, W. J. (1997). *When Work Disappears: The World of the New Urban Poor*. New York: Alfred A. Knopf.

Wirth, L. (1956). *The Ghetto*. Chicago: University of Chicago Press.

Woods, M., McDonald, R. A. and Harris, S. (2003). Predation of wildlife by domestic cats *Felis catus* in Great Britain. *Mammal Review*, **33**, 174–88.

Zhao, S. Q., Da, L. J., Tang, Z. Y. et al. (2006). Ecological consequences of rapid urban expansion: Shanghai, China. *Frontiers in Ecology and the Environment*, **4**, 341–6.

CHAPTER NINE

Urban ecology and human health and wellbeing

JO BARTON AND JULES PRETTY

The importance of urban greening

Westernised societies are becoming more and more urbanised, and throughout the twentieth and twenty-first centuries the number of people living in urban settings has steadily increased. More than half of the world's population currently live in urban areas (UNFPA 2007) and this proportion is still set to increase (Pretty 2007). Urban environments expose people to many stressors, such as traffic noise and congestion, crowding and fear of crime, and are often a source of continual demands prohibiting restoration from mental fatigue (van den Berg et al. 2007). Everyday life revolves around complex information processing activities requiring directed attention (Kaplan & Kaplan 1989). Our capacity for this type of concentrated attention is finite, so it is regularly taxed to its limit, leading to mental fatigue. This is a state characterised by inattentiveness, indecisiveness and increased irritability, and we have fewer cognitive resources available to manage everyday tasks, leading to increased stress (Kaplan 1995). To restore our capacity for directed attention, we need to spend time in settings that utilise involuntary attention requiring no cognitive effort. Having contact with nature and green spaces promotes this type of attention restoration, alleviates fatigue and reduces stress. Thus, with ongoing urban and suburban sprawl, the importance of access to nearby nature is paramount, especially for those regularly exposed to the pressures of urban life.

The type of green space close to where people live and work is important for the quality of life of urban citizens and for the sustainability of towns and cities (Chiesura 2004). Green spaces are defined as 'open, undeveloped land with natural vegetation' (CDC 2009) and there are many types of urban green space, ranging from larger parks and gardens (community, formal and private), city farms and urban agriculture to smaller-scale communal squares, allotments and green roofs. Other types include canals and riverbanks, tree-lined streets, cemeteries, woods and grasslands, cycling routes, disused railways,

Urban Ecology, ed. Kevin J. Gaston. Published by Cambridge University Press.

school playing fields and pitches, golf courses, informal recreation areas and amenity green spaces. Some of these urban green spaces join to form continuous green corridors or networks linking towns, cities and the countryside. The rural–urban fringe consists of new and reinstated areas of woodland, wetland, meadow, nature reserves, parks and many other diverse natural habitats. It can also transport countryside biodiversity to the urban doorstep, thus increasing opportunities for urban dwellers to encounter rarer flora and fauna as well as having regular contact with nature (Countryside Agency and Groundwork 2004). All of these sources provide an important *direct* link to nature for many people and often represent their sole exposure to nature.

The benefits of urban greening

The presence of urban nature is important for a number of reasons including improving human health and wellbeing (Kaplan & Kaplan 1989; Frumkin 2001; Irvine & Warber 2002; Health Council of the Netherlands 2004), improving behaviour and cognitive functioning (Wells 2000; Taylor *et al.* 2001), facilitating social networking (Kuo *et al.* 1998; Ward Thompson 2002) and exercise (Giles-Corti & Donovan 2002; Giles-Corti *et al.* 2005), reducing levels of crime, aggression and violence (Kuo & Sullivan 2001a, 2001b), providing an outdoor classroom (Kaplan & Kaplan 1989; Kahn & Kellert 2002) and improving its aesthetic value (Sheets & Manzer 1991).

Improving human health and wellbeing

The presence of green spaces and vegetation in the built environment can influence human health and wellbeing (Judd *et al.* 2002; Frumkin 2003; Frumkin *et al.* 2004). These terminologies are often used interchangeably, but the term 'health' incorporates physical health, mental or emotional health, social health, spiritual health, lifestyle and functionality. The World Health Organization (WHO) definition of health is still the most widely cited and states that 'health is a state of complete physical, mental and social wellbeing, and not merely the absence of disease or infirmity' (WHO 1948). A universal definition of 'wellbeing' is not available, as many sources interpret and define it differently. However, wellbeing is generally considered in a broader context, and the UK Department for Environment, Food and Rural Affairs has collaborated with other government departments and stakeholders to develop a shared understanding of the meaning of wellbeing within a policy context (Box 9.1; DEFRA 2007).

Attempts to establish a potential link between the urban environment and an individual's mental health have shown that the prevalence of psychiatric morbidity is greater in urban areas and less common in rural domains, after adjusting for confounding variables (Lewis & Booth 1994; White & Heerwagen 1998; Galea *et al.* 2005). For example, Lewis and Booth (1994) found that urban residents' prevalence of psychiatric morbidity (33.7%) was higher than that of

---

Box 9.1 Shared understanding of wellbeing

'Wellbeing is a positive physical, social and mental state; it is not just the
absence of pain, discomfort and incapacity. It requires that basic needs are
met, that individuals have a sense of purpose, that they feel able to achieve
important personal goals and participate in society. It is enhanced by conditions
that include supportive personal relationships, strong and inclusive communities,
good health, financial and personal security, rewarding employment, and a
healthy and attractive environment. Government's role is to enable people
to have a fair access now and in the future to the social, economic and
environmental resources needed to achieve wellbeing. An understanding
of the effect of policies on the way people experience their lives is important
for designing and prioritising them.' *Source:* DEFRA (2007).

---

their rural counterparts (24.8%), after controlling for socioeconomic and other
extraneous variables. Income-related inequalities in health also depend on
exposure to green space. People who live in greener areas reported lower levels
of health inequality relating to income deprivation for both all-cause mortality
and mortality from circulatory diseases (Mitchell & Popham 2008).

A direct link between the amount of accessible local green space and health has
also been evidenced using large-scale epidemiological studies (Takano *et al.* 2002;
de Vries *et al.* 2003; Grahn & Stigsdotter 2003). In one, tree-lined streets, parks and
other green spaces were found to play a key role in longevity and decreased risk of
mental ill-health (Takano *et al.* 2002). This longitudinal study compared access to
local walkable green spaces and mortality rates in elderly residents of Tokyo,
Japan, over a period of five years. After controlling for demographic and socioeconomic variables, they found that out of 3100 Tokyo citizens born between 1903
and 1918, 71% were still alive in 1992 and the probability of living for an
additional five years was linked to their ability to walk in a local park or tree-
lined street (Takano *et al.* 2002). However, although the study asked respondents to
assess the availability of walkable green spaces in their neighbourhood, they did
not establish how frequently these spaces were actually used for walking.

Self-reported health data from over 10 000 Dutch respondents was correlated
with national environmental data characterising the type and quantity of blue
(e.g. rivers, lakes, canals) and green spaces present in their neighbourhood.
Socioeconomic and demographic characteristics were controlled for selection
effects and the study reported that people living in greener neighbourhoods
enjoyed better general health (de Vries *et al.* 2003). The type of green space
did not seem to alter effectiveness, but the total amount of green space in
the living environment seemed to be the most relevant predictor. A criticism
of the study is that the environmental characteristics were separated into

neighbourhoods and all individuals within that particular area were classed as having equal access to green spaces. This crude measure does not acknowledge that the exposure to green space may vary considerably between residents of the same neighbourhood and that durations of exposure may also differ.

In a separate study, one in ten residents felt unhealthy when the majority of the space surrounding their home was green (90%). In contrast, when only 10% of the environment was green, 16% of the residents felt unhealthy (Maas *et al.* 2006). Groenewegen *et al.* (2006) have set up the ongoing 'Vitamin G' project, aiming to build on previous research analysing the relationship between the amount and type of green space and health and wellbeing. The project has three different levels: (i) national Dutch data; (ii) green spaces in urban environments; and (iii) allotment gardens. It combines land-use data and self-reported health states. The findings are going to inform policy development, aid urban planning and design, and raise awareness of the importance of local green pockets.

Perceived neighbourhood greenness is also strongly associated with better mental and physical health. Respondents who perceived their neighbourhood as highly green were 1.37 and 1.60 times more likely to have better physical and mental health respectively, in comparison with those who perceived it as low in greenery (Sugiyama *et al.* 2008). The degree of species richness in urban green spaces has also been positively associated with psychological wellbeing of visitors (Fuller *et al.* 2007), emphasising the importance of locally managed biodiversity for sense of place and reflection.

In terms of overall health, local park users reported fewer visits to a physician for purposes other than routine check-ups in comparison with non-park users (Godbey *et al.* 1998). This difference was apparent even when controlling for the effects of age, income, education level, health status and other potential confounding variables. Frequently active park users also scored better on self-reported health indices and perceived their health states to be better than passive users and non-park users (Godbey *et al.* 1992). Thus, people engaging in leisure recreation in local parks seem to be in disproportionately better health than non-users and are also less likely to be obese than the general population (Ho *et al.* 2003).

Godbey and Blazey (1983) also investigated the leisure behaviour of adults participating in light to moderate aerobic activity in urban parks and found that over half reported better moods after visiting the park. More and Payne (1978) also found that participants' negative moods improved and that park users reported lower levels of anxiety and depression. Often visitors started their recreation experiences in a better mood and their positive moods remained on leaving, implying that outdoor recreation and park use might enhance positive moods, reduce negative ones and alleviate stress.

In a Swedish study, Grahn and Stigsdotter (2003) examined the relationship between use of urban green spaces and health. They found that the level of self-reported stress experienced showed significant relationships with the proximity of urban green spaces, visiting frequency and duration of stay. The findings implied that the more frequent the visits, the lower the incidence of stress-related illnesses. Having access to a public or privately owned garden adjacent to their place of residence was another principal factor, which has implications for both policy and urban landscape planning.

Improving behaviour and cognitive functioning

Parents of children experiencing Attention Deficit Hyperactivity Disorder (ADHD) reported that participating in activities such as fishing or camping in green spaces improved behaviour in 85% of cases (Taylor et al. 2001). When watching television or playing video games in indoor environments, behaviour improved in only 43% of activities. In some indoor activities, behaviour actually deteriorated (57%) and made the children less manageable. Following on from the ideas of attention restoration described earlier, Wells (2000) conducted a longitudinal study with children of low-income urban families and assessed the effects of nature on their cognitive functioning. When the families were relocated to houses with more nature in the window view, they had higher levels of cognitive functioning and their ability to direct attention continued for several months after moving. However, these findings should be treated with caution because it could be argued that these types of families were able to select these types of preferred homes. Therefore, cause and effect can be difficult to disentangle and decipher.

Facilitating social networking

Green space in the form of parks, streets, squares and allotments can be valuable in urban areas for facilitating social contact (Coley et al. 1997; Ward Thompson 2002) and giving rise to stronger neighbourhood ties (Kuo et al. 1998). Green spaces can also foster social inclusion, community development, citizenship and local pride by allowing local residents to assist in the design, management and care of local spaces (DLTR 2002). Activities in green places often occur in social groups, or indeed people undertake activities in order to interact with others. Social capital is thus a component whereby relations of trust and reciprocity tied together by social norms and institutions can help people engage in activities, link to particular places and remain mentally and physically healthy. Social capital is also closely tied to capacity for collective environmental management (Pretty & Smith 2004).

Several key studies researching the link between nature and social contact have all involved the same USA study population (Coley et al. 1997; Kuo et al. 1998; Kweon et al. 1998). The 'Robert Taylor Homes' (RTH) are located in public social housing communities in an underprivileged area of Chicago. They constitute a naturally occurring field experiment as residents are randomly assigned to apartments, which are identical, with the exception of the quantity of surrounding vegetation and greenness of common spaces. All other environmental, cultural and social variables are held constant, and residents are socio-economically homogeneous.

The presence of trees significantly increased the utilisation of public green space by both adults and youths (Coley *et al.* 1997). The communal green spaces provided opportunities for more face-to-face contacts and encouraged social interaction. This led to stronger neighbourhood social ties, which were assessed by the amount of socialising, contact with nearby neighbours and local sense of community (Kuo *et al.* 1998). The greener areas promoted the strongest neighbourhood social ties, although the reasons for visiting the green spaces were not reported (e.g. requiring shade from trees in hot summer months). In addition, the exposure time to communal green spaces was positively linked to social integration of elderly residents in the community (Kweon *et al.* 1998). Outdoor spaces dominated by trees and grass had the greatest effect, and active use of these spaces predicted the strength of neighbourhood social ties and sense of community.

Facilitating green exercise activities

The term 'green exercise' stems from a programme of research which aims to investigate the synergistic benefits of engaging in physical activities whilst simultaneously being directly exposed to nature (Pretty *et al.* 2003; Barton 2009). Access to urban green spaces has been shown to promote healthy living by encouraging participation in green exercise activities, such as walking, jogging and cycling (Ross 2000; Berrigan & Troiano 2002; Craig *et al.* 2002; Handy *et al.* 2002; Parks *et al.* 2003; Wendel-Vos *et al.* 2004; Bedimo-Rung *et al.* 2005; Godbey *et al.* 2005). However, the state and design of the surrounding environment can either be conducive or restrictive to activity participation. Socioeconomic variants in health outcomes are often determined by the immediate environment and thus the individual's behavioural preferences (Owen *et al.* 2000). Typical physically active behaviours may include walking for exercise or recreation, jogging or participating in a sporting activity, whereas sedentary behaviours comprise sitting, socialising, spectating or dining. Therefore, behaviour settings can potentially influence the level of activity experienced and can either encourage or prohibit participation.

Large-scale Australian studies found that accessible public open spaces, such as parks, were used more frequently for physical activity, although this effect was significantly moderated by attractiveness and size (Giles-Corti & Donovan 2002; Giles-Corti *et al.* 2005). People with easy access to an attractive and large public open space were 50% more likely to exceed physical activity recommendations. Open spaces incorporating trees, water features and birdlife were commonly used for walking or jogging (64%) and cycling (12%) activities.

Other studies have reported the physically active behaviour of visitors to parks (Scott 1997; Godbey *et al.* 1998; Raymore & Scott 1998; Tinsley *et al.* 2002). The average visitor spends about half of their time walking when visiting a park, and seven out of ten park visitors engage in moderate to vigorous levels of physical activity (e.g. brisk walking, cycling, jogging; Godbey *et al.* 1998). Participating in vigorous physical exercise was reported to be the most important outcome for many park users (Tinsley *et al.* 2002). Other studies estimate that 7% of urban park users in England visit parks to engage in sporting activities, such as football, bowls, golf and cycling (CABE Space 2004), which represents 7.5 million visitors per annum (Woolley 2003). Walking increases if the local area is attractive and scenic, and if there are safe footpaths and pavements, a diversity of land-use, easy access to public transport, a friendly neighbourhood and ease of accessibility (Humpel *et al.* 2002; Bird 2004; Owen *et al.* 2004). Landowners and managing agencies therefore have a role to play in delivering these requirements to increase visitor numbers and activity levels.

Parks *et al.* (2003) demonstrated a dose–response relationship between the number of places available for exercise within a neighbourhood and the probability of meeting physical activity recommendations. The majority of people who exercise choose to do so in their local park, thus indicating that exercise frequency and park use are both associated with park proximity (Cohen *et al.* 2007). Bedimo-Rung *et al.* (2005) propose a conceptual model to describe the relationships between park characteristics, such as the number of visits and physical activity levels within the park, park use and overall benefits, including physical and psychological health, social, economic and environmental.

Residential environments with large amounts of green and minimal graffiti and litter have been found to be associated with increased physical activity and a reduced incidence of obesity (Ellaway *et al.* 2005). Residents of greener environments were 3.3 times as likely to participate in regular exercise compared with those living in areas with nominal greenery. Although the study was cross-sectional, the findings supported other research studies which also reported improved physical health when good-quality, well-maintained public spaces were easily accessible (CABE Space 2004). Thus, access to nature and green spaces seem to play an important role in increasing physical activity levels in urban communities.

Grahn and Stigsdotter (2003) suggest that urban citizens who live 50 m or less from their nearest green space visit it three to four times per week. If the distance is increased to 300 m, the number of visits reduces to an average of 2.7 times per week. However, a distance of 1000 m reduces the number of visits to only once a week. Residents of communities lacking greenery in their local area do not compensate for this by visiting public parks or urban forests more frequently, which highlights the importance of restoring and conserving nearby local green space. There are also physical, social or cultural barriers which may restrict usage even when green space is available. Perceptions of personal safety and fear of crime may discourage visits, along with cultural barriers where people feel that they are not allowed to use the spaces.

Reducing levels of crime, aggression and violence

Kuo and Sullivan (2001a, 2001b) have made compelling links between small amounts of green in the urban environment of Chicago's poorest public housing neighbourhood and crime, aggression and domestic violence. Vegetation levels surrounding RTH homes were assessed by an independent panel using a combination of photographs from a number of vantages. The assessment took place in June when the grass was green and the tree canopy was in full leaf. Residents living in greener surroundings reported lower levels of fear, fewer incivilities and less generalised aggressive and violent behaviour. Buildings with more vegetation also had 52% fewer property and violent crimes than those with minimal vegetation. There is often a perception that well-vegetated places offer more opportunities for criminals and drug-dealers to hide, so these findings raise some interesting questions.

There was also a greater difference between buildings in non-green and moderately green surroundings than between moderately and very green, suggesting more of a benefit would accrue from a light greening of all urban spaces rather than a dark greening of just a few. Indeed, well-maintained vegetation may instigate new ways of thinking as local people start to care for their environment and so are more vigilant (Kuo & Sullivan 2001a). Levels of aggression and violence were also significantly higher for residents in conditions lacking vegetation than for those who had access to nearby nature (Kuo & Sullivan 2001b). Access to greener surroundings reduced aggression by increasing concentration, which mediates the relationship between mental fatigue and aggressive behaviour. Therefore, neighbourhood settings lacking nearby nature and greenery had a significant impact on human social functioning (Sullivan 2005).

Aesthetic value

Urban environments incorporating green spaces are perceived to be more attractive than urban areas lacking vegetation (Sheets & Manzer 1991; Kuo et al. 1998).

Visitors report a deep sense of personal satisfaction from experiencing the aesthetic pleasures, such as 'enjoying the changing seasons, feeling the sun, the wind or the rain, being able to walk, run or just sit down and enjoy the view' (Burgess et al. 1988). Environmental aesthetics are also positively associated with walking for exercise. Compared with people who had access to a very pleasing aesthetic environment, those reporting a moderately aesthetic environment were 16% less likely to walk for exercise, and those who only had access to a low-rated aesthetic environment were 41% less likely to walk for exercise (Ball *et al.* 2001). Research figures report that 85% of people questioned believe that the quality of public space and the built environment directly affects their lives and the way they feel (CABE Space 2002).

Providing an 'outdoor classroom'

Urban green spaces also provide 'outdoor classrooms' to facilitate learning and enhance knowledge of the natural world and local environment. Outdoor classrooms are important for children (Moore & Wong 1997; Kahn & Kellert 2002; Rickinson *et al.* 2004), with creative social play, concentration and motor ability positively influenced by play in green space (Taylor *et al.* 2001). However, this evidence has not yet been strong enough to change the design of schools, though the emergence of the forest school movement is an indication that green space is being seen as a contributor to positive cognitive outcomes (Bishops Wood Centre 2005).

Levels of engagement with nature

We posit three levels of engagement with nature leading to potentially different outcomes: viewing nature, functional engagement and active participation (Figure 9.1).

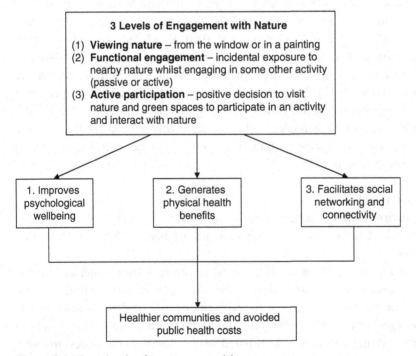

Figure 9.1 Three levels of engagement with nature.

Viewing nature

Viewing nature through a window or in a photograph is an important way of engaging with nature, when direct access is restricted. This type of passive engagement allows the mind to digress and stimulates reflection and recovery, aids recovery from illness, improves mood, reduces stress and improves mental wellbeing.

From the home

Kaplan and Austin (2004) and Kaplan (2001) used black and white photographs of views from the home to explore the importance of differing levels of vegetation for residential satisfaction. Participants were asked how closely the characteristics in the view represented the view from their own home and rated their preferences. The most preferred scenes were predominantly nature views, and trees were strongly associated with feelings of relaxation. Views of gardens, flowers and landscaped areas played an important role in participants' residential satisfaction but characteristics that were most favoured were also rated as least available. Woodlands enhanced residents' satisfaction with their surroundings and contributed to the community and sense of peacefulness (Kaplan & Austin 2004). However, there may be an element of response bias as the participants were aware of the aim of the study and were self-selecting so results may have been skewed.

The value of the view from a window is also reflected in monetary terms as various studies have demonstrated increased economic value for housing and hotels (Peiser & Schwann 1993). The presence of green space affects room pricing policy in hotels in Zurich, Switzerland (Lange & Schaeffer 2001), and increases the value of homes with gardens overlooking lakes and paths in the Netherlands by 25% (Luttick 2000). Street trees in Berlin, Germany, increase real-estate value by 17% (Luther & Gruehn 2001), and the value of housing near to water is greater in Merseyside, UK (Wood & Handley 1999; Lindsey et al. 2004).

From the workplace

Windows present in the workplace can buffer the stresses of work and reduce the frequency of illness, headaches and frustration, and improve patience and enthusiasm for work (Leather et al. 1998; Kaplan 2001). Workers with views of trees and flowers have been shown to feel less stressed in their jobs and to derive greater job satisfaction than workers overlooking built environments (Kaplan 1993). A similar study also reported that natural views buffered the negative impact of job stress on intention to resign (Leather et al. 1998). People working in windowless workplaces were four times more likely to compensate by displaying pictures of landscapes and outdoor natural scenes or indoor plants (Heerwagen & Orians 1993).

From institutions

Two classic and widely cited studies from the 1980s (Moore 1982; Ulrich 1984) suggest that prolonged exposure to window views of nature can have important health-related consequences. The first found that prisoners in Michigan, USA, whose cells overlooked farmland and forests reported 24% fewer sick cell visits compared with those in cells facing the prison yard. Cells were randomly allocated so the findings implied a stress reduction effect. The second 'classic' study exploited the configuration of a hospital in Pennsylvania, USA, where rooms in the surgical section overlooked deciduous trees or a brick wall. It formed a 10-year comparative study of post-operative patients who had undergone identical surgical procedures. Patients were randomly assigned a room which only differed in its view from the window. The hospital stay for those patients with tree views was significantly shorter (7.96 days per patient compared with 8.70); they also required fewer painkillers and took less strong or moderate pain medication. Nursing staff also reported fewer negative comments in the medical records for those with the tree views (1.13 per patient compared with 3.96).

Whilst travelling

The view during the commute to work (e.g. the type and quality of roadside verges) can also influence levels of stress. Participants in one study were exposed to four different simulated types of roadside corridors: rural, urban, golf course or mixed (lots of vegetation but visible commercial buildings; Parsons et al. 1998). There was some evidence that those on the urban drive, dominated by human artefacts, had higher levels of skin conductance, facial muscle tension and blood pressure compared with other settings. Restoration of standard heart rate readings was faster and more complete after viewing scenes of golf courses. Viewing nature-prevalent drives of forests or golf courses facilitated stress recovery quicker than urban settings. In addition, the nature drives offered a protective effect against the negative consequences of future stresses that might have arisen during the working day, implying that there might be an immunisation effect (Parsons et al. 1998).

Using simulated scenes of being in nature

A number of studies have explored the effects of viewing simulated scenes of being in nature by conducting experiments in laboratories. This setting allows the limitation of potentially confounding variables and often participants have been attentionally fatigued, using a variety of demanding mental tests, prior to viewing a series of slides. Photographic simulations have predominantly compared natural settings (e.g. green spaces, forests, woods, open countryside) with contrasting urban scenes lacking nature. Stress reduction qualities of these differential environments have been assessed using a combination of physio-logical (blood pressure, heart rate, cortisol, muscle tension) and psychological measures (e.g. self-reports of emotion, concentration or mood).

Nature slides incorporating water or vegetation were consistently preferred to grey urban scenes lacking greenery and had a more positive effect on emotional states (Ulrich 1981; Hartig *et al.* 1996). The natural settings sustained attention more effectively, and for stressed or excessively aroused participants they offered the greatest opportunity for stress reduction, with blood pressure, muscle tension and skin conductance levels all reducing during recovery (Ulrich *et al.* 1991). Participants' heart rate recordings decreased whilst viewing the nature video, whereas participants viewing the urban environments did not report any change (Laumann *et al.* 2003). Therefore, this implies that natural settings have a relaxing effect on autonomic functions. Recovery was also faster when exposed to natural environments, which supports the theory that nature exposures of short duration are important in urbanised societies. When coloured slides depicting cityscapes with designated green spaces were introduced, positive effect scores were significantly higher compared with no greenery (Honeyman 1992).

Van den Berg *et al.* (2003) analysed the relationship between restorative potential and environmental preference. Participants viewed a frightening movie followed by either a video simulating a nature-based walk or a walk in a built environment. Greater improvements in mood and concentration were reported after viewing natural settings compared with viewing built environ-ments, and the natural settings were perceived as more beautiful and restora-tive. Higher levels of stress were associated with stronger preferences for natural scenes and less liking for urban settings. Berto (2005) explored restora-tive environments' ability to facilitate recovery from mental fatigue by analys-ing the relationship with increased attention performance. Mentally fatigued subjects were exposed to either nature scenes (restorative) or urban streets (non-restorative), and only subjects who viewed the natural scenes regained sufficient attention capacity to perform well in a secondary task.

A study at the University of Essex, UK, tested the physiological and psycho-logical health benefits of exercising on a treadmill whilst being exposed to a series of rural or urban photographic scenes (Pretty *et al.* 2005). Each of these was subdivided into pleasant and unpleasant categories in order to explore the effect of rural scenes compromised with pollutants or other visual impedi-ments (e.g. rubbish, abandoned cars, billboards or pipes carrying effluents) and urban scenes enhanced by the presence of nearby nature in the form of green space. A control group was included which involved exercising without exposure to images. Only those subjects viewing rural pleasant scenes experi-enced significant decreases in mean arterial blood pressure (Figure 9.2). The urban pleasant pictures had no effect on mean arterial blood pressure, whilst

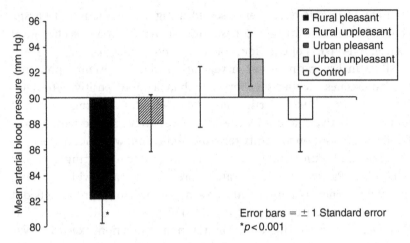

Figure 9.2 Change in mean arterial blood pressure (MABP) after treadmill exercise whilst viewing different scenes on a projector (change in MABP normalised to the starting average for all five groups). See text for details.

the urban unpleasant slightly increased it. All participants viewing rural pleasant pictures experienced a reduction in their mean arterial blood pressure compared with only 60% of people when exposed to the other picture conditions. As control subjects experienced a slight decrease in blood pressure, it is clear that both pleasant and unpleasant urban scenes increased blood pressure relative to the controls. The urban scenes therefore appear effectively to negate the marginal, but potentially beneficial impact of exercise on blood pressure.

Self-esteem significantly improved in all five groups, but the control group produced a greater improvement in self-esteem than the two unpleasant treatments (rural and urban), implying that the latter have a depressive effect on self-esteem relative to exercise alone. Both pleasant treatments, however, produced the greatest increases in self-esteem (Figure 9.3). When viewing both urban and rural pleasant scenes, levels of self-esteem increased by 10% and 9% respectively. This was in comparison with a 7.5% increase after viewing no pictures and only a 4% increase after viewing unpleasant pictures.

For the six measures of mood, viewing rural pleasant scenes during exercise produced consistent, though not always significant, improvements relative to viewing other scenes. Viewing urban pleasant scenes also resulted in improvements in all six mood measures. Unexpectedly, exercise whilst viewing urban unpleasant scenes produced significant improvements for anger-hostility, confusion-bewilderment and tension-anxiety. However, the rural unpleasant scenes had the most differentiated effect on mood measures. There were negative effects on three mood states, the most for any type of scene. This suggests that views embodying threats to the countryside have a greater negative effect on mood than already urban unpleasant scenes.

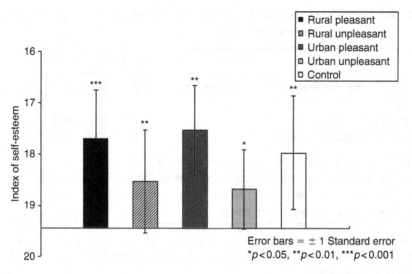

Figure 9.3 Change in self-esteem after treadmill exercise whilst viewing different scenes on a projector (change in self-esteem normalised to the starting average for all five groups). NB: High score = low self-esteem. See text for details.

Functional engagement

The second category of engagement is more functional and often involves incidentally being in the presence of nearby nature, whilst primarily engaging in another activity. The nature setting acts as background scenery for social activities such as walking the dog, cycling to work through an urban park, reading on a garden seat or talking to friends in a park (Hayashi et al. 1999; Ulrich 1999). One of the first longitudinal studies conducted over a decade targeted poorly functioning impoverished neighbourhoods undergoing environmental improvements (Dalgard & Tambs 1997). The mental health states of the residents were initially quite poor, but significant improvements were reported post-intervention for those residents still present. Therefore, selective migration was not the reason for this outcome, indicating that health parameters could be positively influenced by changes in environmental features.

A study at the University of Essex assessed changes in local people's behaviour and health measures following ecological restoration of local green spaces at three urban sites in the UK (urban park, canal and coastal path; Peacock et al. 2005, 2006). Prior to the restoration process all three sites were often unused, but transforming these areas created new opportunities for outdoor recreation and contact with nature and green space. Ecological restoration processes involved re-landscaping grassland, creating a wetland environment, improving biodiversity and improving canal towpaths and coastal paths. A higher proportion of users visited the locations for all of the reasons listed post-restoration

Table 9.1 *A comparison of the number and duration of visits before and after restoration at three sites (see text for details).*

	Before	After	Increase (%)
No. of visitors (who completed question re visitation patterns)	133	150	12.8
No. of visits per month	1535	2007	30.8
No. of visits per person per month	11.5	13.4	15.9
No. of visitors (who completed question re duration of visit)	129	144	11.6
Total time spent at site for all users during one visit (mins)	4580	6068	32.5
Average time spent at site per person (mins)	35.5	42.1	18.7
Total time spent at site per person per month (mins)	410	564	37.6

Note: A month refers to a 4-week period; the numbers of visitors do not match as calculations do not include those who had never visited before the improvements or who had not completed both sections of the questionnaire.

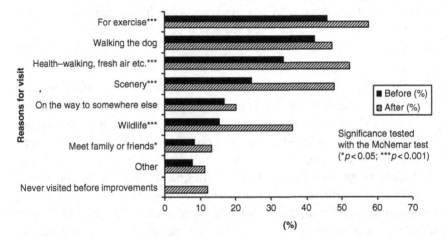

Figure 9.4 Primary reasons users visited sites before and after completion of ecological restoration projects. See text for details.

(Figure 9.4). However, significant differences were found for exercise, meeting family or friends, health, scenery and wildlife. The restored sites attracted new visitors and many more users were now choosing to visit the sites to view the scenery and wildlife and interact with the environment.

Although the environmental improvements encouraged more people to visit the sites, they also visited more frequently and spent longer engaging with nature on each visit (Table 9.1). During a 4-week period before restoration, a total of 133 users visited the sites 1535 times, which equates to an average of 11.5 visits per person per month. However, following restoration of the sites, a total of 150 people visited the sites 2007 times during an equivalent period of

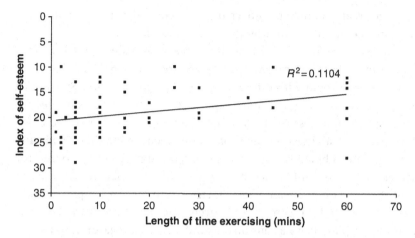

Figure 9.5 Relationship between the length of time exercising at restored sites and self-esteem scores. NB: High score = low self-esteem. See text for details.

time. This equates to an average of 13.4 visits per person per month and represents a significant increase ($p < 0.0001$).

On average, users were spending 35.5 minutes exploring the sites on each visit prior to restoration. Following the environmental improvements, this figure increased to 42.2 minutes and the total time spent at the sites per person per month increased by 154.2 minutes, representing a significant increase of 37.6% ($p < 0.0001$).

A significant improvement in self-esteem was found when comparing arrival scores with departing scores ($p < 0.05$). Thus, those individuals who had been participating in physical activity for longer within the green surroundings reported an improved self-esteem score. A positive correlation between reported self-esteem scores and the length of time exercising within the environment was found ($p < 0.0001$) (Figure 9.5). This implies that the longer people spend exercising in urban green areas, the more their self-esteem improves.

Active participation

The third category of engagement with nature is referred to as 'active participation'. This differs from the second category as it implies a positive decision to visit nature and green spaces and directly participate in an activity, such as gardening, walking, mountaineering, running, cycling or water-based activities. Studies often use psychophysiological measures to compare walking in urban nature parks to city streets. Findings have suggested that nature groups report more positively toned emotional states, higher happiness scores, higher ratings of positive effect, markedly better cognitive performance scores (Hartig *et al.* 1991), a greater ability to reflect on problems (Mayer *et al.* 2006) and improved directed attention abilities (Berman *et al.* 2008).

Other studies have included ambulatory blood pressure measurements during walks in both nature reserves and urban areas with minimal landscaping after sitting in a room with different views (Hartig *et al.* 2003). Sitting in a room with tree views promoted more rapid diastolic blood pressure decline than sitting in a windowless room. Walking in a nature reserve reduced blood pressure more than walking along an urban, non-green street. After 20 minutes of walking, the mean blood pressure values differed significantly between the two settings, before the difference converged. This is probably because the benefits of the exercise (the walk itself) started to surpass any unpleasantness of the urban street. In both contexts, the green room and green walk, people recovered more rapidly from attention-demanding tasks, regardless of antecedent condition. There were no significant differences in blood pressure readings post environmental treatment, but there were visible effects on emotion. The walk in the natural setting increased positive effect and reduced feelings of anger/aggression, while the opposite occurred in the urban setting.

Research has also analysed runners' cognitions and moods whilst exercising in different settings (Bodin & Hartig 2003; Butryn & Furst 2003). A 1-hour run through a nature reserve dominated by greenery, water and pleasant views was compared with an urban route through sidewalks and streets with varying traffic volumes and many buildings (Bodin & Hartig 2003). Running in a nature reserve promoted emotional restoration more effectively than exercising in the urban environment, although there were no significant differences in the reduction in anxiety/depression and anger between the two conditions. The low level of statistical power may have contributed to this, as there were only 12 runners. However, runners did state their preference for the nature reserve as they perceived it to be more psychologically restorative. Butryn and Furst (2003) also examined the effects of natural parks and urban settings on mood, feeling states and cognitive strategies of non-elite female runners. Mood and feeling states significantly increased following both runs, irrespective of setting, but the natural vegetation park setting was overwhelmingly preferred (93%) by the runners. Pre-run moods were elevated, perhaps in anticipation of the run or from accumulated benefits from participation in regular runs. Therefore, there was minimal room for improvement.

Mood change and stress during recreation in an urban park was also compared to indoor leisure activities (Hull & Michael 1995). Findings suggested that anxiety and energy levels decreased with time spent at the park but changes in fatigue and calmness were minimal. A more recent study conducted by the University of Essex assessed the role the environment plays in the effectiveness of exercise for mental wellbeing with members of local Mind groups. Mind is a mental health charity based in England and Wales and provides support for

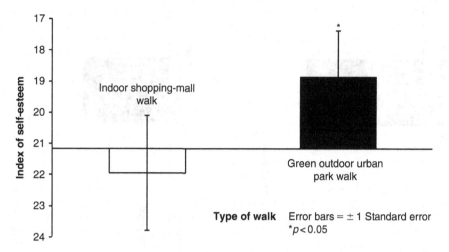

Figure 9.6 The change in self-esteem following participation in both walks (change in self-esteem normalised to the starting average for both groups). NB: High score = low self-esteem. See text for details.

individuals experiencing mental distress. It provides numerous services to all of its members through a network of local Mind associations. Local Mind members participated in both a green outdoor urban walk (Belhus Woods Country Park) and an indoor shopping-mall walk (Lakeside shopping centre), both of which were of the same duration and intensity and led by the same person, to ensure consistency of personality.

Improvements in self-esteem ($p < 0.05$) and overall mood ($p < 0.01$) were significantly greater following the green outdoor urban walk than following the equivalent indoor walk (Mind 2007; Peacock *et al.* 2007). Figure 9.6 compares the significant changes in self-esteem after both walks, highlighting the positive improvement after the green outdoor urban walk and the negative effect after the indoor walk. Figure 9.7 illustrates the significant changes in the subscale mood factors after both of the walks (anger, confusion, depression and tension). Participants felt significantly less angry ($p < 0.05$), confused ($p < 0.05$), depressed ($p < 0.05$) and tense ($p < 0.01$) after the outdoor green urban walk and felt more confused and tense after the indoor walk.

The findings show that exercising outdoors in a green environment is a lot more effective in enhancing mood and improving self-esteem than the equivalent amount of exercise indoors. The enjoyment of engaging in green exercise activities in groups was a valuable part of the experience, as well as the opportunity to breathe in fresh air, admire the scenery and enjoy the wildlife. The findings add significant value to the ever expanding green exercise research programme as they focus on individuals experiencing mental health issues and separate the elements that constitute the green exercise experience.

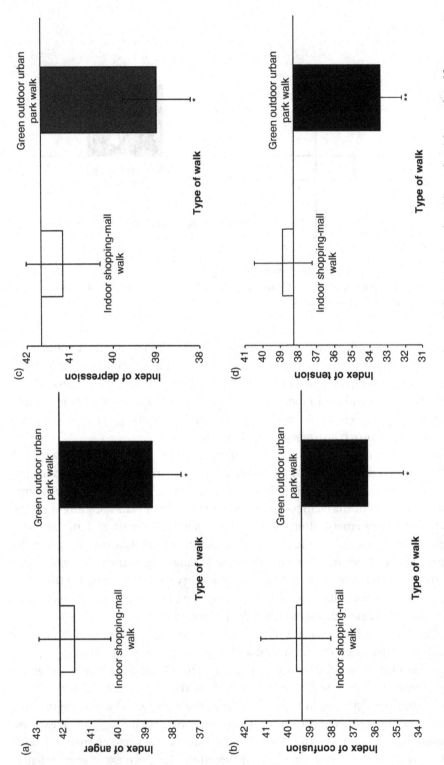

Figure 9.7 The change in feelings following participation in walks. (a) anger; (b) confusion; (c) depression; (d) tension (change in self-esteem normalised to the starting average for both groups). Error bars = ± 1 standard error; *$p < 0.05$, ** $p < 0.01$. See text for details.

Being part of a social group contributes to the green exercise experience, so it is important to compare different group activities to ascertain the importance of the exercise and the contact with nature. A study at the University of Essex therefore evaluated the effectiveness of introducing a 6-week green exercise programme (a series of short countryside and urban park walks) for individuals experiencing a range of mental health problems. The walking group was compared to two other Mind group programmes which were already in existence, including a swimming group (indoor exercise) and a social club (met indoors, but did not participate in any form of exercise). Key findings were that all groups experienced a significant improvement in self-esteem, but the change in self-esteem was significantly greater in the green exercise group than in the social group. All three groups also experienced a significant improvement in overall mood, and feelings of anger, confusion, depression, fatigue and tension all decreased more in the green exercise and swimming groups compared with the social club. These findings imply that participating in exercise is the primary driver in positively enhancing mood, although engaging in sedentary social activities can also still contribute to an improved mood, highlighting the importance of the social contact. Thus, encouraging people to interact with green space and be active outdoors has a potentially therapeutic role in positively influencing emotional and physical wellbeing (Barton *et al.*, unpublished data).

Concluding implications

The central themes emerging from the research are that regular contact with urban green spaces improves human health and wellbeing. It improves physical health by providing opportunities for recreation and green exercise activities. Provision of green spaces encourages cycling and walking to move in and between spaces, instead of relying predominantly on cars, which has numerous health and environmental implications. It enhances psychological health by creating a restorative environment which helps to reduce stress and encourage relaxation. It affords spiritual connections which ensure people start to care about their surrounding environment and influences behaviour by providing a space people will choose to visit more frequently to engage in healthy activities. Urban green spaces also improve social cohesion by encouraging greater social and cultural interaction, community empowerment and sense of place, leading to better community spirit and neighbourhood social ties. Green areas promote social inclusion, generate citizenship and local pride and contribute towards reduced levels of crime and violence.

However, many of the daily settings to which people are regularly exposed are highly dissimilar to the landscapes that shaped human evolution (Sullivan 2005). So, what happens to the health of residents in urban areas lacking access to nature and green spaces? The United Kingdom, for example, is becoming an increasingly urbanised society, and by definition urban areas enjoy less access to nature and green space than rural environments. Some of this will be by choice, as urban areas have more services and jobs concentrated together, with better access to schools, hospitals, recreational facilities and other services. However, disconnection from nature can impose new health costs by affecting psychological wellbeing and reducing the opportunity for recovery from mental stresses or physical tensions (Pretty et al. 2004).

The benefits of urban green spaces extend beyond ecosystem service provision and biodiversity conservation to have real impact on the physical and psychological health and wellbeing of local residents. Yet the development of 'urban sprawl' is compromising public health (Frumkin et al. 2004) as housing rapidly diffuses into green areas and competes for land. Expanding populations as well as changing family demographics are also putting pressure on land for housing developments within the urban envelope and on its fringes. Recent research has found that provision of urban green space is more dependent on city area than on the number of inhabitants (Fuller & Gaston 2009). This implies that the green space provision per capita is very low in small cities of high density, which affects residents' quality of life.

Grahn and Stigsdotter (2003) found that urban citizens who live very close to green space visit it three to four times per week. But at a distance of 1 km the number of visits falls to only one per week. Natural England's Accessible Natural Greenspace Standards (ANGSt) model requires that no one should live more than 300 m from their nearest green space (or 5 minutes' walk) to ensure they have the opportunity for exercise, relaxation and wellbeing (Handley et al. 2003), but what percentage of the population actually meet this recommendation? Harrison et al. (1995) recommend that local nature reserves should be provided in every urban area, with a minimum of 1 hectare per 1000 people. Local accessible green spaces should also be at least 2 hectares in size (Handley et al. 2003), but how far are we from reaching this target?

The findings relating to accessibility to local green spaces across different social groups vary. Some report inaccessibility issues for young people, low-income groups, ethnic minorities and disabled people (CRN 2001; Natural England et al. 2006), whose participation in outdoor recreation remains low. Bedimo-Rung et al. (2005) also report that older adults, ethnic minorities, women and lower-income families are more likely to visit parks infrequently or be non-users. But other evidence suggests that the most deprived groups and older people enjoy the greatest access (Barbosa et al. 2007). However, the importance of locally accessible green spaces remains essential for all cohorts, especially those enjoying the least access. In today's modern society characterised by high stress, sedentary lifestyles, rising obesity problems, poor mental health states and disconnection from nature, there is a new challenge: to find ways to develop more local green spaces and improve accessibility to encourage people to visit them on a more regular basis.

A recent report by CABE indicated that '91% of people believe that public parks and open spaces improve their quality of life'. However, one in five people think that it is 'not worth investing money in the upkeep and maintenance of local parks and public open spaces because they will just get vandalised' (CABE Space 2005). A major concern is that neglected parks attract antisocial behaviour, yet case studies show that often those marginalised and perceived as being a social problem (e.g. disaffected young people, homeless) have become positively involved in transforming space and managing its upkeep. Research suggests that regular park use may result in long-term benefits for health and wellbeing, as the combined benefits of engaging in physical activities and being exposed to the green environment may have a cumulative effect on individuals. Therefore, the approaches that researchers and practitioners use to increase local park and recreation usage to achieve optimal health are becoming increasingly important.

Exposure to urban green spaces also enhances education by providing an 'outdoor classroom' for children to learn about nature and the culture and heritage of communities. It provides them with the opportunity to learn about the natural world in their local environment, engage in creative play and improve their ecological consciousness. Research has indicated that childhood exposure to nature and the frequency of visits to green places at a young age correlate with adult patterns of behaviour. Infrequent woodland or green space experiences as a child correlate with a lower frequency of visits during adulthood. 'Interestingly, not visiting as a child is more predictive of not visiting as an adult than vice versa' (Ward Thompson et al. 2008). The lack of outdoor experiences during childhood may hinder desires to visit such places as adults to engage in physical activity or benefit from its emotional restorative qualities.

Green spaces are often associated with an increased likelihood of exercise, even without actively promoting the health benefits. The outdoor environment therefore exerts a direct influence on the probability of taking leisure-based physical activity and other specific sporting activities. The infrastructure of vehicle-free routes and green open spaces allows more people, irrespective of their social and economic circumstances, easily and safely to enjoy a countryside experience close to their home. A more accessible and attractive rural–urban fringe provides respite from the daily stresses of urban living and affords opportunities for recreational activities such as walking, cycling, horse riding or just relaxing. It also meets a demand for more adventurous sport and water-based leisure activities on rivers and canals as well as making use of disused gravel pits and quarries for fishing and angling. Encouraging more people to engage in a range of recreational activities and interact with nature in the rural–urban fringe will improve physical and mental health states, resulting in reduced costs for both society and the economy. Activities in these areas can serve the needs of both the rural and urban communities contributing towards a more sustainable environment (Countryside Agency and Groundwork 2004).

References

Ball, K., Bauman, A., Leslie, E. and Owen, N. (2001). Perceived environmental aesthetics and convenience and company are associated with walking for exercise among Australian adults. *Preventive Medicine*, **33**, 434–40.

Barbosa, O., Tratalos, J. A., Armsworth, P. R. *et al.* (2007). Who benefits from access to green space? A case study from Sheffield, UK. *Landscape and Urban Planning*, **83**, 187–95.

Barton, J. (2009). The effects of green exercise on psychological health and wellbeing. Unpublished Ph.D. thesis, University of Essex, Colchester.

Bedimo-Rung, A. L., Mowen, A. J. and Cohen, D. A. (2005). The significance of parks to physical activity and public health: a conceptual model. *American Journal of Preventive Medicine*, **28**, 159–68.

Berman, M. G., Jonides, J. and Kaplan, S. (2008). The cognitive benefits of interacting with nature. *Psychological Science*, **19**, 1207–12.

Berrigan, D. and Troiano, R. P. (2002). The association between urban form and physical activity in US adults. *American Journal of Preventive Medicine*, **23**, 74–9.

Berto, R. (2005). Exposure to restorative environments helps restore attentional capacity. *Journal of Environmental Psychology*, **25**, 249–59.

Bird, W. (2004). *Natural Fit. Can Green Space and Biodiversity Increase Levels of Physical Activity?* Sandy, UK: Royal Society for the Protection of Birds.

Bishops Wood Centre (2005). *Worcestershire Forest Schools*. Available at http://www.bishopswoodcentre.org.uk/schools/forest.html

Bodin, M. and Hartig, T. (2003). Does the outdoor environment matter for psychological restoration gained through running? *Psychology of Sport and Exercise*, **4**, 141–53.

Burgess, J., Harrison, C. M. and Limb, M. (1988). People, parks and the urban green: a study of popular meanings and values for open spaces in the city. *Urban Studies*, **25**, 455–73.

Butryn, T. M. and Furst, D. M. (2003). The effects of park and urban settings on the moods and cognitive strategies of female runners. *Journal of Sport Behaviour*, **26**, 335–55.

CABE Space (2002). *Streets of Shame. Summary of Findings from 'Public Attitudes to Architecture and the Built Environment'*. London: CABE.

CABE Space (2004). *The Value of Public Space. How High Quality Parks and Public Spaces Create Economic, Social and Environmental Value*. London: CABE.

CABE Space (2005). *Decent Parks? Decent Behaviour? The Link between the Quality of Parks and User Behaviour*. London: CABE.

CDC (2009). Healthy places terminology. Available at http://www.cdc.gov/healthyplaces/terminology.htm (accessed 28 May 2009).

Chiesura, A. (2004). The role of urban parks for the sustainable city. *Landscape and Urban Planning*, **68**, 129–38.

Cohen, D. A., McKenzie, T. L., Sehgal, A. *et al.* (2007). Contribution of public parks to physical activity. *American Journal of Public Health*, **97**, 509–14.

Coley, R. L., Kuo, F. E. and Sullivan, W. C. (1997). Where does community grow? The social context created by nature in urban public housing. *Environment and Behaviour*, **29**, 468–94.

Countryside Agency and Groundwork (2004). *Unlocking the Potential of the Rural Urban Fringe*. London: CA.

Craig, C. L., Brownson, R. C., Cragg, S. E. and Dunn, A. L. (2002). Exploring the effect of the environment on physical activity: a study examining walking to work. *American Journal of Preventive Medicine*, **23**, 36–43.

CRN (2001). *Removing Barriers Creating Opportunities: Social Inclusion in the Countryside*. 2001 Conference Proceedings of the Countryside Recreation Network, Guildhall, London, Countryside Recreation Network.

Dalgard, O. S. and Tambs, K. (1997). Urban environment and mental health: a longitudinal study. *British Journal of Psychiatry*, **171**, 530–6.

DEFRA (2007). http://www.defra.gov.uk/sustainable/government/what/priority/wellbeing/common-understanding.htm de Vries, S., Verheij, R. A., Groenewegen, P. P. and Spreeuwenberg, P. (2003). Natural environments – healthy environments? An exploratory analysis of the relationship between greenspace and health. *Environment and Planning A*, **35**, 1717–31.

DLTR (2002). *Green Spaces Better Places*. Final Report of the Urban Green Spaces Taskforce. UK: Office of the Deputy Prime Minister.

Ellaway, A., Macintyre, S. and Bonnefoy, X. (2005). Graffiti, greenery and obesity in adults: secondary analysis of European cross-sectional survey. *British Medical Journal* **doi** 10.1136/bmj.38575.664549.F7: 1–2.

Frumkin, H. (2001). Beyond toxicity – human health and the natural environment. *American Journal of Preventive Medicine*, **20**, 234–41.

Frumkin, H. (2003). Healthy places: exploring the evidence. *American Journal of Public Health*, **93**, 1451–6.

Frumkin, H., Frank, L. and Jackson, R. (2004). *Urban Sprawl and Public Health: Designing, Planning and Building for Healthy Communities*. Cambridge, MA: MIT Press.

Fuller, R. A. and Gaston K. J. (2009). The scaling of green space coverage in European cities. *Biology Letters*, **5**, 352–5.

Fuller, R. A., Irvine, K. N., Devine-Wright, P., Warren, P. H. and Gaston, K. J. (2007). Psychological benefits of greenspace increase with biodiversity. *Biology Letters*, **3**, 390–4.

Galea, S., Ahern, J., Rudenstine, S., Wallace, Z. and Vlahov, D. (2005). Urban built environment and depression: a multilevel analysis. *Journal of Epidemiology and Community Health*, **59**, 822–27.

Giles-Corti, B., Broomhall, M. H., Knuiman, M. et al. (2005). Increasing walking: how important is distance to, attractiveness, and size of public open space? *American Journal of Preventive Medicine*, **28**, 169–76.

Giles-Corti, B. and Donovan, R. J. (2002). The relative influence of individual, social and physical environment determinants of physical activity. *Social Science and Medicine*, **54**, 1793–812.

Godbey, G. and Blazey, M. (1983). Old people in urban parks: an exploratory investigation. *Journal of Leisure Research*, **15**, 229–44.

Godbey, G., Caldwell, L., Floyd, M. and Payne, L. (2005). Contribution of leisure studies and recreation and park management research to the active living agenda. *American Journal of Preventive Medicine*, **28**(Supplement 2), 150–8.

Godbey, G., Grafe, A. and James, W. (1992). *The Benefits of Local Recreation and Park Services – A Nationwide Study of the Perceptions of the American Public*. Pennsylvania State University, PA: College of Health and Human Development.

Godbey, G., Roy, M., Payne, L. and Orsega-Smith, E. (1998). *The Relation Between Health and Use of Local Parks*. National Recreation Foundation.

Grahn, P. and Stigsdotter, U. A. (2003). Landscape planning and stress. *Urban Forestry and Urban Greening*, **2**, 1–18.

Groenewegen, P. P., van den Berg, A. E., de Vries, S. and Verheij, R. A. (2006). Vitamin G: effects of green space on health, well-being and social safety. *BMC Public Health*, **6**, **doi** 10.1186/1471-2458-6-149.

Handley, J., Pauleit, S., Slinn, P. *et al.* (2003). *Accessible Natural Green Space Standards in Towns and Cities: A Review and Toolkit for their Implementation*. Peterborough: English Nature.

Handy, S. L., Boarnet, M. G., Ewing, R. and Killingsworth, R. E. (2002). How the built environment affects physical activity: views from urban planning. *American Journal of Preventive Medicine*, **23**, 64–73.

Harrison, C., Burgess, J., Millward, A. and Dawe, G. (1995). *Accessible Natural Greenspace in Towns and Cities: A Review of Appropriate Size and Distance Criteria*. Peterborough: English Nature, pp. 1–18.

Hartig, T., Book, A., Garvill, J., Olsson, T. and Garling, T. (1996). Environmental influences on psychological restoration. *Scandinavian Journal of Psychology*, **37**, 378–93.

Hartig, T., Evans, G., Jamner, L. D., Davis, D. S. and Garling, T. (2003). Tracking restoration in natural and urban field settings. *Journal of Environmental Psychology*, **23**, 109–23.

Hartig, T., Mang, M. and Evans, G. W. (1991). Restorative effects of natural environment experiences. *Environment and Behaviour*, **23**, 3–26.

Hayashi, T., Tsumura, K., Suematsu, C. *et al.* (1999). Walking to work and the risk for hypertension in men: the Osaka Health Survey. *Annals of Internal Medicine*, **130**, 21–6.

Health Council of the Netherlands (2004). *Nature and Health. The Influence of Nature on Social, Psychological and Physical Well-being*. The Hague, the Netherlands: Health Council of the Netherlands and Dutch Advisory Council for Research on Spatial Planning, Nature and the Environment.

Heerwagen, J. H. and Orians, G. H. (1993). Humans, habitats and aesthetics. In S. R. Kellert and E. O. Wilson, eds., *The Biophilia Hypothesis*. Washington, DC: Island Press.

Ho, C.-H., Payne, L., Orsega-Smith, E. and Godbey, G. (2003). *Parks & Recreation – Research Update from April 03: Parks, Recreation and Public Health, The Benefits are Endless*. Pennsylvania: Pennsylvania State University's College of Health and Human Development.

Honeyman, M. C. (1992). Vegetation and stress: a comparison study of varying amounts of vegetation in countryside and urban scenes. In D. Relph, ed., *The Role of Horticulture in Human Well-being and Social Development: A National Symposium*. Portland: Timber Press, pp. 143–5.

Hull, R. B. and Michael, S. E. (1995). Nature-based recreation, mood change, and stress restoration. *Leisure Sciences*, **17**, 1–14.

Humpel, N., Owen, N. and Leslie, E. (2002). Environmental factors associated with adults' participation in physical activity: a review. *American Journal of Preventive Medicine*, **22**, 188–99.

Irvine, K. N. and Warber, S. L. (2002). Greening healthcare: practicing as if the natural environment really mattered. *Alternative Therapies in Health and Medicine*, **8**, 76–83.

Judd, F. K., Jackson, H. J., Komiti, A. *et al.* (2002). High prevalence disorders in urban and rural communities. *Australian and New Zealand Journal of Psychiatry*, **36**, 104–13.

Kahn, P. H. and Kellert, S. R. (2002). *Children and Nature: Psychological, Sociocultural and Evolutionary Investigations*. Cambridge, MA: MIT Press.

Kaplan, R. (1993). The role of nature in the context of the workplace. *Landscape and Urban Planning*, **26**, 193–201.

Kaplan, R. (2001). The nature of the view from home. *Journal of Environment and Behaviour*, **33**, 507–42.

Kaplan, R. and Austin, M. E. (2004). Out in the country: sprawl and the quest for nature nearby. *Landscape and Urban Planning*, **69**, 235–43.

Kaplan, R. and Kaplan, S. (1989). *The Experience of Nature: A Psychological Perspective*. Cambridge, UK: Cambridge University Press.

Kaplan, S. (1995). The restorative benefits of nature: toward an integrative framework. *Journal of Environmental Psychology*, **15**, 169–82.

Kuo, F. E. and Sullivan, W. C. (2001a). Aggression and violence in the inner city: effects of environment via mental fatigue. *Environment and Behaviour*, **33**, 543–71.

Kuo, F. E. and Sullivan, W. C. (2001b). Environment and crime in the inner city – does vegetation reduce crime? *Journal of Environment and Behaviour*, **33**, 343–67.

Kuo, F. E., Bacaicoa, M. and Sullivan, W. C. (1998). Transforming inner-city landscapes: trees, sense of safety and preference. *Environment and Behaviour*, **30**, 28–59.

Kuo, F. E., Sullivan, W. C., Coley, R. L. and Brunson, L. (1998). Fertile ground for community: inner-city neighbourhood common spaces. *American Journal of Community Psychology*, **26**, 823–51.

Kweon, B.-S., Sullivan, W. C. and Wiley, A. R. (1998). Green common spaces and the social integration of inner-city older adults. *Environment and Behaviour*, **30**, 832–58.

Lange, E. and Schaeffer, P. (2001). A comment on the market value of a room with a view. *Landscape and Urban Planning*, **55**, 113-20.

Laumann, K., Gärling, T. and Stormark, K. M. (2003). Selective attention and heart rate responses to natural and urban environments. *Journal of Environmental Psychology*, **23**, 125-34.

Leather, P., Pyrgas, M., Beale, B., Kweon, B. and Tyler, E. (1998). Plants in the workplace: the effects of plant density on productivity, attitudes and perceptions. *Environment and Behaviour*, **30**, 261-82.

Lewis, G. and Booth, M. (1994). Are cities bad for your mental health? *Psychological Medicine*, **24**, 913-15.

Lindsey, G., Man, J., Payton, S. and Dickson, K. (2004). Property values, recreation values, and urban greenways. *Journal of Park and Recreation Administration*, **22**, 69-90.

Luther, M. and Gruehn, D. (2001). Putting a price on urban green spaces. *Landscape Design*, **303**, 23-5.

Luttick, M. (2000). The value of trees, water and open space as reflected by house prices in the Netherlands. *Landscape and Urban Planning*, **48**, 161-7.

Maas, J., Verheij, R. A., Groenewegen, P. P., de Vries, S. and Spreeuwenberg, P. (2006). Green space, urbanity, and health: how strong is the relation? *Journal of Epidemiology and Community Health*, **60**, 587-92.

Mayer, F. S., Frantz, C., Bruehlman-Senecal, E. and Doliver, K. (2006). Why is nature beneficial? The role of connectedness to nature. *Environment and Behaviour*, **41**(5), 607-43.

Mind (2007). *Ecotherapy: The Green Agenda for Mental Health. Mind Week Report*, May 2007. London: Mind.

Mitchell, R. and Popham, F. (2008). Effect of exposure to natural environment on health inequalities: an observational population study. *The Lancet*, **372**, 1655-60.

Moore, E. O. (1982). A prison environment's effect on health care service demands. *Journal of Environmental Systems*, **11**, 17-34.

Moore, R. and Wong, H. (1997). *Natural Learning: Creating Environments for Rediscovering Nature's Way of Teaching*. Berkeley: MIG Communications.

More, T. and Payne, B. (1978). Affective responses to natural areas near cities. *Journal of Leisure Research*, **10**, 7-12.

Natural England (in collaboration with Department for Environment, Food and Rural Affairs, Environment Agency, Forestry Commission, The Broads Authority, Dartmoor National Park Authority, Exmoor National Park Authority, Lake District National Park Authority, North York Moors National Park Authority, Northumberland National Park Authority, Peak District National Park Authority and Yorkshire Dales National Park Authority) (2006). *England Leisure Visits Survey, 2005*.

Owen, N., Humpel, N., Leslie, E., Bauman, A. and Sallis, J. F. (2004). Understanding environmental influences on walking – review and research agenda. *American Journal of Preventive Medicine*, **27**, 67-76.

Owen, N., Leslie, E., Salmon, J. and Fotheringham, M. J. (2000). Environmental determinants of physical activity and sedentary behaviour. *Exercise and Sport Sciences Reviews*, **28**, 153-8.

Parks, S. E., Housemann, R. A. and Brownson, R. C. (2003). Differential correlates of physical activity in urban and rural adults of various socioeconomic backgrounds in the United States. *Journal of Epidemiology and Community Health*, **57**, 29-35.

Parsons, R., Tassinary, L. G., Ulrich, R. S., Hebl, M. R. and Grossman-Alexander, M. (1998). The view from the road: implications for stress recovery and immunization. *Journal of Environmental Psychology*, **18**, 113-39.

Peacock, J., Hine, R. and Pretty, J. (2006). *The Health Benefits of Environmental Improvements to a Circular Route at Easington Coastal Path*. Report for the Environment

Agency and Durham Heritage Coast by the University of Essex.

Peacock, J., Hine, R. and Pretty, J. (2007). *Got the Blues, Then Find Some Greenspace: The Mental Health Benefits of Green Exercise Activities and Green Care.* Report for Mind by the University of Essex.

Peacock, J., Hine, R., Willis, G., Griffin, M. and Pretty, J. (2005). *The Physical and Mental Health Benefits of Environmental Improvements at Two Sites in London and Welshpool.* Report for the Environment Agency by the University of Essex.

Peiser, R. B. and Schwann, G. M. (1993). The private value of public open space within subdivisions. *Journal of Architectural and Planning Research*, **10** (Summer), 91–104.

Pretty, J. (2007). *The Earth Only Endures: On Reconnecting with Nature and Our Place in It.* London: Earthscan.

Pretty, J. and Smith, D. (2004). Social capital in biodiversity conservation and management. *Conservation Biology*, **18**, 631–8.

Pretty, J. N., Griffin, M. and Sellens, M. (2004). Is nature good for you? *ECOS – Quarterly Journal of the British Association of Nature Conservation*, **24**, 2–9.

Pretty, J. N., Griffin, M., Sellens, M. H. and Pretty, C. (2003). *Green Exercise: Complementary Roles of Nature, Exercise and Diet in Physical and Emotional Well-being and Implications for Public Health Policy.* CES Occasional Paper 2003-1, University of Essex.

Pretty, J., Peacock, J., Sellens, M. and Griffin, M. (2005). The mental and physical health outcomes of green exercise. *International Journal of Environmental Health Research*, **15**, 319–37.

Raymore, L. and Scott, D. (1998). The characteristics and activities of older visitors to a metropolitan park district. *Journal of Park and Recreation Administration*, **16**, 1–21.

Rickinson, M., Dillon, J., Teamey, K. *et al.* (2004). *A Review of Research on Outdoor Learning.* National Foundation for Educational Research and King's College London.

Ross, C. E. (2000). Walking, exercising and smoking: does neighbourhood matter? *Social Science and Medicine*, **51**, 265–74.

Scott, D. (1997). Exploring the patterns in people's use of a metropolitan park district. *Leisure Sciences*, **19**, 159–74.

Sheets, V. L. and Manzer, C. D. (1991). Affect, cognition and urban vegetation: some effects of adding trees along city streets. *Environment and Behaviour*, **23**, 285–304.

Sugiyama, T., Leslie, E., Giles-Corti, B. and Owen, N. (2008). Associations of neighbourhood greenness with physical and mental health: do walking, social coherence and local social interaction explain the relationships? *Journal of Epidemiology and Community Health*, **62**, e9.

Sullivan, W. C. (2005). Urban place: reconnecting with the natural world. In P. Bartlett, ed., *Forest, Savanna, City: Evolutionary Landscapes and Human Functioning.* Cambridge, MA: MIT Press, pp. 237–52.

Takano, T., Nakamura, K. and Watanabe, M. (2002). Urban residential environments and senior citizens' longevity in megacity areas: the importance of walkable green spaces. *Journal of Epidemiology and Community Health*, **56**, 913–18.

Taylor, A. F., Kuo, F. E. and Sullivan, W. C. (2001). Coping with ADD: the surprising connection to green play settings. *Environment and Behaviour*, **33**, 54–77.

Tinsley, H., Tinsley, D. and Croskeys, C. (2002). Park usage, social milieu and psychological benefits of park use reported by older urban park users from four ethnic groups. *Leisure Sciences*, **2**, 199–218.

Ulrich, R. S. (1981). Natural versus urban scenes: some psychophysiological effects. *Journal of Environment and Behaviour*, **13**, 523–56.

Ulrich, R. S. (1984). View through a window may influence recovery from surgery. *Science*, **224**, 420–1.

Ulrich, R. S. (1999). Effects of gardens on health outcomes: theory and research. In C. Cooper Marcus and M. Barnes, eds., *Healing Gardens. Therapeutic Benefits and Design*

Recommendations. Marni, New York: John Wiley & Sons, pp. 27–86.

Ulrich, R. S., Simons, R. F., Losito, B. D. *et al.* (1991). Stress recovery during exposure to natural and urban environments. *Journal of Environmental Psychology*, **11**, 201–30.

UNFPA (2007). *State of the World Population 2007: Unleashing the Potential of Urban Growth*. New York, US: United Nations Population Fund.

van den Berg, A. E., Hartig, T. and Staats, H. (2007). Preference for nature in urbanised societies: stress, restoration and the pursuit of sustainability. *Journal of Social Issues*, **63**, 79–96.

van den Berg, A. E., Koole, S. L. and van der Wulp, N. Y. (2003). Environmental preference and restoration: (how) are they related? *Journal of Environmental Psychology*, **23**, 135–46.

Ward Thompson, C. (2002). Urban open space in the 21st century. *Landscape and Urban Planning*, **60**, 59–72.

Ward Thompson, C., Aspinall, P. and Montarzino, A. (2008). The childhood factor: adult visits to green places and the significance of childhood experience. *Environment and Behaviour*, **40**, 111–43.

Wells, N. M. (2000). At home with nature: effects of 'greenness' on children's cognitive functioning. *Environment and Behaviour*, **32**, 775–95.

Wendel-Vos, G. C. W., Schuit, A. J., De Niet, R. *et al.* (2004). Factors of the physical environment associated with walking and bicycling. *Medicine and Science in Sports and Exercise*, **36**, 725–30.

White, R. and Heerwagen, J. (1998). Nature and mental health: biophilia and biophobia. In A. Lundberg, ed., *The Environment and Mental Health: A Guide for Clinicians*. Mahwah, NJ: Lawrence Erlbaum Associates, pp. 175–92.

WHO (1948). *Preamble to the Constitution of the World Health Organization*. As adopted by the International Health Conference, New York, 19–22 June 1946, and entered into force on 7 April 1948.

Wood, R. and Handley, J. (1999). Urban waterfront regeneration in the Mersey Basin, North West England. *Journal of Environmental Planning and Management*, **42**, 565–80.

Woolley, H. (2003). *Urban Open Spaces*. London: Spon Press.

CHAPTER TEN

Bringing cities alive: the importance of urban green spaces for people and biodiversity

JON SADLER, ADAM BATES, JAMES HALE
AND PHILIP JAMES

Introduction

A plethora of papers exist that trumpet the value of urban green spaces as providers of benefits to both people and wildlife (James *et al.* 2009). This body of work emphasises five means by which such spaces improve the urban environment: (i) shaping the character of the city and its neighbourhoods (Pauleit 2003); (ii) engendering a sense of place for city inhabitants (Frumkin 2003); (iii) providing a range of physical (Maas *et al.* 2006) and psychological (Hartig 2008) health benefits to people; (iv) supporting rich assemblages of wildlife, including many rare and endangered species (Gibson 1998; Mortberg & Wallentinus 2000); and (v) possessing important environmental functions that scale to provide a wide range of ecosystem services (Bolund & Hunhammar 1999; Elmqvist *et al.* 2004).

It is estimated that the number of urban areas with over a million people will grow by over 40% by 2015 (Crane & Kinzig 2005). To accommodate this rapidly increasing population and to reduce the deleterious impact of global sprawling cities (European Environment Agency 2006; Irwin & Bockstael 2007), in many countries regulatory bodies have created a range of policies on urban living, housing provision and city development that appear to be in conflict. On the one hand, policies exist espousing the utilisation of as much open space in cities as possible to meet construction targets for new-build housing (e.g. ODPM 2002a), while on the other hand different policy documents highlight the provision of green space for people and wildlife to enhance quality of life (e.g. EEA 2009). The global move towards compact forms of cities (Dantzing & Saaty 1973) has the potential to reduce urban sprawl, thereby preserving rural food production and habitats. It may also contribute towards wider sustainability goals (Cairns 2006) by reducing car use and fuel emissions, encouraging

Urban Ecology, ed. Kevin J. Gaston. Published by Cambridge University Press.

more walking, cycling and the use of public transport, and revitalising inner city areas (Burton 2000). Pauleit *et al.* (2005) illustrated the flip side to this process in Liverpool, UK, where green space has markedly reduced over the past 25 years. This loss was most acute within residential areas as a result of infill development. This intensification of land-use threatens to compromise the quality of the urban environment for both people and biodiversity (RCEP 2007).

Here we review the issues surrounding: (i) the provision and quantification of green spaces in cities, (ii) the value of these spaces for biodiversity, (iii) their importance for providing ecosystem services, (iv) their significance for people, and (v) their design as multi-functional spaces. The chapter concludes by identifying key challenges where further research is necessary. We focus particularly on the United Kingdom as a case study, although we also draw on material from other regions.

Green spaces: classification and quantification
Policy spaces
We stress here that the policy frameworks surrounding the management and provision of green spaces are heavily geographically contextualised (CABE 2009a), which means that generalisations that have widespread applicability are not easily derived. In this section we use the situation in the UK as an example of national initiatives to identify possible commonalities and issues of importance. In the UK there has been significant interest and numerous policy statements concerning the value of green spaces to citizens in cities. This interest has been driven by concerns over health inequalities in cities in much of the developed world (Smyth 2008), and the so-called 'obesity epidemic' (Joshu *et al.* 2008). Concerns over the quality and management of green space in the UK have been driven by local people, and a wide range of reports by non-governmental organisations and government departments which culminated in the Urban Task Force Report (Urban Task Force 1999). This report strongly mainstreamed the significance of urban green space for enhancing the quality of city life into a national policy context. The government's response was the Urban White Paper (DETR 2000), which recommended the creation of the Green Spaces Taskforce (GST), whose remit was to advise government on proposals for the improvement of the quality of urban parks, play areas and other green spaces. GST's report *Green Spaces, Better Places* (DETR 2002) examined the provision of green spaces and their loss, design and utility for people and wildlife, as well as introducing the concept of green networks into planning policy. The government's response to the recommendations in this report arrived the same year in *Living Places – Cleaner, Safer, Greener* (ODPM 2002b). This acknowledged the need for a clearer management process and greater understanding of the provision, quality and access importance of green spaces in cities, and led to the establishment of the Commission for Architecture and the Built Environment (CABE), whose report, *Open Space Strategies: Best Practice*

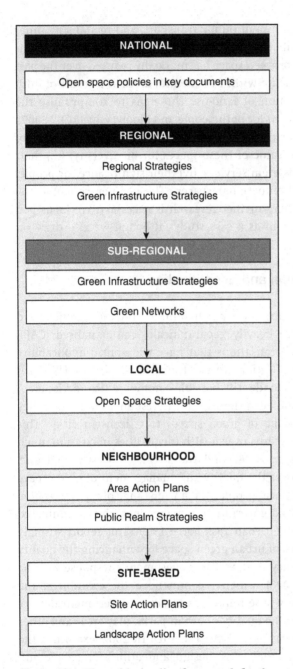

Figure 10.1 Hierarchical policy framework for the management of green spaces in the UK. Reproduced with permission from CABE (2009b).

Guidance (CABE 2009b), called for a hierarchical system (Figure 10.1) that linked national strategies on green space management and provision to policy guidance at local (neighbourhood) scales, a structural pattern that is common to most countries in the world. These policy documents are useful in

Table 10.1 *UK planning policy guidance documents that relate to green spaces and green infrastructure.*

Spatial unit	Planning policy guidance	Other relevant planning policies
Green spaces	PPS 9: Biodiversity and geological conservation PPG 17: Planning for open space, sport and recreation	NI 197: Improved local biodiversity
Green infrastructure	PPS 1: Delivering sustainable development PPS 1 (Supplement): Planning and climate change PPS 12: Local spatial planning	NI 185 & 186: Reduction of CO_2 emissions per capita and via Local Authority operations NI 188: Adapting to climate change NI 189: Flood and coastal risk management NI 194: Level of air quality NI 197: Improved local biodiversity PPS 9: Biodiversity and geological conservation PPS 25: Planning and flood risk PPS 22: Renewable energy

Note: Planning Policy Guidance (PPG) and Planning Policy Statements (PPS) are documents created at a national level which local authorities are required to use as part of their planning processes. National Indicators (NIs) are a sequence of measurable indicators that local authorities have to report to government on an annual basis.

highlighting areas of need, and specifically they identify issues surrounding green space definitions, typology, quantity, quality (including ecology) and the need for standards for access and robust monitoring and management systems. These are issues that are now wired into UK national policy documents and government planning guidance, which is used by local and regional authorities (Table 10.1).

Classification and typology

In the UK, 'open space' is defined in the Town and Country Planning Act 1990 as 'land laid out as a public garden, or used for the purposes of public recreation, or land which is a disused burial ground'. This emphasises land that is public or municipal space, but not the wider range of open green space available in cities. It does, however, include areas of water such as rivers, canals, lakes and reservoirs, which offer important opportunities for sport and recreation and can also act as visual amenity. More recent definitions emphasise a broader portfolio of spaces and include references to 'nearby nature' or 'natural green spaces'. Harrison *et al.* (1995, p. 4), for example, define them as 'Land, water and geological features which have been naturally colonised by plants and animals and which are accessible on foot to large numbers of residents'. In Canada a distinction is drawn between green space and parkland, with the former relating to municipal natural open space (reserves, floodplains etc.) and the latter relating to land set aside by the municipality as part of an established public park (Evergreen 2004).

Green spaces form part of a linked social–economic–ecological system (Yli-Pelkonen & Niemelä 2005) and when scaled up to a landscape or cityscape they form what has been termed green infrastructure. Benedict and McMahon (2002, p. 12) defined this as 'an interconnected network of green space that conserves natural ecosystem values and functions and provides associated benefits to human populations'. The distinction between green spaces and green infrastructure is subtle but worth stressing. Green infrastructure goes beyond the site specific, and provides a means of characterising multifunctional landscapes; it also considers private as well as public assets. This chapter focuses on green spaces but stresses green infrastructure where the issues being discussed demand a focus on a larger spatial scale (e.g. ecosystem services and design and planning).

There is similarly wide variation in the typologies of what constitutes a green space. Table 10.2 illustrates this by examining examples drawn from academic and policy literatures from the UK. It is clear that there is an uneasy mix of classifications that relate to use (e.g. recreational), type (e.g. public/civic space) or habitats (e.g. woodland, wetlands) even within the same scheme. Some typologies include private residential land such as domestic gardens, and most highlight linear features (e.g. greenways, river, road and railway corridors and green networks). This variability in classification has led to inconsistency in adoption of schema in the literature. Handley *et al.* (2003), Tzoulas *et al.* (2007) and Comber *et al.* (2008) used the typology of green spaces provided by the Urban Green Spaces Task Force (DETR 2002), which has been formalised in the UK by Planning Policy Guidance 17 (PPG17): Planning for Open Space, Sport and Recreation (DCLG 2000; Table 10.2).

A recent international comparison of green spaces in 11 cities (CABE 2009a) has highlighted that all had an element of typology in their policies, mostly based around function and size, but that this varied very significantly across the cities examined. For example, in Malmo, Sweden, and Tokyo, Japan, green spaces were classified according to function, and this was used in the planning system to create an even distribution of green space across the city. Some cities had only broad categories, such as Paris, France, where all habitats other than woodlands are classified as gardens, and Minneapolis, USA, where all green space is called parkland. In some countries, such as Japan, the green space

Table 10.2 *Typologies of urban green spaces that exist in UK policy literatures.*

Landscape Institute	Swanwick et al. 2003	DETR	DCLG (PPG17)
Street trees, verges and hedges	Recreation green space	Green roofs	Parks and gardens (not including private)
Green roofs and walls	Incidental green space	Urban parks	Natural and semi-natural spaces
Pocket parks	Private green space	Green corridors	Green corridors
Private gardens	Productive green space	Encapsulated countryside	Outdoor sports facilities
Urban plazas	Burial grounds	Derelict land	Amenity green spaces
Town/village greens and commons	Institutional grounds	Housing green space and domestic gardens	Provision for children and young people
Pedestrian and cycle routes	Wetland	Churchyards, cemeteries and school grounds	Allotments, community gardens and urban farms
Cemeteries and churchyards	Woodland	Open standing and running water	Cemeteries, disused churchyards and other burial grounds
Institutional open spaces	Other habitats (e.g. grassland, disturbed ground)		Public/civic space (e.g. plazas, pedestrianised areas)
Ponds, streams and rivers	Linear green space		
Woodlands			
Nature reserves			
Play areas			
School grounds			
Sports pitches			
Swales, ditches			
Allotments			
Brownfields			

Sources: Landscape Institute (2009); Swanwick et al. (2003); DETR (2002); DCLG (2000).

typology is defined nationally as part of a policy to provide various kinds of green space within walking distance of residential areas.

In this chapter we focus attention on a discussion of the public spaces in the typology of Swanwick *et al.* (2003) and exclude a full discussion of the biodiversity of gardens, even though they comprise a large amount of the green space in many cities (e.g. Loram *et al.* 2007). The importance of urban residential spaces, however, is touched on in our consideration of ecosystem services and green space design.

Quantification

Green space in the UK accounts for 14% of urban space nationally, but this is very variable, with some cities being greener than others. In Leicester, 25% of urban land is green space (Comber *et al.* 2008), in Greater Manchester 32% (Gill *et al.* 2008) and in Sheffield 45% (Fuller & Gaston 2009). Approximately 30% of the total land surface of Greater London is occupied by open green spaces (Wilby & Perry 2006). In Birmingham the proportion of green space is 33.7% not including gardens and land associated with water bodies, roads and railways (unpublished data). Within Europe, available statistics show rather more variability. In a recent and extensive analysis of 386 European cities, Fuller and Gaston (2009) showed that green space coverage at a wider European level was extremely variable, ranging from 1.9% (Reggio di Calabria, Italy) to 46% (Ferol, Spain). Their analyses also illustrated that green space coverage was scaled to city land area so that cities large in area had greater green space provision, even though overall population densities were similar to cities smaller in area. The association showed that green space area increased more than city area, so that cities differing in area by an order of magnitude will have a 15-fold difference in green space area. Perhaps worryingly for UK residents, there the green space provision in 67 cities scales only at the same rate as city area. The implications of the work are concerning, as Fuller and Gaston suggest that access to green space may well decline as cities grow.

Unfortunately, there is a paucity of longitudinal (time series) data on change in urban areas to assess the scale of green space loss, aside from snapshots of individual cities such as Munich, Germany (Pauleit & Duhme 2000), and Stockholm, Sweden (Bolund & Hunhammar 1999), and a European-wide comparative study carried out by the European Environment Agency (EEA 2002). The EEA report suggested that the area of public green spaces in 25 cities had reduced as a proportion of total city land areas between 1950 and 1990. There are no data to show whether this is a result of the limited provision of green spaces in sprawling city margins or whether it relates to real loss of city green spaces due to landscape intensification within cities. Davies *et al.* (2008) showed that the extent of green space is negatively correlated with total area of buildings and the length of the road network in Sheffield, UK, suggesting that urbanisation causes progressive loss of green space via land-use intensification. In their study of Mashad, Iran, Rafiee *et al.* (2009) also found a significant decrease in the extent of urban green spaces had occurred between 1977 and 2006. There is, however, a dearth of global data on this issue, and producing systematic and comparable figures detailing the loss of green space in our cities is an urgent task.

Green space standards

The creation of standards for access to green space is an important part of any policy agenda. In the 1970s, the United Nations suggested a standard of 6 ha per 1000 head of population as a good target figure for urban environments (Wang 2009). Similar standards were subsequently adopted by many countries there-after with figures ranging from 4 ha/1000 people in Japan, 4.05 ha (10 acres)/ 1000 people in the USA and Canada, and 14 ha/1000 people in Germany (Sukopp *et al.* 1995). Within these broad categories there is considerable vari-ation within individual nations. For example, a review of guidelines in Can-adian municipalities showed that they ranged from 0.7 to 6 ha/1000 people (Evergreen 2004). Although such area-based standards have the advantages of being clear and measurable, it is obvious that they do not capture the complex-ity of green space needs in cities. Other approaches are also evident, and they include time spent travelling to the nearest green space, and the nature of those spaces along with area-based guidelines.

The European Environment Agency recommends that accessible green spaces should be within an average 15-minute walk from people's homes. Natural England, the UK's statutory organisation that deals with nature conser-vation and the countryside, created a suite of hierarchical recommendations concerning accessible natural green space standards (ANGSt) (Harrison *et al.* 1995; Handley *et al.* 2003). They recommend at least 2 ha of accessible natural green space per 1000 head of population, but also that:

- every home should be within 300 m (5 minutes' walk) from an accessible natural green space;
- there should be 1 ha of local nature reserves per 1000 people; and
- there should be an accessible 20-ha site, 100-ha site and 500-ha site within 2 km, 5 km and 10 km, respectively, of every home.

Pauleit *et al.* (2001) considered the uptake of the ANGSt model by 30 municipal areas in the UK. Although many (80%) already had in place some kind of green space standard, most (75%) of these related only to 'amenity green space' and ignored other types (brownfields, nature reserves and the like). Some local authorities, such as Birmingham, embraced the model, integrating it, albeit in a modified form, into their Nature Conservation Strategy in 1997, while the majority (90%) had not implemented it into planning policy. The situation has improved substantially since this time with access standards figuring highly in many council and local authority planning documents, including nature conservation strategies that emphasise the importance of green spaces and green infrastructure for local wildlife.

Green spaces, biodiversity and ecosystem services
Green space and biodiversity correlates

Cities are characterised by highly modified and complex landscapes comprising a rich mosaic of relict habitats and designed spaces juxtaposed in combinations that rarely occur in nature (Angold *et al.* 2006). As a result they are home to a wide diversity of organisms (Gilbert 1989), some of which are becoming increasingly rare in the wider countryside (Gibson 1998; Woodward *et al.* 2003), where ecological processes have been stymied by agricultural intensification (Tratalos *et al.* 2007). There is now an extensive literature on the biodiversity of urban areas (e.g. Chapter 5; McKinney 2006, 2008) examining changes in diversity along gradients from the city margins to the city centre (McDonnell & Pickett 1990; McDonnell & Hahs 2008), fragmentation effects (Sadler *et al.* 2006) and studies of total city species pools (e.g. Puth & Burns 2009).

This literature illustrates that the type, structure and location of green spaces are important correlates for the presence, abundance, species richness and persistence of plant and animal species, yet there has not been a systematic study of these effects. Here we draw on a wider literature search aimed at identifying the green space correlates of animal biodiversity in cities. The search identified just under 4000 articles which were subsequently examined to extract information on green space correlates for the target groups (Table 10.3). It should be stressed that the studies revealed a very large variety of approaches to capturing information on both the green space type (target study habitat) and wider urban landscape, so careful interpretation and categorisation were needed to isolate the potential correlates. We divided those correlates into two groups representing landscape and local influences on green space biodiversity. The most evident trend is that there are more local than landscape correlates listed. This was anticipated because researchers carefully monitored and measured the habitats of interest, often generating many local variables, but far fewer landscape variables. Moreover, some authors (e.g. Blair 1999) used multi-variate techniques to analyse their data, and this retained a larger pool of variables than studies that used stepwise and selective regression techniques. There is also considerable variability between correlates, which makes establishing causation problematic. Notwithstanding these complexities, several pervasive themes can be isolated relating to: (i) the importance of local versus landscape influences, and the significance of (ii) local, and (iii) landscape variables. We stress here that because of the scale of the issue our discussion will be selective and focus on very evident and large-scale patterns.

Table 10.3 *Local and landscape biodiversity/green space correlates of animals.*

Taxa group	Green space habitat	Landscape correlates	Local correlates
Vertebrates			
Birds	Urban land-use	Housing density	Green space area
	Brownfield	Building density	Rough grassland
	Wetland	Urban land-use	Supplementary feeding
	Woodland	Isolation	Human disturbance
	Public green space	Adjacent gardens	% Woody vegetation
	Natural green space	Connectivity	Riparian vegetation
	Gardens		Habitat structure
			Size
			% Woodland cover
Bats	Woodland reserve	Distance to water	Shrub density
	Natural relict green space	Density of built land	Tree diameter
	Hedge lines	Woodland fragmentation	Tree density
		Connectivity	Forest cover
			Roads
Amphibians	Ponds	Woodland cover	Pond depth
		Pond isolation	Woodland cover
		Pond density	Pond structure
			Pond size
			Fish presence
			Pond vegetation
			Roads
Reptiles	Natural relict green space	% Urban land	Habitat physiognomy
	Gardens	Distance to nearest similar patch	Vegetation structure
		Housing density	Vegetation type
		Urban sprawl	Grassland presence
			Soil compaction
			Introductions (Vegetation)
			Aspect
			Geology
Invertebrates			
Coleoptera	Brownfield	Site location	Vegetation structure
	Greenways	% Urban land	Disturbance
	Gardens	Amount of similar habitat locally	Site age
	Woodland	Corridors	Urban land-use
	Wetland		Size
			Trampling

Table 10.3 (*cont.*)

Taxa group	Green space habitat	Landscape correlates	Local correlates
Butterflies	Grassland	Isolation	
	Garden	Habitat quality	
	Office park	Amount of similar habitat locally	
	Golf course	% Urban land	
	Park		
	Reserve		
	Amenity space		
Galls	Gardens		Edge effects
	Woodland		Plant richness
	Natural green space		
	Road verge		
Spiders	Woodland	Woodland fragmentation	Prey availability
	Desert park	Isolation	Habitat productivity
	Urban desert remnants		Size
	Industrial		Introduced species
	Agricultural		
	Xeric- and mesic-residential yards		
Bees	Gardens	Roads	Grazing density
	Allotments	% Urban land	Plant diversity
	Grassland	% Suburban land	Tree abundance
	Woodland	Fragmentation	Shrub diversity
Ants	Woodland	% Urban land	Vegetation cover
	Gardens		Tree density
	Brownfields		Nest resources
Orthoptera	Reserve	Road density	Road density
	Brownfield	Landscape context	Road age
	Grassland		Vegetation structure
			Landscape context
			Soil parameters
			Site age
Hemiptera	Brownfield	Landscape context	Vegetation structure
	Traffic roundabouts		Size
			Soil parameters
			Site age
			Mowing
			Floral evenness

Note: Data were derived from searches on the Web of Knowledge using a sequence of keyword queries (e.g. URBAN+BIRDS, URBAN*BATS and so on). The subsequent results were sifted for duplication and then categorised into the groups in the table.

Several researchers modelled variability and partitioned it to either landscape or local factors or combinations of both. In these studies there is a general tendency for local factors to be more important than landscape variables for biodiversity. The pattern is pervasive and cuts across taxonomic groups. Parris (2006) showed that declining pond quality (measured as the proportion of ponds with a surrounding wall) led to a 44–56% decrease in amphibian species in ponds in urban Melbourne, Australia. Evans et al. (2009) state that local correlates such as increased structural complexity, woody vegetation diversity and supplementary feeding influence avian assemblages more strongly than factors operating at larger spatial scales. Many invertebrate taxa such as hemipterans and ground beetles exhibit similar responses, where site-based variables such as age since last disturbance (Small et al. 2006) and management regime (Helden & Leather 2004) account for a greater proportion of the variability in the assemblage pools. These findings are important because they suggest that management at the scale of individual green spaces may provide significant benefits to local biodiversity. This is a strategy that is much more easily operationalised than attempting to manage whole multi-functional cityscapes, especially as planning tends to operate on a site-by-site basis.

Local-scale biodiversity–green space correlates tend to fall under the umbrella of what is frequently termed 'habitat quality' and include variables such as the size of the green space, its vegetation type, structural diversity, disturbance regime, management and so on. Perhaps the most evident pattern is the link between biodiversity metrics and site size which is found in a wide range of studies of birds (Chamberlain et al. 2007), amphibians (Dickman 1987) and insects (Small et al. 2003). This is not altogether surprising as many studies have shown pervasive area effects in habitats (e.g. Buckton & Ormerod 2002), a pattern often attributed to the Theory of Island Biogeography (MacArthur & Wilson 1967). However, several researchers have argued that this is not directly applicable to urban areas where green spaces are not fully isolated islands (Niemelä 1999), rather the matrix habitats are permeable and enhance species movements. Perhaps the best example of this is residential gardens (Davies et al. 2009), which are not only home to a wide range of organisms (e.g. Loram et al. 2008) but are permeable to a great many others. Indeed, the high turnover rate exhibited by many taxonomic groups (Smith et al. 2006a, 2006b) suggests that the presence of gardens may enhance species dispersal through the city. Nonetheless, utilising the area effect relationship may provide a means of gaining rough estimates of likely species richness loss in the face of increasing land intensification in cities.

Other variables illustrate the importance of structural habitat influences (Denoel & Lehmann 2006; Garden et al. 2007), especially vegetation structure (Strauss & Biedermann 2006), vegetation cover (Loeb et al. 2009), vegetation type (Palomino & Carrascal 2006), substrate type (Small et al. 2003), soil compaction

(Grandchamp *et al.* 2000) and site age (Bolger *et al.* 2000). Several papers illustrate the importance of biotic linkages (Miyashita *et al.* 1998; Bolger *et al.* 2008) and their likely influence on ecosystem services (Sandford *et al.* 2008). Most of the physical factors are amenable to management at a local scale, although their importance differs across taxa suggesting caution is needed when integrating them into management plans.

Landscape variables used to contextualise green spaces in cityscapes are usually correlates of urbanisation (e.g. the amount of built land, percentage of concrete per unit area, building densities) and/or relate to fragmentation and isolation (e.g. percentage of habitat per unit area, distance to nearest similar green space). Many studies also highlight the importance of connectivity and linkage between green spaces. Several taxa such as butterflies (Hardy & Dennis 1999), beetles (Magura *et al.* 2004), bats (Gehrt & Chelsvig 2003, 2004), birds (Evans *et al.* 2009), reptiles (Germaine & Wakeling 2001) and amphibians (Pillsbury & Miller 2008) illustrate negative responses to increased urbanisation pressure. Studies of fragmentation effects caused by urbanisation are also widespread in the literature (Table 10.3), as are studies examining the impact of roads in creating barriers to movement for poorly dispersing species, notably amphibians (Forman & Deblinger 2000) and small mammals (Dickman & Doncaster 1987). Recent research on taxa as diverse as bats (Gorrensen *et al.* 2005), bees (Ahrne *et al.* 2009) and beetles (Ishitani *et al.* 2003) has shown that species responses to urbanisation and green space fragmentation are complex, frequently species-specific and often best predicted by using ecological traits (Cane *et al.* 2006). There are few data quantifying the likely loss of species from urban green spaces as a result of fragmentation effects. For example, Parris (2006) showed that the predicted amphibian species richness in the smallest ponds in their Melbourne survey was 2.8–5.5 times lower than predicted in the largest ones. She noted also that the most isolated pond was predicted to have 5–9 times fewer species than the least isolated pond. The potential loss of species richness also has more pervasive functional implications. In a recent paper, Flynn *et al.* (2009) used an extensive meta-analysis to show that intensification in the agricultural landscape has led to widespread loss of functional diversity (FD) in temperate and tropical New World mammal and bird communities. Significantly, in over one-quarter of the bird and mammal communities they analysed, declines in FD were steeper than predicted by species number. It seems likely that the loss of urban green spaces may be leading to similar losses of FD in our cities.

Connective features such as green networks and corridors have been influential in guiding planning policies in many areas of the world (Turner 2006; von Haaren & Reich 2006), but are also subject to considerable debate and confusion (Hess 1994; Hess & Fischer 2001). They have long been seen as providing connectivity, linking green spaces and minimising the potential effects of fragmentation on wildlife (Jongman *et al.* 2004), while providing important recreational, leisure and nature experience possibilities for people (Gobster & Westphal 2004). Only a few empirical studies have illustrated the successful role of corridors as conduits for species movement, and most of these derive from carefully controlled landscape-scale experiments (Beier & Noss 1998; Haddad & Baum 1999; Haddad *et al.* 2003). This is a vastly different situation from most urban areas where green corridors are multi-functional spaces, with a lot of habitat variability, and in these instances they do not appear to provide a strong functional conduit role (Angold *et al.* 2006). Nonetheless, some mobile and generalist species do utilise them as habitat (Mason *et al.* 2007), and even small-scale green features, such as street trees, are thought to enhance move-ments between green spaces in cities by birds (Fernández-Juricic 2000) and bats (Avila-Flores & Fenton 2005). Moreover, the role that green corridors may play as key elements in defining the social-nature landscape in cities is undeniable (Hartig *et al.* 1991; Evans 2007).

Ecosystem function and services

Constanza *et al.* (1997, p. 253) defined ecosystem services as 'the benefits human populations derive, either directly or indirectly, from ecosystem functions'. Although widely studied in broader ecological research, an evaluation of the quantities and flows of these services in cities is lacking (Elmqvist *et al.* 2004). A fuller review of this burgeoning literature is the subject of another chapter in this volume (Chapter 3), so we have limited our discussions to key areas that are of significance to green space planning in cities. Green spaces provide a wide range of locally generated ecosystem services in urban areas (Bolund & Hunhammar 1999) as well as important habitat and refugia functions as discussed above. These include a mix of regulatory and provisioning services such as climate amelioration (Akbari *et al.* 2001; Nowak *et al.* 2001), carbon sequestration and storage (Nowak & Crane 2002; Sahely *et al.* 2003, Pouyat *et al.* 2006), floodwater storage (Pauleit & Duhme 2000), water (Bolund & Hunhammar 1999) and air purification (McPherson *et al.* 1997; Beckett *et al.* 1998). It seems likely that several of these services have the capacity to help future-proof cities against global climate changes (Gill *et al.* 2007), but their quantification and the science underlying their application in this manner require considerable research.

The structure, composition, location and spatial configuration of green spaces influence their ecological performance and hence the services they provide (Pauleit & Duhme 2000; Whitford *et al.* 2001). Tratalos *et al.* (2007) compared the performance of differing landscape units across five cities in the UK and concluded that although ecosystem quality tends to decline con-tinuously as urban density increases, there is variability evident in many of the relationships between landscape type and service provision. This suggests that potential exists for maximising ecological services at a given landscape density by careful consideration of the structure (i.e. proportion and configuration) of landscape units. Currently planning systems still operate by focusing on individual units or development schemes (Niemelä 1999) but to gain the full benefit of ecosystem services, planning of the green space resource needs to occur at larger spatial scales (Andersson 2006).

Green spaces and people
Use and access
The final report of the World Health Organization's Commission on the Social Determinants of Health (CSDH 2008) and the latest European Union report on 'urban living' (EEA 2009) both called for wide-ranging improvements in daily living conditions. They both highlighted the need for improved access to green space for all, especially for those elements of society that live in poverty and are subject to deprivation. Green spaces are certainly well used in many regions of the world (e.g. De Sousa 2003). Comparative data on green space usage are few, but they are available from the UK, and they provide a useful benchmark. The Urban Parks Forum (2001) estimated that there are somewhere in the region of 1.5 billion visits per year to historic parks, using visitor numbers from three-quarters of local authorities. This is similar to an estimate of 2.6 billion visits per annum, based on a detailed questionnaire survey in Sheffield and ten other UK cities (Swanwick *et al.* 2003). There are also relatively few studies that consider access, especially measured against any standards that have been stipulated by governing bodies.

In a comparison of six major European cities, Stanners and Bourdeau (1995) reported that all citizens have access to urban green space within a 15-minute walk from their homes. Similarly, several large UK cities meet Natural England's basic ANGSt criteria of providing at least 2 ha of green space per 1000 head of population, with estimates ranging from 3.5 ha per 1000 in Leicester, 8.8 ha per 1000 in Manchester and 9.3 ha per 1000 in Birmingham to 21.5 ha per 1000 in Greater London. This basic criterion, however, is the lowest base level provision and a more rigorous application of the standard is certainly needed, utilising the tighter, more stringent remaining rules, which are much more discriminatory (Handley *et al.* 2003). The ANGSt rules have been more fully tested recently by studies in both Sheffield and Leicester. Barbosa *et al.* (2007) noted that green spaces are chronically underprovided for in Sheffield. Only 36.5% of Sheffield's households are located within 300 m of their nearest green space, which is the stipulated threshold figure recommended by Natural England. In contrast, 95.6% meet the European guidance, living within a 15-min walk from a green space. When considering the 85 larger municipal spaces (e.g. parks and public gardens) only 18% and 58% of Sheffield households meet the UK and European recommended access thresholds, respectively. A study of access in Leicester showed that only 10% of the population live within 300 m of a local green space and 40% lack access to large (20-ha) sites within 2 km of their homes (Comber *et al.* 2008). On the basis of the few available studies it seems that, when measured against UK and EU standards, access to green space is very limited.

The CSDH (2008) report highlights that access to green spaces should be for all sectors of society. Another important question is who has access to green spaces in cities. Barbosa *et al.* (2007) showed that there was great variability in access by different socioeconomic groups in Sheffield, although deprived social groupings and the aged fared better than the rest. Studies elsewhere have found opposite income-related patterns. Heynen (2006) showed a strong positive correlation between access and median household income. Mitchell and Popham (2008) studied the association between green space and socioeconomic inequalities in England using health statistics data derived from the English Index of Multiple Deprivation (EIMD). They noted strong covariation between green space access and socioeconomic group where poorer groups had less access.

Inequality of access is also a significant issue for people from ethnic groups. Studies in the USA have shown that the use of, and preferences of citizens for, green space vary in relation to ethnicity (Gobster 1998, 2002; Frumkin 2005). This pattern is paralleled in the city of Leicester, UK, where Comber *et al.* (2008) showed that certain ethnic groups (Indian, Hindu and Sikh) had limited access to green space in the city. Although few comparative data and little research exist, the findings are worrying as such patterns could be exacerbating other health equality issues (Smyth 2008).

Health and wellbeing benefits of green space

Green spaces have figured large in studies examining the relationships of 'place to health' (e.g. Frumkin 2003). This work has a strong historical dimension that focuses on extraordinary places of healing (e.g. Lourdes, France) (Smyth 2005) and on the sociopolitical context surrounding the rapid growth and creation of green spaces during the nineteenth century, which aimed at improving the lot of the working classes who were subjected to squalid and overcrowded living conditions (Giles-Corti *et al.* 2005). There is now an influential and substantial body of evidence exploring the links between green space provision in cities and human health (physical and psychological) and wellbeing (Tzoulas *et al.* 2007). The evidence can be grouped into three classes: (i) epidemiological studies linking health benefits to exposure to an improved natural environment, (ii) epidemiological evidence linking green space to behavioural changes leading to increased levels of physical exercise, and (iii) improvements in psychological (mental) health engendered by exposure to natural places and scenes. We examine each in turn.

(i) Epidemiological studies of green space health links

De Vries et al. (2003) showed a strong positive association between self-reported health and available neighbourhood green space. Their results suggest that a 10% increase in green space provision reduced symptoms of ill health in their survey pool of people ($n = 13090$) to a level comparable with a decrease in age of 5 years (p. 1726). A recent evidence-based study of 250 782 people, which was carefully constructed to avoid potential selection errors, highlighted a strong positive association between their local provision of green space (i.e. within 1 and 3 km of their homes) and perceptions of their own health (Maas et al. 2006). Mitchell and Popham (2007) aimed to determine the association between the percentage of green space in an area and the rate of self-reported 'not good' health, using health statistics and census data ($n = 32482$). They report that, generally, higher proportions of green space in an area were associated with better health, but the association was mixed depending on the degree of urbanisation and the level of income deprivation. Although these are all well-controlled multi-level and cross-sectional studies which account for variable socioeconomic status and age, their design does not help to identify or prove causal relationships that underlie the associations.

(ii) Behaviour and physical activity

An increasingly overweight and obese population in many developed countries is strongly correlated with reductions in the numbers of people taking part in physical activity (Joshu et al. 2008). Both issues are linked to major risk factors for a wide range of ailments and potentially fatal diseases (e.g. diabetes, cancers, heart disease, stroke). The many studies that have examined this important health issue fall into two categories: those that use questionnaires and self-reporting to examine levels of physical activity, and those that link indicators of overall physical activity (e.g. data on body mass index, BMI) to green space provision.

The association between green space provision and an increase in physical activity levels has been illustrated in several studies, which controlled for age, gender and educational level. Li et al. (2005) considered aspects of self-reported walking activity in the aged (people over 65 years old; $n = 577$) in Portland, Oregon, USA. The results suggest that neighbourhood building density and green infrastructure provision were important determinants of walking. Takano et al. (2002) observed a positive effect of living in an area that has walkable green spaces on the longevity of elderly people in Japan.

Frank et al. (2004) examined this issue in Atlanta, USA, by associating obesity (i.e. BMI $\geq$ 30 kg m^{-2}) with other factors such as land-use mix, residential density and car use in adults. The analyses indicated that an additional hour spent in a car per day was associated with a 6% increase in the likelihood of obesity, whereas an additional kilometre walked per day was associated with a

4.8% reduction in the likelihood of obesity. Although the variables interacted the authors suggest that land-use mix is an important determinant of walking activity, and thus lower BMI, in their sample populations. In a follow-up study accelerometers were deployed over a 2-day period to capture objective levels of physical activity in 357 adults in the same city (Frank *et al.* 2005). Activity levels were then compared to a walkability index derived from each household's physical location, the adjacent land-use mix, residential density and street connectivity. The walkability index of these urban form factors was related to levels of physical activity. Nielsen and Hansen's (2007) survey of access and use of green areas and their impact on obesity in 1200 Danish adults (18–80 years old) provided similar results, indicating that access to a garden or short distances to green areas are associated with lower likelihood of obesity. The authors suggest that the significance of distance to green areas is mainly derived from its correlation with the character of the neighbourhood and its conduciveness to outdoor activities and 'healthy' lifestyle modes.

The positive association between local green space provision and weight is prevalent for all ages of society. Bell *et al.* (2008) note that African-American economically disadvantaged children and youth exhibited an inverse association between neighbourhood greenness and BMI scores. Their study examined the lasting effect of green space provision by examining changes in the BMI of children and youth (*n* = 3842) in Indianapolis, USA. The results suggest that children and youth in greener neighbourhoods were less likely to increase their BMI scores over the two years of their study than their counterparts in less green neighbourhoods. But associative links such as these are context-dependent, notwithstanding the care and thought that has gone into study designs. Garden and Jalaludin (2009) showed that urban sprawl in Sydney, Australia, was associated with reduced activity levels and increased obesity owing to an overreliance on cars for transport and poor neighbourhood resources for shopping, community activities and the like. It appears that where people are in the cityscape matters as well as their socioeconomic status, gender and ethnicity, highlighting the need for spatially explicit models that factor in geographic locations.

(iii) Improvements in psychological (mental) health and wellbeing
The mechanisms underpinning psychological responses to green spaces derive from psycho-evolutionary theory (Ulrich *et al.* 1991) and relate to the capacity of natural green spaces to act as 'restorative places' (Hartig *et al.* 2003). Attention Restoration Theory (Kaplan & Kaplan 1989; Kaplan 1995) suggests that direct and focused attention can be enhanced by direct contact with nature, which replenishes an individual's attention resource. In contrast, Ulrich's (1984) work on the restorative effects of nature, while in some ways similar to Kaplin's, emphasises an affective rather than cognitive response, and focuses on

emotional and mental responses to taxing/threatening stimuli, not attention deficits due to everyday humdrum activity (Hartig *et al.* 1991). Ulrich (1984) highlighted the positive impact that a 'green view' can have on patients convalescing in hospital after major surgery. In this ground-breaking study, 23 patients whose beds overlooked 'natural scenery' showed enhanced recovery rates when compared to a control group ($n = 23$) whose beds faced a brick wall. Kuo (2001) studied attentional functioning and effectiveness in dealing with major stresses in 149 urban public housing residents randomly assigned to housing with or without nearby nature. Residents placed in housing with little adjacent green space (nearby trees and grass) reported higher levels of procrastination and stress when facing major issues than their counterparts who were inhabiting homes in greener areas, indicating that access to nature is a powerful means of coping with the daily demands of crushing poverty. Wells and Evans (2003) considered whether exposure to nature in a sample of rural children was related to their levels of stress. Their results suggest that children living near nature were buffered against the potential impacts of life stress to a greater extent than those with little nature nearby. Nielsen and Hansen (2007) also show that Danish adults (18–80 years old) who have access to a garden or short distances to green areas in the neighbourhood are associated with lower self-reported stress levels. These studies suffer from the same associative links so cannot be used to prove causality.

Studies of stress recovery are, however, amenable to experimental work that provides stronger links between stress levels and human physiology. Ulrich *et al.* (1991) subjected 120 people to a stressful movie and then exposed them to videotapes of six different natural and urban settings. Data concerning their recovery from the stressor were obtained from self-ratings of affective states and a range of physiological measures including heart period, muscle tension, skin conductance and pulse transit time. Findings from both measures converged, indicating that recovery was faster and more complete in subjects that were exposed to natural rather than urban environments. Hartig *et al.* (1991) carried out quasi-experimental and true experimental work to assess the utility of different theoretical models of restorative experience related to nature and greenness. This illustrated that groups of people who vacationed in wilderness situations showed much improved proofreading ability in comparison to control groups who vacationed elsewhere. In a second study, in the same paper, they report that a group of people situated in a 'natural' rather than an 'urban' environment fared better in a range of attentionally fatiguing tasks.

Parsons *et al.* (1998) examined whether stress and recovery from stress varied as a function of roadside environment in a sample of 160 college participants. The participants were subjected to two stressors while being continuously monitored for facial electromyographic (EMG) and autonomic (electrocardiogram, blood pressure and skin conductance) activity and then played a videotape that simulated a drive in one of four different environments. Participants who viewed nature-dominated roadside environments exhibited less stress and recovered from it more fully and faster than the control group who viewed urban environments. Hartig *et al.* (2003) compared psycho-physiological stress recovery and attention restoration in natural and urban field settings using measurements of blood pressure, emotion and attention collected from 112 young adults. In the two-way experiments urban settings led to increased blood pressure, heightened negative emotional responses and attention deficiencies.

Health inequalities

There is now a growing body of evidence that supports the notion that disadvantaged groups in society can gain more health benefits from green space provision. Mitchell and Popham's (2008) study factored in covariability of access to green space by socioeconomic group and showed that people who were more deprived and had good access to green space exhibited lower incidence rates of all-cause and circulatory mortality than did those in areas where green spaces were fewer. Importantly, this suggests that higher exposure to green space could save 1328 lives per year in people suffering severe deprivation. Other large epidemiological studies provide corroboratory results. De Vries *et al.* (2003) showed that health effects of green space were more significant in lower-income groups, and Maas *et al.* (2006) found that people belonging to lower socioeconomic groups, the elderly, youth and higher-educated (post-school) people seemed to gain more from access to green space.

Green space quality

One of the limitations of health/green space studies is that the analyses fail to examine what it is about the green space that leads to improvements in health. Is it the distance to the green space, its quality or its structure that matters, or all three? Hillsdon *et al.* (2006) examined the relationship between access to high-quality urban green space and the level of recreational activity in middle-aged people (40–70 years). Their results show no clear linkages between the distance to, quality of, or size of the green spaces and the amount of physical activity recorded for the respondents. This is in contrast to other studies that showed that proximity was important (Nielsen & Hansen 2007) and that access to a large attractive public space is associated with increased probability of a high level of walking (Giles-Corti *et al.* 2005). A recent study suggests that people do appear to value and recognise nature in their local green spaces, illustrating additional and important synergies between biodiversity in green space and wellbeing. Fuller *et al.* (2007) analysed the linkage between biodiversity in green spaces and psychological wellbeing. Plants, butterflies and birds were sampled in 15 green spaces in the city of Sheffield during 2005. Linear regression showed that the psychological measures were positively related to species richness and park area, and that local users can perceive species richness reasonably well. Although the links found are illuminating, the authors stress that they are not proof of causality.

In summary, the association of green space provision and usage to health benefits is a positive one although it is not easily generalised. Notwithstanding the attempts of researchers to use innovative sample design and careful modelling of covariates, most studies are correlative and do not indicate underlying mechanisms. Several researchers have argued that the distinctions between people, place and health are contrived, and that there is a need to reconceptualise (or reimagine) the relationship between the variables to emphasise their complexity, history and relationality, and how these manifest themselves in health outcomes and individual behaviours (Tunstall et al. 2004). This sentiment is shared by a growing number of health professionals (e.g. Frumkin 2003) who suggest that more emphasis needs to be placed on researching what kinds of nature are experienced in green spaces. Put another way: what nature is going to be viewed and how are users going to contact it? This should lead to the formulation and definition of more meaningful health endpoints, rather than the endless run of metrics and targets that are pervasive currently. Frumkin (2003) also suggests that design features (e.g. amount and type of vegetation, quiet areas for sitting, recreational amenities and so on) also need to be carefully considered.

Synthesis: designing and managing green spaces for people and wildlife

The use of habitat creation schemes as compensation for ecological features in green spaces that are lost because of (re)development is commonplace, but where the redevelopment schemes replace former derelict or brownfield sites, these heavily managed landscaped open spaces are much less biologically diverse than the pre-development landscapes they replace (Donovan et al. 2005; Figure 10.2). Much more careful thought, testing and design of these schemes is needed, similar to the work carried out examining the biodiversity value and possibilities for enhancement of garden ecosystems (Davies et al. 2009). Clearly, if agricultural land is developed for housing, as part of an increasing urban sprawl, then it is possible that local diversity could increase rather than decrease (Gaston et al. 2005). Where the removal of green space has to occur, and remnants of the former landscape cannot be preserved, potential still exists for 'green engineering' the built environment in a manner that maximises its ecological function. A myriad of green technologies now exist that can be used for this purpose, ranging from mitigation for particular species, such as nest and roost boxes for birds, bees and bats, to broader-scale initiatives such as permeable pavements, living walls and green (biodiversity) roofs which may provide some biodiversity, ecosystem and social benefits

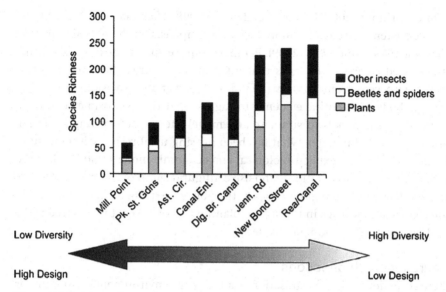

Figure 10.2 Comparison of species richness of designed and derelict open spaces in a redevelopment site in Eastside (Birmingham, UK). Site names on *x* axis are as follows: Mill. Point = Millennium Point; Pk. St. Gdns = Park Street Gardens; Ast. Cir. = Aston Circus; Canal Ent. = Canal Entrance; Dig. Br. Canal = Digbeth Branch Canal; Jenn. Rd = Jennings Road; New Bond Street; Rea/Canal = River Rea and canal junction. Reproduced with permission from Donovan *et al.* (2005).

(Oberndorfer *et al.* 2007). There are, however, no systematic studies that test their effectiveness in urban areas to mitigate the effects of lost green space.

The UK Royal Commission on Environmental Pollution (RCEP 2007) stated that 'the natural environment should be at the heart of urban design and management' (p. 83). The wide range of policy documents at a national (Table 10.1) and international level (EEA 2009) suggests that the political will now exists to move towards a city future that includes more, better-designed and multi-functional green spaces. For this to take place green spaces must be viewed as an important component of the urban complex social–economic–ecological linked system. Pickett *et al.* (2001) provide a conceptual framework which helps contextualise the importance of urban green spaces, but its emphasis is on socio-ecological linkages. Tzoulas *et al.* (2007) add to this greatly by articulating the relationships between green spaces and public and ecosystem health, while others focus their attention on creating models to predict the impact of urbanisation on biodiversity (Williams *et al.* 2009).

While such frameworks are of considerable use in highlighting the complex linkages between the diverse systems, and targeting areas of research need and policy gaps, they do not necessarily provide concrete guidance for action on the ground where there is a clear need for a stronger design framework

(Hunter & Hunter 2008; Pickett & Cadenasso 2008). Niemelä and his colleagues have consistently argued for more focus and emphasis on the planning process (Niemelä 1999; James *et al.* 2009). Planners require spatially explicit data that can be used to assess the roles that different green areas have in providing people with ecosystem services (including biodiversity provision). There is a growing body of scientific evidence to suggest that this is possible, at least in terms of creating tools to simulate and model the impacts of fragmentation (Andersson & Bodin 2009). What has been missing until recently is a means of integrating spatially explicit biological and economic models that can be used to find the most parsimonious land-use configurations (Polasky *et al.* 2008). We suggest that, in some regions at least, the policy instruments needed to facilitate change are in place in the spatial planning system. What is required is the local political will to implement them.

Future research directions

There is no doubt that urbanisation is a pervasive environmental issue that is set to increase as the global population grows (Grimm *et al.* 2008). This will lead to further fragmentation, isolation and degradation of green spaces in cities, especially if urban sprawl is contained. This review has illustrated that green spaces provide numerous benefits in terms of their value for biodiversity, provision of ecosystem services, and benefits for public heath and wellbeing. There is a pressing need therefore to place green spaces high on the political agenda and work towards the use of policies that are already in place to protect and enhance green space provision for people and wildlife. For this to occur, several key challenges must be addressed (cf. James *et al.* 2009). First, the physical science base needs strengthening. There is a need for: (i) more reliable and quantified data on green space provision in cities; (ii) the creation of baseline data from which loss and predicted loss can be estimated; (iii) estimates of the changes in species richness and functional diversity as a result of green space loss; (iv) the extension of landscape modelling to a wider portfolio of ecosystem services; (v) more careful testing and monitoring of the efficacy of designed green spaces that are created as part of ongoing urban changes and development; and (vi) a greater emphasis on species-specific studies, perhaps focusing on indicator species of conservation concern (Evans *et al.* 2009). There are still large gaps in our autecological understanding of a range of species that inhabit urban green spaces, even those that have evident and significant relevance such as insect pollinators (Matteson *et al.* 2008).

Second, in terms of the social science, health and wellbeing and policy arenas, emphasis needs placing on: (i) monitoring and assessing variability of access to green spaces – few comparable studies exist here, but this lies at the heart of current government and international policies concerning the quality of life of urban citizens; (ii) strengthening the qualitative research that views green spaces as green places that are in part socially constructed – this should help in the creation of better-designed spaces that maximise health and wellbeing benefits; and (iii) the creation of models that are easily integrated into planning systems.

Acknowledgements

This work and research has been funded by EPSRC grants EP/E021603 and EP/007426/1 and an award from the Big Lottery Fund via the Open Air Laboratories (OPAL) network. We are grateful for comments and discussions with colleagues within our respective schools and the wider Urban Nature Community of Interest.

References

Ahrne, K., Bengtsson, J. and Elmqvist, T. (2009). Bumble bees (*Bombus* spp) along a gradient of increasing urbanization. *PLoS One*, **4**, e5574.

Akbari, H., Pomerantz, M. and Taha, H. (2001). Cool surfaces and shade trees to reduce energy use and improve air quality in urban areas. *Solar Energy*, **70**, 295–310.

Andersson, E. (2006). Urban landscapes and sustainable cities. *Ecology and Society*, **11**, 23–86.

Andersson, E. and Bodin, O. (2009). Practical tool for landscape planning? An empirical investigation of network based models of habitat fragmentation. *Ecography*, **32**, 123–32.

Angold, P. G., Sadler, J. P., Hill, M. O. *et al.* (2006). Biodiversity in urban habitat patches. *Science of the Total Environment*, **360**, 196–204.

Avila-Flores, R. and Fenton, M. B. (2005). Use of spatial features by foraging insectivorous bats in a large urban landscape. *Journal of Mammalogy*, **86**, 1193–204.

Barbosa, O., Tratalos, J. A., Armsworth, P. R. *et al.* (2007). Who benefits from access to green space? A case study from Sheffield, U.K. *Landscape and Urban Planning*, **83**, 187–95.

Beckett, K. P., Freer-Smith, P. H. and Taylor, G. (1998). Urban woodlands: their role in reducing the effects of particulate pollution. *Environmental Pollution*, **99**, 347–60.

Beier, P. and Noss, R. F. (1998). Do habitat corridors provide connectivity? *Conservation Biology*, **12**, 1241–52.

Bell, J. F., Wilson, J. S. and Liu, G. C. (2008). Neighborhood greenness and 2-year changes in Body Mass Index of children and youth. *American Journal of Preventive Medicine*, **35**, 547–53.

Benedict, M. A. and McMahon, E. T. (2002). Green infrastructure: smart conservation for the 21st century. *Renewable Resources Journal*, **20**, 12–17.

Blair, R. B. (1999). Birds and butterflies along an urban gradient: surrogate taxa for assessing biodiversity? *Ecological Applications*, **9**, 164–70.

Bolger, D. T., Beard, K. H., Suarez, A. V. and Case, T. J. (2008). Increased abundance of native and non-native spiders with habitat fragment. *Journal of Biogeography*, **14**, 655–65.

Bolger, D. T., Suarez, A. V., Crooks, K. R., Morrison, S. A. and Case, T. J. (2000). Arthropods in urban habitat fragments in southern California: area, age, and edge effects. *Ecological Applications*, **10**, 1230–48.

Bolund, P. and Hunhammar, S. (1999). Ecosystem services in urban areas. *Ecological Economics*, **29**, 293–301.

Buckton, S. T. and Ormerod, S. J. (2002). Global patterns of diversity among the specialist birds of riverine landscapes. *Freshwater Biology*, **47**, 695–709.

Burton, E. (2000). The compact city: just or just compact? A preliminary analysis. *Urban Studies*, **37**, 1969.

CABE (2009a). *Is the Grass Greener. . .? Learning from International Innovations in Urban Green Space Management*. London: ODPM.

CABE (2009b). *Open Space Strategy: Best Practice Guidance*. London: Commission for the Built Environment.

Cairns, J. (2006). Designing for nature and sustainability. *International Journal of Sustainable Development and World Ecology*, **13**, 77–81.

Cane, J. H., Minckley, R. L., Kervin, L. J., Roulston, T. H. and Williams, N. M. (2006). Complex responses within a desert bee guild (Hymenoptera: Apiformes) to urban habitat fragmentation. *Ecological Applications*, **16**, 632–44.

Chamberlain, D. E., Gough, S., Vaughan, H., Vickery, J. A. and Appleton, G. F. (2007). Determinants of bird species richness in public green spaces. *Bird Study*, **54**, 87–97.

Comber, A., Brunsdon, C. and Green, E. (2008). Using a GIS-based network analysis to determine urban greenspace accessibility for different ethnic and religious groups. *Landscape and Urban Planning*, **86**, 103–14.

Costanza, R., d'Arge, R., deGroot, R. *et al.* (1997). The value of the world's ecosystem services and natural capital. *Nature*, **387**, 253–60.

Crane, P. and Kinzig, A. (2005). Nature in the metropolis. *Science*, **308**, 1225.

CSDH (2008). *Closing the Gap in a Generation: Health Equity through Action on the Social Determinants of Health*. Final Report of the Commission on Social Determinants of Health. Geneva: World Health Organization.

Dantzing, G. B. and Saaty, T. (1973). *Compact City: A Plan for a Livable Urban Environment*. San Francisco: W. H. Freeman.

Davies, R. G., Barbosa, O., Fuller, R. A. and Tratalos, J. (2008). City-wide relationships between green spaces, urban land use and topography. *Urban Ecosystems*, **11**, 269–87.

Davies, Z. G., Fuller, R. A., Loram, A. *et al.* (2009). A national scale inventory of resource provision for biodiversity within domestic gardens. *Biological Conservation*, **142**, 761–71.

DCLG (2000). *Planning Policy Guidance 17: Planning for Open Space, Sport and Recreation*. HMSO, London: Department for Communities and Local Government.

De Sousa, C. A. (2003). Turning brownfields into green space in the City of Toronto. *Landscape and Urban Planning*, **62**, 181–98.

de Vries, S., Verheij, R. A. and Groenewegen, P. P. (2003). Natural environments-healthy environments? An exploratory analysis of the relationship between greenspace and health. *Environment and Planning A*, **35**, 1717–31.

Denoel, M. and Lehmann, A. (2006). Multi-scale effect of landscape processes and habitat quality on newt abundance: implication for conservation. *Biological Conservation*, **130**, 495–504.

DETR (2000). *Our Towns and Cities: Delivering an Urban Renaissance*. London: Department of the Environment, Transport and the Regions.

DETR (2002). *Green Spaces, Better Places*. London: Department of the Environment, Transport and the Regions.

Dickman, C. R. (1987). Habitat fragmentation and vertebrate species richness in an urban environment. *Journal of Applied Ecology*, **24**, 337–51.

Dickman, C. R. and Doncaster, C. P. (1987). The ecology of small mammals in urban habitats. I. Populations in a patchy environment. *Journal of Animal Ecology*, **56**, 629–40.

Donovan, R., Sadler, J. P. and Bryson, J. R. (2005). Urban biodiversity and sustainable development. *Engineering Sustainability*, **158**, 105–14.

EEA (2002). Towards an urban atlas: assessment of spatial data on 25 European cities and urban areas. *Environmental Issues Report No. 30*. Copenhagen: European Commission, Joint Research Centre, pp. 117.

EEA (2006). *Urban Sprawl in Europe: The Ignored Challenge*. Copenhagen: European Commission, Joint Research Centre, pp. 56.

EEA (2009). *Ensuring Quality of Life in Europe's Cities and Towns*. Copenhagen: European Commission, pp. 108.

Elmqvist, T., Colding, J., Barthel, S. *et al.* (2004). The dynamics of social-ecological systems in urban landscapes – Stockholm and the National Urban Park, Sweden. *Urban Biosphere and Society: Partnership of Cities*, **1023**, 308–22.

Evans, J. (2007). Wildlife corridors: an urban political ecology. *Local Environment*, **12**, 129–52.

Evans, K. L., Newson, S. E. and Gaston, K. J. (2009). Habitat influences on urban avian assemblages. *Ibis*, **151**, 19–39.

Evergreen (2004). *Green Space Acquisition and Stewardship in Canada's Urban Municipalities*. Toronto: Evergreen.

Fernandez-Juricic, E. (2000). Avifaunal use of wooded streets in an urban landscape. *Conservation Biology*, **14**, 513–21.

Flynn, D. F. B., Gogol-Prokurat, M., Nogeire, T. et al. (2009). Loss of functional diversity under land use intensification across multiple taxa. *Ecology Letters*, **12**, 22–33.

Forman, R. T. T. and Deblinger, R. D. (2000). The ecological road-effect zone of a Massachusetts (USA) suburban highway. *Conservation Biology*, **14**, 36–46.

Frank, L. D., Andresen, M. A. and Schmid, T. L. (2004). Obesity relationships with community design, physical activity, and time spent in cars. *American Journal of Preventive Medicine*, **27**, 87–96.

Frank, L. D., Schmid, T. L., Sallis, J. F., Chapman, J. and Saelens, B. E. (2005). Linking objectively measured physical activity with objectively measured urban form – findings from SMARTRAQ. *American Journal of Preventive Medicine*, **28**, 117–25.

Frumkin, H. (2003). Healthy places: exploring the evidence. *American Journal of Public Health*, **93**, 1451–6.

Frumkin, H. (2005). Health, equity, and the built environment. *Environmental Health Perspectives*, **113**, A290–1.

Fuller, R. A. and Gaston, K. J. (2009). The scaling of green space coverage in European cities. *Biology Letters*, **5**, 352–5.

Fuller, R. A., Irvine, K. N., Devine-Wright, P., Warren, P. H. and Gaston, K. J. (2007). Psychological benefits of greenspace increase with biodiversity. *Biology Letters*, **3**, 390–4.

Garden, F. L. and Jalaludin, B. B. (2009). Impact of urban sprawl on overweight, obesity, and physical activity in Sydney, Australia. *Journal of Urban Health*, **86**, 19–30.

Garden, J. G., McAlpine, C. A., Possingham, H. P. and Jones, D. N. (2007). Habitat structure is more important than vegetation composition for local-level management of native terrestrial reptile and small mammal species living in urban remnants: a case study from Brisbane, Australia. *Austral Ecology*, **32**, 669–85.

Gaston, K. J., Smith, R. M., Thompson, K. and Warren, P. H. (2005). Urban domestic gardens (II): experimental tests of methods for increasing biodiversity. *Biodiversity and Conservation*, **14**, 395–413.

Gehrt, S. D. and Chelsvig, J. E. (2003). Bat activity in an urban landscape: patterns at the landscape and microhabitat scale. *Ecological Applications*, **13**, 939–50.

Gehrt, S. D. and Chelsvig, J. E. (2004). Species-specific patterns of bat activity in an urban landscape. *Ecological Applications*, **14**, 625–35.

Germaine, S. S. and Wakeling, B. F. (2001). Lizard species distributions and habitat occupation along an urban gradient in Tucson, Arizona, USA. *Biological Conservation*, **97**, 229–37.

Gibson, C. W. D. (1998). *Brownfield: Red Data. The Values Artificial Habitats Have for Uncommon Invertebrates*. Peterborough: English Nature.

Gilbert, O. (1989). *The Ecology of Urban Habitats*. London: Chapman and Hall.

Giles-Corti, B., Broomhall, M. H., Knuiman, M. et al. (2005). Increasing walking – how important is distance to, attractiveness, and size of public open space? *American Journal of Preventive Medicine*, **28**, 169–76.

Gill, S. E., Handley, J., Ennos, A. R. and Pauleit, S. (2007). Adapting cities for climate change: the role of the green infrastructure. *Built Environment*, **33**, 115–33.

Gill, S. E., Handley, J. F., Ennos, A. R. et al. (2008). Characterising the urban environment of

U.K. cities and towns: a template for landscape planning. *Landscape and Urban Planning*, **87**, 210–22.

Gobster, P. H. (1998). Urban parks as green walls or green magnets? Interracial relations in neighborhood boundary parks. *Landscape and Urban Planning*, **41**, 43–55.

Gobster, P. H. (2002). Managing urban parks for a racially and ethnically diverse clientele. *Leisure Sciences*, **24**, 143–59.

Gobster, P. H. and Westphal, L. M. (2004). The human dimensions of urban greenways: planning for recreation and related experiences. *Landscape and Urban Planning*, **68**, 147–65.

Gorrensen, P. M., Willig, M. R. and Strauss, R. E. (2005). Multivariate analysis of scale-dependent associations between bats and landscape structure. *Ecological Applications*, **15**, 2126–36.

Grandchamp, A.-C., Niemelä, J. and Kotze, J. (2000). The effects of trampling on assemblages of ground beetles (Coleoptera, Carabidae) in urban forests in Helsinki, Finland. *Urban Ecosystems*, **4**, 321–32.

Grimm, N. B., Faeth, S. H., Golubiewski, N. E. et al. (2008). Global change and the ecology of cities. *Science*, **319**, 756–60.

Haddad, N. M. and Baum, K. A. (1999). An experimental test of corridor effects on butterfly densities. *Ecological Applications*, **9**, 623–33.

Haddad, N. M., Bowne, D. R., Cunningham, A. et al. (2003). Corridor use by diverse taxa. *Ecology*, **84**, 609–15.

Handley, J., Pauleit, S., Slinn, P. et al. (2003). *Accessible Natural Green Space Standards in Towns and Cities: A Review and Toolkit for their Implementation*. Peterborough: English Nature Report No. 526.

Hardy, P. B. and Dennis, R. L. H. (1999). The impact of urban development on butterflies within a city region. *Biodiversity and Conservation*, **8**, 1261–79.

Harrison, C., Burgess, J., Millward, A. and Dawe, G. (1995). *Accessible Natural Greenspace in Towns and Cities: A Review of Appropriate Size and*

Distance Criteria. Peterborough: English Nature Research Report No. 153.

Hartig, T. (2008). Green space, psychological restoration, and health inequality. *The Lancet*, **372**, 1614–15.

Hartig, T., Evans, G. W., Jamner, L. D., Davis, D. S. and Gärling, T. (2003). Tracking restoration in natural and urban field settings. *Journal of Environmental Psychology*, **23**, 109–23.

Hartig, T., Mang, M. and Evans, G. W. (1991). Restorative effects of natural environment experiences. *Environment and Behavior*, **23**, 3–26.

Helden, A. J. and Leather, S. R. (2004). Biodiversity on urban roundabouts – Hemiptera, management and the species area. *Basic and Applied Ecology*, **5**, 367–77.

Hess, G. R. (1994). Conservation corridors and contagious-disease – a cautionary note. *Conservation Biology*, **8**, 256–62.

Hess, G. R. and Fischer, R. A. (2001). Communicating clearly about conservation corridors. *Landscape and Urban Planning*, **55**, 195–208.

Heynen, N. (2006). Green urban political ecologies: toward a better understanding of inner-city environmental change. *Environment and Planning A*, **38**, 499–516.

Hillsdon, M., Panter, J., Foster, C. and Jones, A. (2006). The relationship between access and quality of urban green space with population physical activity. *Public Health*, **120**, 1127–32.

Hunter, M. R. and Hunter, M. D. (2008). Designing for conservation of insects in the built environment. *Insect Conservation and Diversity*, **1**, 189–96.

Irwin, E. G. and Bockstael, N. E. (2007). The evolution of urban sprawl: evidence of spatial heterogeneity and increasing land fragmentation. *Proceedings of the National Academy of Sciences of the USA*, **104**, 20672–7.

Ishitani, M., Kotze, D. J. and Niemelä, J. (2003). Changes in carabid beetle assemblages across an urban-rural gradient in Japan. *Ecography*, **26**, 481–9.

James, P., Tzoulas, K., Adams, M. D. *et al.* (2009). Towards an integrated understanding of green space in the European built environment. *Urban Forestry and Urban Greening*, **8**, 65–75.

Jongman, R. H. G., Kulvik, M. and Kristiansen, I. (2004). European ecological networks and greenways. *Landscape and Urban Planning*, **68**, 305–19.

Joshu, C. E., Boehmer, T. K., Brownson, R. C. and Ewing, R. (2008). Personal, neighbourhood and urban factors associated with obesity in the United States. *Journal of Epidemiology and Community Health*, **62**, 202–8.

Kaplan, R. and Kaplan, S. (1989). *The Experience of Nature*. New York: Cambridge University Press.

Kaplan, S. (1995). The restorative benefits of nature: towards an integrated framework. *Journal of Environmental Psychology*, **15**, 169–82.

Kuo, F. E. (2001). Coping with poverty: impacts of environment and attention in the inner city. *Environment and Behavior*, **33**, 5–34.

Landscape Institute (2009). *Green Infrastructure: Connected and Multifunctional Landscapes*. London: Landscape Institute.

Li, F. Z., Fisher, K. J., Brownson, R. C. and Bosworth, M. (2005). Multilevel modelling of built environment characteristics related to neighbourhood walking activity in older adults. *Journal of Epidemiology and Community Health*, **59**, 558–64.

Loeb, S. C., Post, C. J. and Hall, S. T. (2009). Relationship between urbanisation and bat community structure in national parks of the southern U.S. *Urban Ecosystems*, **12**, 197–214.

Loram, A., Thompson, K., Warren, P. H. and Gaston, K. J. (2008). Urban domestic gardens (XII): the richness and composition of the flora in five U.K. cities. *Journal of Vegetation Science*, **19**, 321–30.

Loram, A., Tratalos, J., Warren, P. H. and Gaston, K. J. (2007). Urban domestic gardens (X): the extent & structure of the resource in five major cities. *Landscape Ecology*, **22**, 601–15.

Maas, J., Verheij, R. A., Groenewegen, P. P., de Vries, S. and Spreeuwenberg, P. (2006). Green space, urbanity, and health: how strong is the relation? *Journal of Epidemiology and Community Health*, **60**, 587–92.

MacArthur, R. H. and Wilson, E. O. (1967). *The Theory of Island Biogeography*. Princeton, NJ: Princeton University Press.

Magura, T., Tothmeresz, B. and Molnar, T. (2004). Changes in carabid beetle assemblages along an urbanisation gradient in the city of Debrecen, Hungary. *Landscape Ecology*, **19**, 747–59.

Mason, J., Moorman, C., Hess, G. and Sinclair, K. (2007). Designing suburban greenways to provide habitat for forest-breeding birds. *Landscape and Urban Planning*, **80**, 153–64.

Matteson, K. C., Ascher, J. S. and Langellotto, G. A. (2008). Bee richness and abundance in New York city urban gardens. *Annals of the Entomological Society of America*, **101**, 140–50.

McDonnell, M. J. and Hahs, A. K. (2008). The use of gradient analysis studies in advancing our understanding of the ecology of urbanizing landscapes: current status and future directions. *Landscape Ecology*, **23**, 1143–55.

McDonnell, M. J. and Pickett, S. T. A. (1990). Ecosystem structure and function along urban–rural gradients: an unexploited opportunity for ecology. *Ecology*, **71**, 1232–7.

McKinney, M. L. (2006). Urbanization as a major cause of biotic homogenization. *Biological Conservation*, **127**, 247–60.

McKinney, M. L. (2008). Effects of urbanization on species richness: a review of plants and animals. *Urban Ecosystems*, **11**, 161–76.

McPherson, E. G., Nowak, D. and Heisler, G. (1997). Quantifying urban forest structure, function, and value: the Chicago Urban Forest Climate Project. *Urban Ecosystems*, **1**, 49–61.

Mitchell, R. and Popham, F. (2007). Greenspace, urbanity and health: relationships in England. *Journal of Epidemiology and Community Health*, **61**, 681–3.

Mitchell, R. and Popham, F. (2008). Effect of exposure to natural environment on health inequalities: an observational population study. *The Lancet*, **372**, 1655–60.

Miyashita, T., Shinkai, A. and Chida, T. (1998). The effects of forest fragmentation on web spider communities in urban areas. *Biological Conservation*, **86**, 357–64.

Mortberg, U. and Wallentinus, H. G. (2000). Red-listed forest bird species in an urban environment – assessment of green space corridors. *Landscape and Urban Planning*, **50**, 215–26.

Nielsen, T. S. and Hansen, K. B. (2007). Do green areas affect health? Results from a Danish survey on the use of green areas and health indicators. *Health and Place*, **13**, 839–50.

Niemela, J. (1999). Ecology and urban planning. *Biodiversity and Conservation*, **8**, 119–31.

Nowak, D. J. and Crane, D. E. (2002). Carbon storage and sequestration by urban trees in the USA. *Environmental Pollution*, **116**, 381–9.

Nowak, D. J., Noble, M. H., Sisinni, S. M. and Dwyer, J. F. (2001). People & trees – assessing the US urban forest resource. *Journal of Forestry*, **99**, 37–42.

Oberndorfer, E., Lundholm, J., Bass, B. *et al.* (2007). Green roofs as urban ecosystems: ecological structures, functions, and services. *BioScience*, **57**, 823–33.

ODPM (2002a). *Planning Policy Guidance 3: Housing*. London: Office of the Deputy Prime Minister.

ODPM (2002b). *Living Places – Cleaner, Safer, Greener*. London: Office of the Deputy Prime Minister.

Palomino, D. and Carrascal, L. M. (2006). Urban influence on birds at a regional scale: a case study with the avifauna of northern Madrid province. *Landscape and Urban Planning*, **77**, 276–90.

Parris, K. M. (2006). Urban amphibian assemblages as metacommunities. *Journal of Animal Ecology*, **75**, 757–64.

Parsons, R., Tassinary, L. G. and Ulrich, R. S. (1998). The view from the road: implications for stress recovery and immunization. *Journal of Environmental Psychology*, **18**, 113–39.

Pauleit, S. (2003). Perspectives on urban greenspace in Europe. *Built Environment*, **29**, 89–93.

Pauleit, S. and Duhme, F. (2000). Assessing the environmental performance of land cover types for urban planning. *Landscape and Urban Planning*, **52**, 1–20.

Pauleit, S., Ennos, R. and Golding, Y. (2005). Modeling the environmental impacts of urban land use and land cover change – a study in Merseyside, U.K. *Landscape and Urban Planning*, **71**, 295–310.

Pauleit, S., Slinn, P., Handley, J. and Lindley, S. (2001). Promoting the natural green structure of towns and cities: English Nature's *Accessible Natural Greenspace Standards* model. *Built Environment*, **29**, 157–70.

Pickett, S. T. A. and Cadenasso, M. L. (2008). Linking ecological and built components of urban mosaics: an open cycle of ecological design. *Journal of Ecology*, **96**, 8–12.

Pickett, S. T. A., Cadenasso, M. L., Grove, J. M. *et al.* (2001). Urban ecological systems: linking terrestrial ecological, physical, and socioeconomic components of metropolitan areas. *Annual Review of Ecology and Systematics*, **32**, 127–57.

Pillsbury, F. C. and Miller, J. R. (2008). Habitat and landscape characteristics underlying anuran community structure along an urban–rural gradient. *Ecological Applications*, **18**, 1107–18.

Polasky, S., Nelson, E., Camm, J. *et al.* (2008). Where to put things? Spatial land management to sustain biodiversity and economic returns. *Biological Conservation*, **141**, 1505–24.

Pouyat, R. V., Yesilonis, I. D. and Nowak, D. J. (2006). Carbon storage by urban soils in the United States. *Journal of Environmental Quality*, **35**, 1566–75.

Puth, L. M. and Burns, C. E. (2009). New York's nature: a review of the status and trends in species richness across the metropolitan region. *Diversity and Distributions*, **15**, 12–21.

Rafiee, R., Salman Mahiny, A. and Khorasani, N. (2009). Assessment of changes in urban green spaces of Mashad city using satellite data. *International Journal of Applied Earth Observation and Geoinformation*, **11**, 431–8.

RCEP (2007). *The Urban Environment*. Norwich: Royal Commission on Environmental Pollution.

Sadler, J. P., Small, E. C., Fiszpan, H., Telfer, M. G. and Niemela, J. (2006). Investigating environmental variation and landscape characteristics of an urban–rural gradient using woodland carabid assemblages. *Journal of Biogeography*, **33**, 1126–38.

Sahely, H. R., Dudding, S. and Kennedy, C. A. (2003). Estimating the urban metabolism of Canadian cities: Greater Toronto Area case study. *Canadian Journal of Civil Engineering*, **30**, 468–83.

Sandford, M. P., Manley, P. N. and Murphy, D. D. (2008). Effects of urban development on ant communities: implications for ecosystem services and management. *Conservation Biology*, **23**, 131–41.

Small, E., Sadler, J. P. and Telfer, M. (2006). Do landscape factors affect brownfield carabid assemblages? *Science of the Total Environment*, **360**, 205–22.

Small, E. C., Sadler, J. P. and Telfer, M. G. (2003). Carabid beetle assemblages on urban derelict sites in Birmingham, U.K. *Journal of Insect Conservation*, **6**, 233–46.

Smith, R. M., Thompson, K., Hodgson, J. G., Warren, P. H. and Gaston, K. J. (2006a). Urban domestic gardens (IX): composition and richness of the vascular plant flora, and implications for native biodiversity. *Biological Conservation*, **129**, 312–22.

Smith, R. M., Warren, P. H., Thompson, K. and Gaston, K. J. (2006b). Urban domestic gardens (VI): environmental correlates of invertebrate species richness. *Biodiversity and Conservation*, **15**, 2415–38.

Smyth, F. (2005). Medical geography: therapeutic places, spaces and networks. *Progress in Human Geography*, **29**, 488–95.

Smyth, F. (2008). Medical geography: understanding health inequalities. *Progress in Human Geography*, **32**, 119–27.

Stanners, D. and Bourdeau, P. (1995). *Europe's Environment: The Dobris Assessment*. Copenhagen: The European Environment Agency.

Strauss, B. and Biedermann, R. (2006). Urban brownfields as temporary habitats: driving forces for the diversity of phytophagous insects. *Ecography*, **29**, 928–40.

Sukopp, H., Numata, M. and Huber, A. (1995). *Urban Ecology as the Basis of Urban Planning*. The Hague: SPB Academic Publications.

Swanwick, C., Dunnett, N. and Woolley, H. (2003). Nature, role and value of green space in towns and cities: an overview. *Built Environment*, **29**, 94–106.

Takano, T., Nakamura, K. and Watanabe, M. (2002). Urban residential environments and senior citizens' longevity in mega-city areas: the importance of walkable green space. *Journal of Epidemiology and Community Health*, **56**, 913–16.

Tratalos, J., Fuller, R. A., Warren, P. H., Davies, R. G. and Gaston, K. J. (2007). Urban form, biodiversity potential and ecosystem services. *Landscape and Urban Planning*, **83**, 308–17.

Tunstall, H. V. Z., Shaw, M. and Dorling, D. (2004). Places and health. *Journal of Epidemiology and Commununity Health*, **58**, 6–10.

Turner, T. (2006). Greenway planning in Britain: recent work and future plans. *Landscape and Urban Planning*, **76**, 240–51.

Tzoulas, K., Korpela, K., Venn, S. et al. (2007). Promoting ecosystem and human health in urban areas using green infrastructure: a literature review. *Landscape and Urban Planning*, **81**, 167–78.

Ulrich, R. S. (1984). View through a window may influence recovery from surgery. *Science*, **224**, 420–1.

Ulrich, R. S., Simons, R. F., Losito, B. D. and Fiorito, E. (1991). Stress recovery during exposure to natural and urban environments. *Journal of Environmental Psychology*, **11**, 201–30.

Urban Parks Forum (2001). *A Survey of Local Authority Owned Parks of Historic Interest.* Reading: DLTR.

Urban Task Force (1999). *Towards an Urban Renaissance, Report of the Urban Task Force.* London: Spon.

von Haaren, C. and Reich, M. (2006). The German way to greenways and habitat networks. *Landscape and Urban Planning*, **76**, 7–22.

Wang, X. (2009). Analysis of problems in urban green space system planning in China. *Journal of Forestry Research*, **20**, 79–82.

Wells, N. M. and Evans, G. W. (2003). Nearby nature: a buffer of life stress among rural children. *Environment and Behavior*, **35**, 311–30.

Whitford, V., Ennos, A. R. and Handley, J. F. (2001). City form and natural processes: indicators for the ecological performance of urban areas and their application to

Merseyside, U.K. *Landscape and Urban Planning*, **20**, 91–103.

Wilby, R. L. and Perry, G. L. W. (2006). Climate change, biodiversity and the urban environment: a critical review based on London, U.K. *Progress in Physical Geography*, **30**, 73–98.

Williams, N. S. G., Schwartz, M. W., Vesk, P. A. *et al.* (2009). A conceptual framework for predicting the effects of urban environments on floras. *Journal of Ecology*, **97**, 4–9.

Woodward, J. C., Eyre, M. D. and Luff, M. L. (2003). Beetles (Coleoptera) on brownfield sites in England: an important conservation resource? *Journal of Insect Conservation*, **7**, 223–31.

Yli-Pelkonen, V. and Niemelä, J. (2005). Linking ecological and social systems in cities: urban planning in Finland as a case. *Biodiversity and Conservation*, **14**, 1947–67.

CHAPTER ELEVEN

Integrating nature values in urban planning and design

ROBBERT SNEP AND PAUL OPDAM

Introduction

In several chapters of this book it is emphasised that urban ecosystems are under constant threat by ongoing urbanisation, by biological invasions and by human use. Other chapters, however, highlight the specific character of urban ecosystems (that by definition is indicated by the dominant role of urban dynamics) and their importance to human wellbeing and quality of life. This divergence illustrates two fundamentally different views of urban ecosystems: as remnants of (previously extensive) natural ecosystems threatened by ongoing human pressure, or as green patches with high use value to people who work and live in the city landscape. From the first perspective urban ecosystems are separated as much as possible from the urban system, and from the second they are an essential part of that system.

These divergent views on urban ecosystems illustrate the type of debate that can be expected from the variety of participants in an urban planning process. They may hold opposing views and economic stakes, but have to develop a common position on the desired future of the urban landscape, and what it would take to achieve this. In this chapter we consider how both the conservation (in terms of defending pieces of highly valued ecosystems against undesired changes) and the enhancement (in terms of creating additional ecological, social or economic value) of urban ecosystems can be integrated into decision making on urban development.

We consider urban ecosystems at two spatial scales. At the site level we distinguish a patch of a particular ecosystem type, interacting with its urban context. At the level of the urban landscape we distinguish a pattern of ecosystem sites, which may interact and thereby build an ecosystem network embedded in the urban matrix. This network may also interact, at a regional scale, with ecosystems situated outside the urban context. This distinction between scales is essential, because decision making at the local level is different from

Urban Ecology, ed. Kevin J. Gaston. Published by Cambridge University Press.
© British Ecological Society 2010.

that at the city level, occurring on a different timescale, with different actors using different criteria when identifying goals and measures. For example, biodiversity in a local park is affected by decisions on local management taken by the park manager. The size of the park, and the configuration of park sites across the city, are the result of city planning by the local authority.

How is nature in the urban context conceptualised?

Urban ecosystems differ from the rest of the urban landscape because their structure as well as their functioning is the result of an interplay between human and natural processes. Green elements dominate the site of an urban ecosystem, instead of stone, glass and concrete, and a significant part of the site has established without the direct action of humans. So from this perspective, nature is the opposite of human culture and technique. Planning for urban ecosystems is then about deciding on the amount, character and functioning of sites where natural processes dominate. The basis for this decision is value, attributed by citizens, entrepreneurs or politicians.

In this chapter we distinguish two essentially different definitions of value: the intrinsic value of natural assets, for example the value attributed by humans to the occurrence of rare species, and the use value, the benefit humans experience from urban ecosystems. Most current nature conservation takes intrinsic value as a goal. From that perspective, conservation can be defined as being a philosophy of managing the environment in a manner that does not despoil, exhaust or extinguish (Jordan 1995). Urban environments, however, are so much altered that only few remnant patches may resemble the natural situation. The major part of the urban landscape consists of built-up areas or designed urban green spaces in which natural mechanisms (e.g. ground and surface water flows, succession of vegetation) are predominantly replaced by anthropogenic structures and management (e.g. sewer systems, gardening). This affects the quality of the soil, water and air (e.g. Sukopp & Starfinger 1999; Li *et al.* 2001; Penttinen *et al.* 2001) and the local climate in terms of temperature and rainfall (Landsberg 1981; Oke 1982; Arnfield 2003). The functioning of the urban ecosystem is therefore much more influenced by human activities than in other ecosystems. Consequently, the prediction of, for example, the distribution of plant and animal species cannot be achieved predominantly from knowledge of natural driving-forces and disturbances. Therefore, the planning and management of urban ecosystems cannot simply draw on the experiences obtained in conservation planning in rural or natural landscapes, and much effort should be put into defining what is conserved or restored, with what purpose and against what reference.

Conservation of urban ecosystems may be legitimised by the occurrence of endangered plant and animal species (e.g. Mörtberg & Wallentinus 2000), and conservation measures may be aimed at sustaining the local populations.

Compared with rural and natural areas, the inherent value of urban environments is often considered to be low, based on the assumption that only a few urban species are listed as endangered. However, when urban areas have been thoroughly mapped for their biodiversity, it has become apparent that numerous endangered species can be found and that urban species richness is much higher than anticipated (e.g. Araújo 2003; Hope *et al.* 2003; Kühn *et al.* 2004). Nevertheless, urban ecosystems are less appreciated for their inherent value by policy- and decision-makers, and by the general public (e.g. Tucker *et al.* 2005).

A second justification for conserving urban ecosystems is that they bring nature into the mindset of urban people. People may learn the importance of nature and natural processes at first hand, and get to appreciate the value of nature. According to Murphy (1988), 'Our urban centers can be viewed as bellwethers of our global environmental fate. Our success at meeting the challenges of protecting biological diversity in urban areas is a good measure of our commitment to protect functioning ecosystems worldwide. If we cannot act as responsible stewards in our own backyards, the long-term prospects for biological diversity in the rest of this planet are grim indeed.' So, even if restoring the natural situation is physically impossible, urban environments can act as a demonstration area for nature conservation and restoration practices (see also McKinney 2002; Tucker *et al.* 2005).

The third justification for urban ecosystems is the services that they provide to the quality of the physical environment for human life. As described in earlier chapters (Chapters 9 and 10), urban ecosystems directly and indirectly increase the quality level for people to live and work in cities in terms of, for example, health, wellbeing, recreation opportunities, water retention and air pollution filtering.

In the remainder of this chapter, we will summarise current conservation practices, and subsequently present our views as to how nature conservation can become part of the integrated planning of urban landscapes.

A review of conservation practices
Current support for conservation in urban societies
Because of the close contact between urban ecosystems and citizens, one would expect these ecosystems to be a popular and broadly discussed issue in city planning. However, based on a review of 217 papers published in the journal *Conservation Biology* between 1995 and 1999, Miller and Hobbs (2002) state that so far little attention has been paid to conservation in human settlements. They argue that, for a variety of reasons, conservation has tended to focus on lands with a relatively small human presence, often dominated by resource extraction and agriculture. They therefore call for extra attention to areas where people live and work (Miller & Hobbs 2002). Their highly cited paper provides a view on the conservation of wildlife and other natural values in urban

environments. Opposing views have been put forward in response to this paper. Adams (2005), for example, illustrates that wildlife conservation in cities and villages has a long history, and includes a range of research and conservation activities taken in different countries throughout the world. Among other initiatives, Adams refers to the UNESCO-MAB (Man and Biosphere) programme which included the Urban-Forum in the United Kingdom, the Urban Wildlife Trusts in several UK cities, the pioneer work on urban vegetation by Sukopp in Berlin, Germany (Sukopp 1986), Durban's Metropolitan Open Space System (South Africa), the multiple-use green corridors in Singapore, the work of the Australian Research Centre for Urban Ecology (ARCUE) in Melbourne, Australia, and the Center for Urban Ecology of the US National Park Service. Adams shows that over time numerous institutions, both public and private, have recognised the value of urban nature, and therefore the need for conservation. This notion was confirmed at the international mayors' conference 'Local action for bio-diversity' (Bonn, Germany, May 2008), where the importance of paying attention to biodiversity conservation in cities, the place where most people live, was stressed. Adam's review also suggests that urban nature conservation has a distinct character. Education, human–wildlife interactions and the multi-functional use of wildlife habitat seem to play a more prominent role in urban nature conservation than in the conservation of other areas. Also, the 'urban' aspect means that many conservation initiatives are taken at the city level, often in close cooperation with local governments and citizens. Urban conservation practices are therefore more of a city than a regional or national event, are on a highly applied level, and most times include a range of societal and non-governmental stakeholders. This could explain why these initiatives are rarely described in journals like *Conservation Biology*. Miller and Hobbs (2002) do, however, bring up a good point by stressing the lack of interest by most ecologists, and conservation biologists in particular, in the urban environment.

Current strategies

Traditionally, conservation and development of urban nature may start from the perspective of water management, forest management, urban green management or wildlife management. For all, the perception of citizens towards urban nature plays an important role. Based on literature and practical experience we here distinguish three types of nature conservation and development practices in urban areas: traditional nature conservation, conservation and development of multi-functional urban green and water, and conservation-inclusive architecture (Table 11.1). This classification is based on the relation between the zoning of the area, how citizens perceive that zoning, and – as a result – the applied conservation or development practice. All three types of practice can take place both at the city and site level, but in general only the conservation and development of multi-functional urban green takes place at both.

Table 11.1 *Three types of nature conservation and development practices in urban areas.*

	Traditional nature conservation	Conservation and development of multi-functional green and water	Conservation-inclusive architecture
Zoning	Conservation of biodiversity and natural resources is a main goal of the area	Conservation of biodiversity and natural resources is only one of the goals of the area	Conservation of biodiversity and natural resources is not a goal of the area, but only an extra value (on top of the targeted (socio)economic or other values)
Value	The intrinsic value of nature is emphasised; use value has less priority	The use value of nature is emphasised; intrinsic value has less priority	Both use and intrinsic value can be important, depending on the actual practice
Type of action	Conservation measures are similarly applied as in other, 'more natural', areas outside the city	Conservation practices are tuned with the other green functions of the zoning area	Conservation practices are tuned with the architectural functions of the zoning area
Appearance	The zoning area is (looks like) a remnant of a natural area	The zoning area looks like 'well-managed urban green or water'	The zoning area is a built-up area with extra attention for biodiversity conservation
Location	These areas are often located at the margin of the urban landscape	These areas are often located in between the built-up areas within the urban landscape	These areas are located within the built-up area of the urban landscape

The other two types of practice are implemented at the site level. In Table 11.1 we describe the three classes in terms of zoning (how are the areas labelled?), value (is the emphasis on intrinsic or user value?), type of action (what is the character of the conservation or development actions?), appearance (what does the area look like?) and location (in what part of the city the action is taken?).

We now further examine these urban nature conservation and development practices, using existing cases to illustrate what type of activities are included.

Traditional nature conservation practices applied in urban landscapes

Although cities contain 'typical urban habitats' (e.g. Sukopp & Weiler 1988) with associated species, such as the well-known feral pigeon *Columba livia* (Murton *et al.* 1972), these species occur in cities because urban environments may resemble their natural habitats (Wheater 1999). Plant and animal species that have colonised urban areas have recognised the habitat opportunities that cities offer and have adapted their lifestyle to the urban situation (e.g. Moller 2009). In terms of conservation, this means that urban environments – although unique as habitat – are not the only places where a species is occurring. If conservation of a particular species is targeted, this can also be done in rural areas where implementing conservation measures is probably easier and cheaper (e.g. more land is available for conservation). Conservation in urban environments may therefore be much more based upon the demand from citizens to protect local nature than considered from a nationwide perspective on species conservation.

Nature conservation in cities and towns is often focused on (i) preserving native vegetation, or (ii) conservation and restoration of wildlife habitat. We here describe some examples of these practices.

(i) Preservation of native vegetation

Preserving native vegetation requires considerable efforts from green managers, because urban environments often appear to favour exotic above native species. Intensive management to protect native vegetation may only be possible (or fundable) if protecting this vegetation is of interest to citizens and other parties. Here we provide some examples.

In a fragmented forest in Nishinomiya Shrine, southeastern Hyogo Prefecture, Japan, active vegetation management was used to control the abundant invasive plant species (e.g. *Trachycarpus fortunei*). In 2005, restoration measures were undertaken to remove all individuals of *T. fortunei* from the forest, resulting in increased canopy openness in the understorey. However, as the understorey lacked seedlings and saplings of native species, re-vegetation may additionally be necessary to accelerate re-establishment of native species in this and other urban forest areas that are heavily invaded by exotic species (Ishii & Iwasaki 2008).

In Wisconsin, USA, Wilcox *et al.* (2007) investigated whether the use of herbicide with burning, clipping or seeding could reduce the cover of the invasive perennial grass *Phalaris arundinacea* (reed canarygrass) and increase the cover of native species in an urban wet prairie that receives stormwater runoff. Although initially the use of herbicides decreased the abundance of *P. arundinacea* and led to an increased cover of native species, two years after treatment the exotic plant species recovered and its abundance was no different from control plots. The researchers conclude that land managers should be selective in allocating efforts to control monotypes of invasive species. Where wetlands continually receive stormwater, the replacement of invasive monotypes by native species will be difficult and the effort required might be prohibitively expensive and ineffective in the long term. Highest priority should go to sites that can best support native species restoration.

(ii) Conservation and restoration of wildlife habitat

Practices of wildlife conservation in urban areas resemble those in rural environments. The conservation of specific species and their habitats is the main goal of the measures taken, and 'urban' characteristics such as housing development and intensive human presence are considered as threats rather than as part of the dynamics of the urban landscape. These practices put the emphasis on 'conservation' and 'restoration'. For example, Brawn and Stotz (2001) describe the case of the Chicago Wilderness initiative in which 60 public and private organisations undertake efforts to conserve and restore habitats for breeding and migratory birds within the Chicago metropolitan region, USA. They conclude that habitat protection in otherwise developed metropolitan areas may offer significant conservation opportunities in regions where natural habitat is scarce. Similarly, the London Biodiversity Action Plan (LBAP) details plans to enhance the survival of rare and valued species associated with some 15 London habitats such as woodland, heathland, railway line sides, private gardens, cemeteries and wastelands (Harrison & Davies 2002). This action plan is part of the UK-wide Biodiversity Action Plans (BAP), in which conservation of endangered plant and animal species and their habitats are targeted. For the Greater London area, species like the peregrine falcon *Falco peregrinus*, house sparrow *Passer domesticus* and black redstart *Phoenicurus ochruros* are listed, and specific conservation measures such as habitat restoration are defined (see the website http://www.lbp.org.uk/londonap.html). These measures include contacting owners of suitable buildings for construction of nesting sites, education of the public in the species' habitat requirements, and monitoring and studying the London populations.

In South Africa, Gauteng's last mountain wildlife ecosystem is being decimated by ongoing urbanisation from cities like Johannesburg. Environmental NGOs aim to protect the natural ecosystems on which the locally present birds of prey (including the black eagle *Aquila verreauxii*) and other wildlife species depend for their continued survival (http://www.blackeagles.co.za). In 2008, several plans (including an 'Urban Wildlife Reserve') were developed to conserve raptors occurring at the edge of the cities. The proposed Urban Wildlife Reserve aims to protect fauna and flora currently threatened by development, by creating a unique sanctuary within the Johannesburg Metropolitan Area.

Conservation and development of multi-functional green and water

In urban green planning, design and management, there are two concepts that address the conservation of natural values as part of multi-functional green or water at the city level, those being the greenway concept and the concept of urban stream rehabilitation. Both take the use value of the urban green or water as a starting point.

Greenways are networks of land containing linear elements that are planned, designed and managed for multiple purposes including ecological, recreational, cultural, aesthetic or other purposes compatible with the concept of sustainable land-use (Ahern 1995). 'Multi-objective' greenways address needs of wildlife, flood damage reduction, water quality, education and other infrastructure in addition to urban beautification and recreation (Searns 1995). These greenways may also include streams and other water bodies (Asakawa *et al.* 2004). In urban environments, greenways can support the occurrence of wildlife. Bryant (2006) demonstrated the critical role of ecological greenways and parks in addressing the need to conserve biodiversity. More specifically, Mason *et al.* (2007) illustrated that, if appropriately designed, greenways may provide habitat for neotropical migrants, insectivores and forest-interior specialist birds that decrease in diversity and abundance as a result of suburban development. They recommend that 'landscape and urban planners can facilitate the conservation of development-sensitive birds in greenways by minimising the width of the trail and associated mowed and landscaped surfaces adjacent to the trail, locating trails near the edge of greenway forest corridors, and giving priority to the protection of greenway corridors at least 100 m wide with low levels of impervious surface (pavement, buildings) and bare earth in the adjacent landscape'. In general, the extent to which urban open space corridors meet recreational, habitat and wildlife needs is highly dependent on the degree to which the interactive effects of these uses are identified and incorporated into planning, design and management decisions (Briffet 2001). For managers who aim to enhance compatibility between people and wildlife, the challenge is to bridge the gap between established practice and a more multi-functional approach. Ultimately the aim should be to put in place an enlightened and sympathetic management regime that supports indigenous biodiversity maintainable at an acceptable cost and is compatible with providing sufficient access to cater for a wide range of recreational activities.

Whereas greenways are networks of land planned, designed and managed for multi-functional purposes, stream rehabilitation is more focused on existing urban water networks with ecological roles that have over the years been neglected. If the rehabilitation of ecological functioning is the goal, urban streams may better support the intrinsic and use value of the urban ecosystem. For example, Larned *et al.* (2006) illustrated the effects on invertebrates and non-native macrophytes of transplanting patches of native macrophytes into a 230-m-long stream section. In this experimental urban water restoration project, native macrophytes seemed to limit the spread of non-natives, which were absent in the planted section by the second spring. Native macrophyte establishment did not enhance invertebrate communities as predicted; few invertebrate metrics differed significantly between planted and unplanted sections. Pollution- and sediment-tolerant invertebrate taxa were abundant in both sections, suggesting that invertebrate colonisation was limited by water quality or sedimentation, not macrophyte composition. Survey respondents considered the stream to be visually and ecologically improved after rehabilitation, and macrophyte establishment was generally considered positive or neutral. According to Booth *et al.* (2004), successful stream rehabilitation requires coordinated diagnosis of the causes of degradation and integrative management to treat the range of ecological stressors within each urban area, and it depends on appropriate remedies at scales from backyards to regional stormwater systems. Recent studies of urban impacts on streams in Melbourne, Australia, suggested that the primary degrading process in many urban areas is effective imperviousness, the proportion of a catchment covered by impervious surfaces directly connected to the stream by stormwater drainage pipes (Walsh *et al.* 2005). This implies that even small rainfall events can produce sufficient surface runoff to cause frequent disturbance through regular delivery of water and pollutants; where impervious surfaces are not directly connected to streams, small rainfall events are intercepted and infiltrated. Walsh *et al.* (2005) showed, in a sample catchment, that it is possible to redesign the drainage system to reduce effective imperviousness to a level at which the models predict detectable improvement in most ecological indicators. Distributed, low-impact design measures are required that intercept rainfall from small events and then facilitate its infiltration, evaporation, transpiration or storage for later in-house use.

Conservation-inclusive architecture

In discussions of the opportunities for conservation and development of biodiversity and other ecosystem services in city environments, the focus has mostly been on the non-built-up parts. It is only over the past 20 years that the conservation value of buildings and other built-up structures present in urban areas has begun to be recognised. This increasing attention has originated from several directions, from architects aiming for ecological design (Todd & Todd 1994), from the emergence of ecological engineering as a discipline (Shijun 1985; Gattie *et al.* 2003) and from biodiversity conservation actions in urban environments (e.g. the EU Countdown 2010 programme).

Over several decades, architects have developed an interest in integrating ecology in the design of the buildings. This movement was called 'green design' (1980s), later 'ecological design' (1990s) and these days 'sustainable design', representing an increasing broadening of scope in theory and practice (Madge 1997). Well-known examples of the 'greening of buildings' are 'Le Mur Vegetal' by Patrick Blanc (complex vegetation structures growing vertically, attached to several buildings in Paris, France) and the green roof of the Chicago City Hall (an impressive rooftop garden in the centre of Chicago; Velasquez 2005).

Whereas architects focus on buildings and combine visual appearance with functional quality, engineers are more interested in innovative technical systems that improve both human experiences and wellbeing, and environmental qualities. According to Bergen *et al.* (2001), ecological engineering is 'the design of sustainable systems, consistent with ecological principles, which integrate human society with its natural environment for the benefit of both. It recognises the relationship of organisms (including humans) with their environment and the constraints on design imposed by the complexity, variability and uncertainty inherent to natural systems.' In the built-up environment engineers search for new ways to deal with water retention, air quality, energy saving, noise reduction and other issues to improve their liveability. As the ecological design of buildings may increase their sustainability, engineers have developed new building techniques that are inspired by ecosystem functioning. Just as for the architects, the concept of green roofs and living walls (Dunnett & Kingsbury 2004) is an important part of that, but here functionality is much more important than visual quality.

A third group that explores how buildings can support biodiversity and other ecosystem services is wildlife conservationists. As in recent years the conservation of biodiversity in the urban environment has received increasing attention, so now buildings and other built-up structures are – apart from their human functionality – also considered from a wildlife conservation point of view. Because of their size, shape, material and location, conventional buildings may mimic habitat conditions for particular species that originate from rocky and cliff environments. In addition, when 'green design' measures (green roofs, living walls) are added to buildings, the range of habitat opportunities is broadened, and more plant and animal species may find their habitat in or on these man-made structures (Brenneisen 2003, 2006; Grant *et al.* 2003). In more and more cities, wildlife conservationists actively encourage owners to integrate specific conservation measures for wildlife (e.g. nest boxes for breeding birds or bats) in their buildings (Baines 2000).

Incorporating nature into urban planning
Legitimacy of ecosystems as part of the city landscape

Spatial landscape planning is the organised process of attributing significance to landscape functions, making decisions on values to be created or preserved, and spatially organising and designing the physical structure of landscapes to ensure the provision of landscape services in the short and long term. City planning which includes ecological ('green') functions can be considered as urban landscape planning. It is fundamentally different from conservation planning, in that it considers a wide range of functions and values of multi-functional landscapes. Because urban development is guided by human values (Andersson 2006), urban landscapes are not planned primarily for rare species and habitats (such as in conservation planning).

If remnant patches of former large natural landscapes in cities are considered as threatened sites to be protected from the devastating influence of humankind, any conservation action is opposed to and conflicts with the human socioeconomic system (as for example in Breuste 2004). In this chapter, we take a different point of view, and consider ecosystems as part of an ecological–social system. This implies that ecosystems (the 'green' elements in the urban landscape) are considered as an intrinsic part of the urban system, and that planning of ecosystems in the urban landscape is about the values they could provide to the urban system as a whole. Therefore, although ecological principles determined in rural landscapes are applicable to urban landscapes (Savard et al. 2000), conservation planning rules as discussed in the literature (Cabeza & Moilanen 2001; Margules & Sarkar 2007; Pressey et al. 2007) cannot simply be transferred to urban landscapes. We propose that a transition from traditional conservation towards ecology as a cornerstone of sustainable urban development (as described by Wu 2008) creates opportunities to better position nature and biodiversity in urban planning.

The consequence of this view is that urban ecosystems are not planned and designed for protected species or native vegetation, but for the ecosystem services that the urban society cares for, at a level of intensity valued by that society, and on which government and private enterprises choose to spend money (Boland & Hunhammar 1999). Urban ecosystems are known to provide a series of values to inhabitants, including human health, social cohesion, regulation of air quality and temperature, stormwater regulation, nature perception and leisure activities (Jackson 2003; Chiesura 2004; Tzoulas et al. 2007; Miller 2008). Similar benefits were found in relation to employees and enterprises (Kaplan 2007; Snep et al. 2009). Biodiversity can be valued as a key factor in well-functioning ecosystems which provide ecosystem services in the long term (Hooper *et al.* 2005), as well as for the significance that humans attribute to observing wild animals and plants in their living environment.

Planning at three levels of scale

Planning of ecosystems within the urban landscape occurs at multiple spatial scales driven by the scale of influence of the decision-maker (Andersson 2006). We distinguish three levels of spatial scale: national or regional scale (depending on the scale at which governmental responsibilities for spatial planning are established), the scale of the city and that of the site. At the first level, strategic decisions are made about the density of buildings and infrastructure in comparison to the area of green structures. As shown by Tratalos *et al.* (2007), the contribution of ecosystems to the urban system is a matter of the proportion of green area within the city landscape: the more ecosystem area, the higher the level of ecosystem services. However, increasing the proportion of the green component within the city limits means that the urban zone expands further into the rural landscape (causing potentially detrimental impacts on rural values, including nature reserves). It also implies that humans have to cover greater distances between functional sites, which is potentially detrimental to environmental quality, increases travel costs and the emission of greenhouse gases, and may cause loss of social cohesion. Balancing these trade-offs across different spatial scales requires a fundamental debate at the highest level of spatial organisation, and might be organised by the government and national planning authorities.

At the city level, the amount and spatial pattern of ecosystems (parks, greenways, amenities and watercourses) is determined. Ecosystems in urban landscapes are often small and narrow and embedded in a stony environment with high-density road networks. Biodiversity and related ecosystem functioning is limited by the impact of fragmentation effects. An effective strategy to improve the values of ecosystems under pressure of fragmentation is to ensure that the individual sites are linked in a network in which they support each other (Opdam & Steingröver 2008) and increase the adaptive capacity of the ecosystems (Opdam *et al.* 2006). Key characteristics of such networks include the total network area, network density and the permeability of the urban space between the sites (Opdam *et al.* 2003). These spatial characteristics are largely determined at the level of city planning (Sukopp *et al.* 1995). This is also the case for the overall pattern of urban expansion: for example, along transport corridors or in satellite suburbs, or in concentric circles. The shape of the urban zone is significant for the level of biodiversity in the city centre. Ecosystem sites in the peri-urban zone may be functionally connected to the urban ecosystem network (Snep *et al.* 2006), which supposedly increases the species diversity within the city. An irregular shape with green wedges intruding into the urban zone ensures a better connectivity between rural and urban ecosystem sites than does a perfect circular shape. At this level of planning, the main actors are the city planners and the governmental authorities, in dialogue with NGOs and well-organised pressure groups.

At the site level, planning involves the identification of local targets and the design of the site, including decisions on how to combine incompatible functions spatially. The main players are governmental bodies, real estate enterprises, the local citizens and other interested parties. Here, the challenge is to incorporate higher-level interests into the local decision making, for example in the type of ecosystem which is developed. For example, does the site functionally contribute to the larger network, or is it so different that it is functionally isolated? A key issue is whether private partners are motivated to invest in public services.

Scale levels are also relevant in funding conservation measures. In many countries, nature conservation funding agencies do not consider the urban environment as an area of interest. This means that urban nature projects may not meet the requirements for (inter)national subsidies (e.g. because they lack the minimum area size, the right label ('conservation area') or the presence of sufficient target species). Urban nature measures are therefore often dependent on local funding sources that are not really equipped to support biodiversity conservation and development in the long term. This will limit opportunities to insert wildlife conservation targets in local city planning, because creating or conserving urban ecosystems will be considered as expensive extra costs that inhibit the opportunities for economic expansion. However, if urban ecosystem services are appreciated for their contribution to the economic and social values of the urban system, local funding could become available. This opportunity is illustrated by the study of Snep et al. (2009), who showed that biodiversity was not among the functions of the green office parks which were most preferred by stakeholders, but that plant and animal species could profit from the high preference for other urban green functions related to quality of human life.

Goal setting

The chances for optimising the spatial pattern of ecosystem sites are much greater if target setting is done early in the urban planning and design processes. If conservation and development actions are undertaken once major decisions concerning the layout of a specific area have already been completed, it is more difficult to tune the conservation actions with the surroundings and the (a)biotic conditions of the area, and consequently the urban ecosystem's functioning will be less effective. At the city level, a strategic vision of the desirable role of ecosystems and the level of biodiversity and ecosystem services is an important cornerstone in positioning ecosystems in negotiations about,

Table 11.2 *Different starting points for integrating biodiversity values in urban planning and design.*

	Ecological values are addressed early in the planning and development process	Ecological values are introduced afterwards, in existing situations
Emphasis on species targets	Conditions for required persistence of species are inserted into the planning process, and balanced with other targeted urban values	Conservation measures for biodiversity depend fully on the opportunities offered by the existing green space and building conditions, e.g. in parks
Emphasis on ecosystem services	Minimal requirements for chosen provision level of ecosystem services are inserted into the planning process; different services can be combined in the same green structure and unite different interest groups	Which ecosystem services can be provided and at what level fully depend on the opportunities offered by the existing conditions; no balancing of ecological, social and economic values possible

for example, how to structure new suburbs, and how to restructure office parks. Examples have been published in the scientific literature (e.g. Conine *et al.* 2004; Li *et al.* 2005), and many more can be found on internet websites. A vision may include a choice for target ecosystem types, target ecosystem services and aspiration levels. For example, an important decision is whether ecosystem types should be preferred that occur in the surrounding rural countryside, to create an opportunity to link the urban ecosystem network to the rural network and create a higher level of urban biodiversity. The vision also emphasises where weak links in the network exist, so that any opportunity that might occur in the dynamic urban landscape to place an ecosystem patch in a reconstruction plan can be used. If ecosystem conservation and development are not a part of the planning and design process, the emphasis is on what is still possible within the constraints set by the design, and there is no drive and no space to create more effective solutions (Table 11.2).

Legal aspects need to be considered at an early stage of planning. City environments are multi-stakeholder centres, with highly dynamic property ownership and land-use. Developments, even if they aim at creating opportunities for ecosystems, may include destruction of habitat, and the presence of protected plants and animals in the target area will provide environmental non-governmental organisations with an argument to stop or delay development projects. Again, a strategic vision on the quality of life that is well established in the urban society may facilitate the debate about why and how to adapt the metropolitan landscape to future needs.

Planning green and blue structures for ecosystem networks

Urban biodiversity and ecosystems are best conserved and developed in the city's green structure, because ecosystems are allowed more time to develop and because the spatial configuration of ecosystem sites can better be optimised. A key feature in urban ecosystems is spatial dynamics: urban development may destroy sites, but at the same time may offer new opportunities. Ecosystem networks (and greenways) offer a spatial structure that can better deal with such dynamics. In efforts to optimise the functioning of this green structure for urban biodiversity and ecosystem services, several key principles based on metapopulation ecology need to be considered (Opdam & Steingröver 2008).

Size matters

Habitat size is an important factor determining urban biodiversity (Cornelis & Hermy 2004; Donnelly & Marzluff 2004; Opdam & Steingröver 2008; Palmer *et al.* 2008). Larger patches mean reduced edge effects (disturbances by adjacent land-use), opportunities for species with larger area requirements, and larger local populations of species. Species groups may differ in their response to changing patch size (Godefroid & Koedam 2003). Mason *et al.* (2007) distinguished optimal width of greenways for the occurrence of different groups of bird species, for example urban adaptors, forest interior and ground-nesting species. Again, this emphasises the importance of goal setting. Also the functional requirements of ecosystem services which are not based on biodiversity may depend on the size of green elements, for example for climate regulation (see Goméz *et al.* 2001 for a quantitative approach to relate human comfort to the size of green zones in Valencia, Spain).

Large patches increase value for money

Verboom *et al.* (2001) showed that in an ecosystem network, a relatively large patch allowed smaller total network areas to offer sustainable conditions for metapopulations. Larger remnants of (semi-)native vegetation are the obvious focal elements in any spatial green or blue structure. Measures to increase their supportive role include enlargement of these sites and improved connectivity to other sites (Opdam & Steingröver 2008; Palmer *et al.* 2008). While this may be a difficult task in many older cities where space is limited (e.g. Fernández-Juricic & Jokimäki 2001), it can be a valid approach within newer cities, such as those in Australia, or within recently developed peri-urban areas.

Connectivity

Connectivity is a popular concept among planners and landscape architects, but its application is still too often based on structural considerations, instead of a functional analysis of ecosystem network structure. Improving connectivity may be a solution to the total network area not being adequate for a persistent population of a target species, or for raising the biodiversity level in a park area in the city centre. Structural measures, such as a green corridor or stepping stones that facilitate dispersal, may improve the spatial cohesion of the network. Alternatively, measures that increase the carrying capacity of the network patches (through improving habitat quality and enlarging sites) may contribute to achieving the same end (Opdam & Steingröver 2008). In the city landscape, road and water infrastructure may be used as a physical template to improve connectivity for aquatic, marsh and terrestrial ecosystems (Adams & Dove 1989).

City edge
The city edge is the most dynamic part of the urban landscape, with a constant change in land-use. Owing to these dynamics temporary habitats often occur that can have unique biodiversity levels, suggesting that a spatial development vision for the city edge from the point of view of pioneer ecosystems may be important. Also, in the process of urban sprawl, today's city edge will be well within the urban zone in a couple of decades. Therefore, any decision on incorporating parts of today's city edge into a future urban ecosystem network has to be anticipated at an early stage of development.

Peri-urban areas
Rural habitat patches located adjacent to the city (peri-urban) may act as source areas for inner city nature (e.g. ants in urban sites are more abundant where natural areas are near; Pacheco & Vasconcelos 2007). Key features are size and ecosystem type, as well as connectivity, for example supported by (rail)road verges (Snep et al. 2006). Green wedges protruding into the city may increase the functional connectivity between the inner city part of the ecosystem network and the surrounding rural landscape. They also improve the liveability of city environments for human beings (see for example Li et al. 2005).

Incorporating land-use dynamics in the ecosystem network
Because of its functional cohesion, an ecosystem network can stand a certain level of turnover of patches without losing sustainability. However, there is a relation between the turnover rate and the total network area required for sustainability (A. J. A. Van Teeffelen et al., unpublished results). In ecosystem types with a long development time, compensating loss of area by developing an equal amount elsewhere is not sufficient as a compensation measure, since it takes a long time before the newly developed patch is a functional part of the network. If turnover is foreseen, the total area of the network should be expanded to ensure that the network stays sustainable. This suggests that

within city landscapes it is efficient to keep part of the urban network stabilised, and to connect this network with networks in the peri-urban zone. Pioneer species often colonise new urban development lots. By including those opportunities in the urban planning process and connecting the pioneer habitats with existing networks, added value for urban biodiversity conservation can be obtained. For business parks, industrial estates and port areas, such an approach has been elaborated into the 'habitat backbone' strategy, as applied in the case of the natterjack toad *Bufo calamita* in the Port of Antwerp, Belgium (Snep & Ottburg 2008).

Making use of the opportunities for temporary habitats
When urban green is addressed in city planning, only those green structures are discussed that will last for many years. City environments, however, are highly dynamic in land-use, creating numerous temporary opportunities for the conservation and development of natural values. Locations within or at the edge of the urban matrix that are not yet (re)developed may – for several months or years – offer habitat for a wide range of pioneer plants and animals, including threatened species (Breuste 2004). Typical areas for temporary nature are located at older industrial sites, along (rail)road tracks and at planned but not yet developed sites on the city edge (Straus & Biedermann 2006). By including those temporary opportunities in the city planning process, an urban green structure with permanent and temporary elements can be defined such that both types of green may strengthen each other. By doing so, the opportunities for early-successional habitats are better utilised, which will lead to an increase in the overall conservation value of the city environment.

Planning and management of urban ecosystem sites
The effectiveness of measures at the site scale fully depends on how the spatial pattern at a higher level supports the functional connectivity with other ecosystem patches elsewhere. Local targets and measures have to be consistent with the higher-level vision and strategies. We highlight here several issues frequently discussed at the site level with opportunities to enhance nature values in the planning and design process.

Business sites: contributing to common societal values
Business sites (e.g. business parks, industrial estates and distribution centres) are currently not designed and managed to provide added values in terms of sustainability to their (urban) surroundings. Snep (2009) explored options and opportunities in planning, design and management of business sites to accommodate ecosystems as a source of ecosystem services to their surroundings. His findings illustrate the significance that business sites can have for biodiversity

Figure 11.1 An office site design in Groningen, the Netherlands, in which specific biodiversity conservation measures (such as a bat cave in the building and specific vegetation for birds and butterflies in the green surrounding the building) are incorporated in the site design. This site is actually (2008–11) being developed, including the measures for plant and animal conservation. Artist's impression: UNStudio and Lodewijk Baljon Landscape architects, by order of DUO[2].

conservation, based on their location, land-use and urban design. Snep shows that by implementing different 'green design' measures, business sites may be able to support biodiversity in adjacent areas, improve the liveability of neighbouring residential areas and create a better quality for work at the business sites (e.g. Figure 11.1). He demonstrates that, in terms of conservation value, business sites are able to support the viability of threatened species, especially those linked with early-successional vegetation (e.g. Snep & Ottburg 2008). Implementing green roofs and opportunities for temporary habitats, and managing the existing business site green in a more ecological way are among the green design measures that could help to make use of the opportunities that business sites potentially offer for the conservation and development of urban ecosystems.

Buildings: green architecture
Built structures offer different opportunities for the conservation and development of natural values. Green roofs are capable of supporting a whole range of ecosystem services including water retention, energy savings, wildlife habitat,

Figure 11.2 Like the well-known Hanging Gardens of Babylon, this urban green design in the inner city of Eindhoven, the Netherlands, illustrates how highly built-up areas can still offer opportunities for conservation. The mix of plant species used at this location are selected on both visual quality and habitat suitability for bees, butterflies and birds. Photograph used with permission of Soontiëns Gardening Eindhoven.

sustainable use of building materials, and diminishing air pollution and urban heat effects (Oberndorfer *et al.* 2007). Concerning their conservation value for biodiversity, the substrate depth (not too shallow), soil origin (aim for natural soils from the direct surroundings of the roof location) and vegetation structure (aim for variation in species mix and height of vegetation) of green roofs are important factors (Kadas 2002; Brenneisen 2006; Dunnett *et al.* 2008; Emilsson 2008). Under the right conditions, green roofs can have a significant value for the conservation of invertebrates (Jones 2002) and breeding birds (Baumann 2006). Additionally, in some cases they can also increase wildlife experience in urban environments (Figure 11.2).

With living walls (planted facades of buildings) there are also habitat opportunities for wildlife, according to Köhler (2008). He specifically mentions wood-adapted birds and spiders, beetles and others preferring thermophile or synanthropic vegetation (species adapted to urban settings, including those preferring warm temperatures; Klausnitzer 1993).

Finally, by including specific spaces in the actual building, birds, bats and other wildlife may find nesting and hibernation opportunities. These days there is a small industry that produces all types of 'building blocks' for housing development (e.g. custom-made roof tiles with nesting places for birds) so that these opportunities can be created on a large scale.

Figure 11.3 Small parcel of urban green in a highly built-up setting (business site). Traditionally such an area will be developed and managed as lawn with some trees (left), but with ecological management the same area could be designed in such a way that both wildlife and citizens might much better appreciate this green site (potential design, right). Reproduced with permission from Soontiëns Gardening Eindhoven.

Urban green management

Citizens appreciate wildlife in their direct living environment (with some exceptions like gulls, starlings, feral pigeons), with diversity being more important than density (Clergeau *et al.* 2001). To support a wide variety of urban wildlife, great variation in the management of urban vegetation is required. Offering a broad range of urban green types and using different methods to manage the urban green will attract a high diversity of plant and animal species. However, the management practice for a city's public green is usually rather straightforward, with cost-effectiveness and traditions in green management being the main drivers. This results in a dominance by lawns and street trees, and a shortage of shrub and brushwood vegetation. The latter two, especially, include many opportunities for wildlife. If in the management of the urban green particular attention is being paid to habitat opportunities for wildlife, not only plant and animal species will benefit but also the people who like to enjoy the city's wildlife (e.g. Figure 11.3).

Gardens: private management for common values

Some recent visions of urban development focus on expansion within the city limits, sacrificing existing urban green spaces (like gardens) for new housing or business developments. Although a careful handling of the remnant open space in peri-urban areas is in itself a good idea, one should also consider the value of existing green for ecosystems and nature experience in cities (Figure 11.4). Among urban green, gardens 'constitute a considerable proportion of "green space" in urban areas and are therefore of potential significance for maintaining biodiversity and ecosystem service provision in such areas' (Loram *et al.* 2007). For example, Wheater (1999) addresses the benefits of

Figure 11.4 Communal garden in 'eco' neighbourhood (De Kersentuin, Leidschenrijn, the Netherlands): citizens share part of their gardens with their neighbours, thereby creating semi-private green spaces that offer great places for both ecosystems and nature experience.

garden plants (including many non-native plants) for producing nectar and food (even in early spring and autumn) for bees, bumblebees, hoverflies and birds. Baines (2000) mentions that gardens can optimally contribute to the city's biodiversity if managed in an ecological way, by so-called 'wildlife gardening'. Wildlife gardening can be broadly defined to encompass any actions conducted in private or domestic gardens to increase their suitability for wildlife, and thus includes the provision of a diversity of resources (e.g. substrates, food, breeding and overwintering sites; Davies *et al.* 2009). Concerning these actions, Gaston *et al.* (2005) observed that 'whilst some methods for increasing the biodiversity of garden environments may be very effective, others have a low probability of success on the timescales and spatial scales likely to be acceptable to many garden owners. If one of the functions of small scale biodiversity enhancement is to develop and encourage awareness of biodiversity and its conservation, then encouragement to conduct particular activities must be balanced with a realistic appraisal of their likely success.'

Integrating nature values in urban planning and design

So far, city development and design has mainly been the territory of urban planners, project developers and architects. Consequently, the attention and budget have been focused on the built-up part of cities, with 'green' considered as 'accessories' in architecture and as 'necessary recreation area' in between the housing blocks. However, as we think of humans being part of the global ecosystem, one could see the city as a local ecosystem where people, plants and animals coexist, offering each other 'ecosystem services'. From the viewpoints of climate change, sustainability and urban 'liveability', such an idea would offer humans a whole range of physical and social advantages (e.g. the impact of urban green in city temperature buffering). Unfortunately, most of us have probably not yet linked ecosystem functioning with the quality of human city life. In particular, getting urban planners, project developers and architects interested in the ecosystem perspective seems the largest challenge (besides informing the general public). If in the future these urban professionals are willing to broaden their current point of view with the ecosystem approach, new ways to combine 'red' and 'green' in the city will be asked for. For now, this chapter provides a whole array of clues as to how to integrate nature values in current urban planning and design.

References

Adams, L. W. (2005). Urban wildlife ecology and conservation: a brief history of the discipline. *Urban Ecosystems*, **8**, 139–56.

Adams, L. W. and Dove, L. E. (1989). *Wildlife Reserves and Corridors in the Urban Environment: A Guide to Ecological Landscape Planning and Resource Conservation*. Columbia, MD: National Institute of Urban Wildlife.

Ahern, J. (1995). Greenways as a planning strategy. *Landscape and Urban Planning*, **33**, 131–55.

Andersson, E. (2006). Urban landscapes and sustainable cities. *Ecology and Society*, **11**, 34. Available at http://www.ecology and society.org/vol11/iss1/art34/

Araújo, M. B. (2003). The coincidence of people and biodiversity in Europe. *Global Ecology and Biogeography*, **12**, 5–12.

Arnfield, A. J. (2003). Two decades of urban climate research: a review of turbulence, exchanges of energy and water, and the urban heat island. *International Journal of Climatology*, **23**, 1–26.

Asakawa, S., Yoshida, K. and Yabe, K. (2004). Perceptions of urban stream corridors within the greenway system of Sapporo, Japan. *Landscape and Urban Planning*, **68**, 167–82.

Baines, C. (2000). *How to Make a Wildlife Garden*. London: Frances Lincoln.

Baumann, N. (2006). Ground-nesting birds on green roofs in Switzerland: preliminary observations. *Urban Habitats*, **4**, 37–50.

Bergen, S. D., Bolton, S. M. and Fridley, J. L. (2001). Design principles for ecological engineering. *Ecological Engineering*, **18**, 201–10.

Boland, P. and Hunhammar, S. (1999). Ecosystem services in urban areas. *Ecological Economics*, **29**, 293–301.

Booth, D. B., Karr, J. R., Schauman, S. *et al.* (2004). Reviving urban streams: land use, hydrology, biology, and human behaviour. *Journal of the American Water Resources Association*, **40**, 1351–64.

Brawn, J. D. and Stotz, D. E. (2001). The importance of the Chicago region and the Chicago Wilderness initiative for avian conservation. In J. M. Marzluff, R. Bowman and R. Donnelly, eds., *Avian Ecology and Conservation in an Urbanizing World*. New York: Kluwer, pp. 509–22.

Brenneisen, S. (2003). The benefits of biodiversity from green roofs: key design consequences. In *Proceedings of the First Annual Greening Rooftops for Sustainable Communities Conference, Awards and Trade Show*

(Chicago, 2003). Toronto: Green Roofs for Healthy Cities.

Brenneisen, S. (2006). Space for wildlife: designing green roofs for habitat in Switzerland. *Urban Habitats*, **4**, 27–36.

Breuste, J. (2004). Decision making, planning and design for the conservation of indigenous vegetation within urban development. *Landscape and Urban Planning*, **68**, 439–52.

Briffett, C. (2001). Is managed recreational use compatible with effective habitat and wildlife occurrence in urban open space corridor systems? *Landscape Research*, **26**, 137–63.

Bryant, M. M. (2006). Urban landscape conservation and the role of ecological greenways at local and metropolitan scales. *Landscape and Urban Planning*, **76**, 23–44.

Cabeza, M. and Moilanen, A. (2001). Design of reserve networks and the persistence of biodiversity. *Trends in Ecology and Evolution*, **16**, 242–8.

Chiesura, A. (2004). The role of urban parks for the sustainable city. *Landscape and Urban Planning*, **68**, 129–38.

Clergeau, P., Mennechez, G., Sauvage, A. and Lemoine, A. (2001). Human perception and appreciation of birds: a motivation for wildlife conservation in urban environments of France. In J. M. Marzluff, R. Bowman and R. Donnelly, eds., *Avian Ecology in an Urbanizing World*. Norwell, MA: Kluwer, pp. 69–86.

Conine, A., Xiang, W.-N., Young, J. and Whitley, D. (2004). Planning for multipurpose greenways in Concord, North Carolina. *Landscape and Urban Planning*, **68**, 271–87.

Cornelis, J. and Hermy, M. (2004). Biodiversity relationships in urban and suburban parks in Flanders. *Landscape and Urban Planning*, **69**, 385–401.

Davies, Z. G., Fuller, R. A., Loram, A. *et al.* (2009). A national scale inventory of resource provision for biodiversity within domestic gardens. *Biological Conservation*, **142**, 761–71.

Donnelly, R. and Marzluff, J. M. (2004). Importance of reserve size and landscape context to urban bird conservation. *Conservation Biology*, **18**, 733–45.

Dunnett, N. P. and Kingsbury, N. (2004). *Planting Green Roofs and Living Walls*. Portland, OR: Timber Press.

Dunnett, N., Nagase, A. and Hallam, A. (2008). The dynamics of planted and colonising species on a green roof over six growing seasons 2001–2006: influence of substrate depth. *Urban Ecosystems*, **11**, 373–84.

Emilsson, T. (2008). Vegetation development on extensive vegetated green roofs: influence of substrate composition, establishment method and species mix. *Ecological Engineering*, **33**, 265–77.

Fernández-Juricic, E. and Jokimäki, J. (2001). A habitat island approach to conserving birds in urban landscapes: case studies from southern and northern Europe. *Biodiversity and Conservation*, **10**, 2023–43.

Gaston, K. J., Smith, R. M., Thompson, K. and Warren, P. H. (2005). Urban domestic gardens (II): experimental tests of methods for increasing biodiversity. *Biodiversity and Conservation*, **14**, 395–413.

Gattie, D. K., Smith, M. C., Tollner, E. W. and McCutcheon, S. C. (2003). The emergence of ecological engineering as a discipline. *Ecological Engineering*, **20**, 409–20.

Godefroid, S. and Koedam, N. (2003). How important are large vs. small forest remnants for the conservation of the woodland flora in an urban context? *Global Ecology and Biogeography*, **12**, 287–98.

Gómez, F., Tamarit, N. and Jabaloyes, J. (2001). Green zones, bioclimatic studies and human comfort in the future development of urban planning. *Landscape and Urban Planning*, **55**, 151–62.

Grant, G., Engleback, L. and Nicholson, B. (2003). *Green Roofs: Existing Status and Potential for Conserving Biodiversity in Urban Areas*. English Nature Research Report No. 498. Peterborough, UK: English Nature.

Harrison, C. and Davies, G. (2002). Conserving biodiversity that matters: practitioners' perspectives on brownfield development and urban nature conservation in London. *Journal of Environmental Management*, **65**, 95–108.

Hooper, D. U., Chapin, F. S. III, Ewel, J. J. *et al.* (2005). Effects of biodiversity on ecosystem functioning: a consensus of current knowledge. *Ecological Monographs*, **75**, 3–35.

Hope, D., Gries, C., Zhu, W. X. *et al.* (2003). Socioeconomics drive urban plant diversity. *Proceedings of the National Academy of Sciences of the USA*, **100**, 8788–92.

Ishii, H. T. and Iwasaki, A. (2008). Ecological restoration of a fragmented urban shrine forest in southeastern Hyogo Prefecture, Japan: initial effects of the removal of invasive *Trachycarpus fortunei*. *Urban Ecosystems*, **11**, 309–16.

Jackson, L. (2003). The relationship of urban design to human health and condition. *Landscape and Urban Planning*, **64**, 191–200.

Jones, R. A. (2002). *Tecticolous Invertebrates: A Preliminary Investigation of the Invertebrate Fauna on Ecoroofs in Urban London*. London: English Nature.

Jordan, C. F. (1995). *Conservation: Replacing Quantity with Quality as a Goal for Global Management*. Chichester: John Wiley and Sons.

Kadas, G. (2002). Study of invertebrates on green roofs: how roof design can maximize biodiversity in an urban environment. Unpublished M.Sc. thesis, University College London, England.

Kaplan, R. (2007). Employees' reactions to nearby nature at their workplace: the wild and the tame. *Landscape and Urban Planning*, **82**, 17–24.

Klausnitzer, B. (1993). *Ökologie der Großstadtfauna*. G. Fischer.

Köhler, M. (2008). Green facades – a view back and some visions. *Urban Ecosystems*, **11**, 423–36.

Kühn, I., Brandl, R. and Klotz, S. (2004). The flora of German cities is naturally species rich. *Evolutionary Ecology Research*, **6**, 749–64.

Landsberg, H. (1981). *The Urban Climate*. International Geophysics Series 28. New York: Academic Press.

Larned, S. T., Suren, A. S., Flanagan, M., Biggs, B. J. F. and Riis, T. (2006). Macrophytes in urban stream rehabilitation: establishment, ecological effects, and public perception. *Restoration Ecology*, **14**, 429–40.

Li, F., Wang, R., Paulussen, J. and Liu, X. (2005). Comprehensive concept planning of urban greening based on ecological principles: a case study in Beijing, China. *Landscape and Urban Planning*, **72**, 325–36.

Li, X., Poon, C.-S. and Liu, P. S. (2001). Heavy metal contamination of urban soils and street dusts in Hong Kong. *Applied Geochemistry*, **16**, 1361–8.

Loram, A., Tratalos, J., Warren, P. H. and Gaston, K. J. (2007). Urban domestic gardens (X): the extent and structure of the resource in five major cities. *Landscape Ecology*, **22**, 601–15.

Madge, P. (1997). Ecological design: a new critique. *Design Issues*, **13**, 44–54.

Margules, C. and Sarkar, S. (2007). *Systematic Conservation Planning*. Cambridge: Cambridge University Press.

Mason, J., Moorman, C., Hess, G. and Sinclair, K. (2007). Designing suburban greenways to provide habitat for forest-breeding birds. *Landscape and Urban Planning*, **80**, 153–64.

McKinney, M. L. (2002). Urbanization, biodiversity, and conservation. *BioScience*, **52**, 883–90.

Miller, J. R. (2008). Conserving biodiversity in metropolitan landscapes. *Landscape Journal*, **27**, 115–26.

Miller, J. R. and Hobbs, R. J. (2002). Conservation where people live and work. *Conservation Biology*, **16**, 330–7.

Moller, A. P. (2009). Successful city dwellers: a comparative study of the ecological characteristics of urban birds in the Western Palearctic. *Oecologia*, **159**, 849–58.

Mörtberg, U. and Wallentinus, H.-G. (2000). Red-listed forest bird species in an urban environment – assessment of green space corridors, *Landscape and Urban Planning*, **50**, 215–26.

Murphy, D. D. (1988). Challenges to biological diversity in urban areas. In E. O. Wilson, ed., *Biodiversity*. Washington, DC: National Academy Press, pp. 71–6.

Murton, R. K., Thearle, R. J. P. and Thompson, J. (1972). Ecological studies of the feral pigeon *Columba livia var.* I. Population, breeding biology and methods of control. *Journal of Applied Ecology*, **9**, 835–74.

Oberndorfer, E., Lundholm, J., Bass, B. *et al.* (2007). Green roofs as urban ecosystems: ecological structures, functions, and services. *BioScience*, **57**, 823–33.

Oke, T. R. (1982). The energetic basis of the urban heat island. *Quarterly Journal of the Royal Meteorological Society*, **108**, 1–24.

Opdam, P. and Steingröver, E. (2008). Designing metropolitan landscapes for biodiversity: deriving guidelines from metapopulation ecology. *Landscape Journal*, **27**, 69–80.

Opdam, P., Steingrover, E. and van Rooij, S. (2006). Ecological networks: a spatial concept for multi-actor planning of sustainable landscapes. *Landscape and Urban Planning*, **75**, 322–32.

Opdam, P., Verboom, J. and Pouwels, R. (2003). Landscape cohesion: an index for the conservation potential of landscapes for biodiversity. *Landscape Ecology*, **18**, 113–26.

Pacheco, R. and Vasconcelos, H. L. (2007). Invertebrate conservation in urban areas: ants in the Brazilian Cerrado. *Landscape and Urban Planning*, **81**, 193–9.

Palmer, G. C., Fitzsimons, J. A., Antos, M. J. and White, J. G. (2008). Determinants of native avian richness in suburban remnant vegetation: implications for conservation planning. *Biological Conservation*, **141**, 2329–41.

Penttinen, P., Timonen, K. L., Tiittanen, P. *et al.* (2001). Ultrafine particles in urban air and respiratory health among adult asthmatics. *European Respiratory Journal*, **17**, 428–35.

Pressey, R. L., Cabeza, M., Watts, M. E., Cowling, R. M. and Wilson, K. A. (2007). Conservation planning in a changing world. *Trends in Ecology and Evolution*, **22**, 583–92.

Savard, J.-P. L., Clergeau, Ph. and Mennechez, G. (2000). Biodiversity concepts and urban ecosystems. *Landscape and Urban Planning*, **48**, 131–42.

Searns, R. M. (1995). The evolution of greenways as an adaptive urban landscape form. *Landscape and Urban Planning*, **33**, 65–80.

Shijun, M. (1985). Ecological engineering: application of ecosystem principles. *Environmental Conservation*, **12**, 331–5.

Snep, R. P. H. (2009). Biodiversity conservation at business sites – options and opportunities. Unpublished Ph.D. thesis, Wageningen University, the Netherlands. Full text available at http://edepot.wur.nl/92.

Snep, R. P. H. and Ottburg, F. G. W. A. (2008). The 'habitat backbone' as a nature conservation strategy for industrial areas: lessons from the natterjack toad (*Bufo calamita*) in the Port of Antwerp (Belgium). *Landscape Ecology*, **23**, 1277–89.

Snep, R. P. H., Opdam, P. F. M., Baveco, J. M. *et al.* (2006). How peri-urban areas can strengthen animal populations in cities: a modeling approach. *Biological Conservation*, **127**, 345–55.

Snep, R. P. H., Van Ierland, E. C. and Opdam, P. (2009). Enhancing biodiversity at business sites: what are the options, and which of these do stakeholders prefer? *Landscape and Urban Planning*, **91**, 26–35.

Strauss, B. and Biedermann, R. (2006). Urban brownfields as temporary habitats: driving forces for the diversity of phytophagous insects. *Ecography*, **29**, 928–40.

Sukopp, H. (1986). Naturschutzstrategien in der Stadt. In G. Kaule, ed., *Arten- und Biotopschutz*. Stuttgart, pp. 434–9.

Sukopp, H. and Starfinger, U. (1999). Disturbance in human ecosystems. In L. R. Walker, ed., *Ecosystems of Disturbed Ground*. Amsterdam: Elsevier, pp. 397–412.

Sukopp, H. and Weiler, S. (1988). Biotope mapping and nature conservation

strategies in urban areas of the Federal
Republic of Germany. *Landscape and Urban
Planning*, **15**, 39–58.

Sukopp, H., Numata, M. and Huber, A. (eds.)
(1995). *Urban Ecology as the Basis for Urban
Planning*. The Hague, the Netherlands: SPB
Academic.

Todd, N. J. and Todd, J. (1994). *From Eco-cities to
Living Machines: Principles of Ecological Design*.
Berkeley, CA: North Atlantic Books.

Tratalos, J., Fuller, R. A., Warren, P. H., Davies,
R. G. and Gaston, K. J. (2007). Urban form,
biodiversity potential and ecosystem
services. *Landscape and Urban Planning*,
83, 308–17.

Tucker, G., Ash, H. and Plant, C. (2005). *Review of
the Coverage of Urban Habitats and Species
Within the UK Biodiversity Action Plan*. English
Nature Research Reports 651. Northminster
House, Peterborough: English Nature.

Tzoulas, K., Korpela, K., Venn, S. *et al.* (2007).
Promoting ecosystem and human health in
urban areas using green infrastructure: a
literature review. *Landscape and Urban
Planning*, **81**, 167–78.

Velasquez, L. S. (2005). Organic greenroof
architecture: design considerations and

system components. *Environmental Quality
Management (Summer)*. Available at http://
www.greenroofs.com/pdfs/news-
EQM_VelazquezPart2.pdf.

Verboom, J., Foppen, R., Chardon, P., Opdam, P.
and Luttikhuizen, P. (2001). Introducing the
key patch approach for habitat networks
with persistent populations: an example for
marshland birds. *Biological Conservation*,
100, 89–101.

Walsh, C. J., Fletcher, T. D. and Ladson, A. R.
(2005). Stream restoration in urban
catchments through redesigning
stormwater systems: looking to the
catchment to save the stream. *Journal of the
North American Benthological Society*,
24, 690–705.

Wheater, C. P. (1999). *Urban Habitats*. London:
Routledge.

Wilcox, J. C., Healy, M. T. and Zedler, J. B. (2007).
Restoring native vegetation to an urban wet
meadow dominated by Reed Canarygrass
(*Phalaris arundinacea* L.) in Wisconsin. *Natural
Areas Journal*, **27**, 354–65.

Wu, J. (2008). Making the case for landscape
ecology an effective approach to urban
sustainability. *Landscape Journal*, **27**, 41–50.

CHAPTER TWELVE

Urban futures

MICHAEL L. MCKINNEY

In this chapter, I review some of the extensive literature on urbanisation to examine the future of urban growth and its impact on the biosphere. I argue that the relatively few ecological studies on urbanisation thus far have typically focused on the environmental impact of cities at local spatial scales when we also need to consider urban impacts at regional and global scales. By expanding our view of urban impacts beyond describing local environmental impacts, it becomes clear that cities can provide opportunities for biological conservation at many spatial scales. For example, cities concentrate human populations and produce economies of scale that can reduce per capita human impact on the atmosphere and on regional watersheds, as well as on local air quality. A 'sustainable' city is thus not just a city that sustains its own existence. Rather it is a city that contributes to the long-term persistence of the biosphere and indeed the *global* environment (McGranahan & Satterthwaite 2003). Ecologists should therefore find and recommend ways that urban impacts can be reduced on regional and global scales as well as the local scale. To that end, this chapter devotes considerable discussion on ways to create more sustainable cities.

The future growth of cities

The proportion of humanity living in urban areas has consistently increased in recent decades, from just 10% in 1900 to 50% in 2007. Put another way, the twentieth century witnessed a ten-fold increase in urban populations. It has been estimated that by 2025, five billion people will live in urban areas and account for 65% of the global population (United Nations 2008).

More than 95% of this future net increase in the global urban population will be in cities of the developing world. The greatest proportion of this growth is expected to occur in Asia, where 71% of the global rural population currently resides; the urban population of Africa, Asia, Latin America and the Caribbean is now nearly three times the size of the urban population of the rest of the

Urban Ecology, ed. Kevin J. Gaston. Published by Cambridge University Press.

world. Projections by the United Nations suggest that urban populations are growing so much faster than rural populations that 85% of the growth in the world's population between 2010 and 2020 will be in urban areas, and almost all of this growth will be in Africa, Asia and Latin America (United Nations 2008).

This rapid urbanisation of developing nations has enormous consequences for conservation planning because many of the developing nations are disproportionately poor in monetary terms but are disproportionately wealthy in biodiversity, especially of course those containing tropical rainforest habitats. Roughly 72% of the urban population of Africa now lives in slums (Cohen 2006). The slum-dwelling proportion is 43% for Asia and the Pacific, 32% for Latin America and 30% for the Middle East and Northern Africa (Cohen 2006). Unfortunately, in many of the world's poorest countries, the proportion of urban poor is increasing faster than the overall rate of urban population growth (United Nations 2008). This combination of increasing poverty and urban population growth in nations that are rich in biodiversity is perhaps the single greatest challenge to biological conservation today.

Another important urban trend is the growth of megacities, cities containing over 10 million people (Decker et al. 2000). At the beginning of the twentieth century, just 16 cities in the world contained a million people or more. Today, over 400 cities contain over a million people, and at least 25 cities contain over 10 million people (Cohen 2006). Most megacities are found in mid-latitude regions and are located near large river mouths within 100 km of a coastline (Decker et al. 2000). The number of megacities will undoubtedly grow, as will their impact on the environment at many scales. Grimm et al. (2008) discuss the environmental importance of 'megapolitan' regions, consisting of multi-city conglomerates covering hundreds of square kilometres. Because of the enormous volumes of resources consumed by these areas, and the enormous volumes of pollution that they discharge, these regions will have an increasingly powerful influence on climate, watersheds, the oceans and many other regional and global processes. However, urban conservation efforts must also include smaller cities. Indeed, most urban population growth is expected to occur in smaller cities with populations of less than one million. In 2005, cities with populations under 500 000 accounted for 51% of urban dwellers (United Nations 2008).

Finally, many urban areas, especially in the USA, have been trending toward increasingly dispersed settlement patterns. Often referred to as 'suburban sprawl' or 'exurban' growth, this has been driven by mass construction of large residential developments of low-density housing beyond the urban periphery. Low-density rural home development is the fastest-growing form of land-use in the USA since 1950 (Hansen et al. 2005). Housing growth, in the form of both suburban and rural sprawl, has been identified as one of the major threats to ecosystems in the USA (Lepczyk *et al.* 2007). This is also a harmful trend in other regions. According to Liu *et al.* (2003), 'In recent years, the number of households globally, and particularly in countries with biodiversity hotspots (areas rich in endemic species and threatened by human activities), was more rapid than aggregate population growth.' This has increased per capita impacts on biodiversity by increasing per capita consumption of wood for fuel, habitat alteration for home building and associated activities, and greenhouse gas emissions.

Cities as engines of biological loss: extinction, eco-function and homogenisation

Urbanisation is, justifiably, often seen as a major cause of species extinction and ecological damage. For example, urbanisation is the second-leading cause of habitat loss and pollution in Canada (Venter *et al.* 2006), and it has been documented as the second-leading cause of species endangerment in the USA (Czech *et al.* 2000).

The increasingly harmful impacts of cities on the natural environment are rooted in their origins. Cities are artificial habitats created primarily for one species, and this requires greatly modifying, and often obliterating, whatever natural habitats occur in the immediate vicinity. An obvious example is the replacement of a local forest with asphalt and row housing. Such 'constructional' urban impacts arise from the building activities of urbanisation.

Also, the maintenance of high human population densities in any city requires the importation and consumption of massive amounts of resources from many distant areas. As noted long ago by Odum (1963), cities are highly heterotrophic entities which require huge inputs of matter and energy from many distant sources. As these resources are exploited for urban consumption (e.g. logging, mining, agriculture), natural habitats in those areas are inevitably affected in negative ways. Conversely, vast quantities of waste from this consumption are released into the air and water which transport these materials as pollution into surrounding and distant regions. Such 'maintenance' urban impacts occur from the consumption and pollution produced in maintaining large urban human populations.

The accumulation of these constructional and maintenance impacts of cities on the environment is of course increasingly large. Although urban area now occupies less than 3% of the Earth's terrestrial surface (Mills 2007), this area produces roughly 78% of anthropogenic carbon emissions, and accounts for 60% of residential water use and 76% of wood used for industrial purposes (Brown 2001). These numbers exemplify the extremely wide-ranging and diverse impacts of cities on the global and regional atmosphere, the hydrosphere and the biosphere. In addition to these wide-ranging impacts at regional to global scales, urbanisation has exceptionally severe local impacts. As

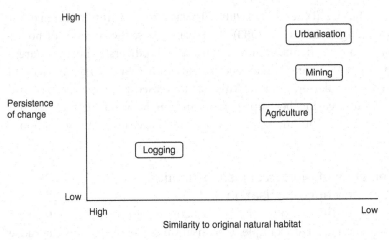

Figure 12.1 The adverse effects of different human activities on natural habitat. In terms of duration and intensity (deviation from the pristine natural condition), urbanisation typically involves the greatest anthropogenic impact. Modified from Marzluff and Ewing (2001).

discussed by Marzluff and Ewing (2001), urbanisation impacts on the local environment are more intense and more enduring than any other common human activity. Thus, the effects of logging, mining, farming and many other anthropogenic human disturbances are typically less intensive and enduring as a cause of habitat loss than urbanisation. Whereas logging, mining, farming and most other land-uses have the goal of intermittent resource extraction, urbanisation represents a more permanent (on ecological and human time-scales) land-use with the goal of long-term habitation. Furthermore, our desire to alter the land drastically, to fit our own aesthetic and functional needs, produces extreme modifications of the pre-existing natural communities. For instance, concrete and pavement, with zero primary productivity, would seem to be a more extreme disturbance than occurs with logging, mining and farming. These other activities at least generally allow soil and vegetation to persist (see Figure 12.1).

In addition to habitat loss and pollution, urbanisation is also strongly associated with many other anthropogenic causes of extinction such as non-native species introductions and overharvesting. As transportation hubs, cities contain a disproportionate number of non-native species which become major sources for the export of non-native species into the surrounding region (von der Lippe & Kowarik 2008). Urban consumers are also major markets for products from illegally harvested species (Robinson 2001). This includes cities in poor nations where local and regional species are often harvested for 'bushmeat' and other animal products which are increasingly sold in urban areas (Edderai & Dame 2006).

Although the impacts of urbanisation on ecosystem function (services) are not nearly as well studied as its impacts on species losses, the existing evidence generally confirms what might be intuitively inferred: that urbanisation degrades ecosystem function, especially with increasing urban density (Chapter 3; Tratalos et al. 2007). For example, a study of an urbanisation gradient in the Lake Tahoe basin of the USA showed that ecosystem services became impaired after moderate amounts (30–40%) of development occurred (Sanford et al. 2009). In this case, decline of ecosystem function was documented via its effect on ant communities, with the loss of decomposer and aerator ants which, in turn, reduced soil productivity and rainwater infiltration. Severe negative impacts of urbanisation are also seen in the ecosystem function of aquatic systems arising from drastic alteration of water flow regimes, sedimentation and nutrient inputs (Roach et al. 2008). Fortunately, there seems to be substantial variation in the negative relationship between urbanisation and ecosystem function, suggesting that there is considerable opportunity to maximise ecological performance (Tratalos et al. 2007).

A common consequence of the natural disruptions caused by urbanisation is biological homogenisation (McKinney 2006). This can have several manifestations. At the taxonomic level, the extirpation of local native species followed by their replacement with widespread non-native species can produce homogenisation. Such a process is promoted by two basic dynamics of urbanisation (McKinney 2006): (i) the creation of similar (artificial, human) habitats in many places that are out of equilibrium with their natural matrix, and (ii) the importation (deliberate and accidental) of non-native species that can survive in these urban habitats. At the ecosystem level, urbanisation can also promote functional homogenisation. A study of bird communities in France, for example, showed that urbanisation induced community homogenisation and that populations of specialist species became increasingly unstable with increasing urbanisation of the landscape, increasing the functional similarity of the bird communities (Devictor et al. 2007).

Cities as future centres of conservation

In view of the extensive environmental impacts caused by urbanisation, it may seem counterintuitive to view cities in terms of conservation. There are, however, several basic reasons why urbanisation, where done in a more sustainably planned way, can make substantial contributions toward nature conservation (McKinney 2002). For one, the very fact that cities currently have such enormous environmental impacts means that any reductions in those impacts are positive steps toward sustainability. There are many ways (discussed below) that the construction and maintenance of cities, ranging from housing to transportation, can be redesigned to reduce their impact. Second, urban settlements may have intrinsic properties that, of themselves, are the best way for humans to minimise their impact on the natural environment. By concentrating humans into small areas, cities reduce the human population density in other locations. More importantly, this density concentration creates economies of scale such that urban inhabitants have lower per capita impacts (e.g. resource consumption, pollution) than rural human populations. This is because of a physical infrastructure (e.g. mass transit, mass housing) that is more efficient on a per capita basis. For example, total road surface, electrical cable length and fuel stations scale more slowly than population size as a city grows (Bettencourt et al. 2007). Similarly, per capita consumption of motor fuel declines exponentially with increasing human population density (Kennedy et al. 2007).

This potential of urbanisation to concentrate human impacts and 'free up' natural areas for preservation has a parallel in a debate about agricultural impacts. Is wildlife preservation more effectively achieved by expanding 'wildlife-friendly' farming techniques or by using more intensive farming methods that will more drastically alter farmland but will also reduce the area of land that is farmed? While empirical data are sparse, available evidence seems to suggest that (as with urbanisation) intensification may be the more effective conservation strategy (Green et al. 2005).

Urbanisation can also promote nature conservation via the social and psychological impacts it has on city residents. Citizens of suburban and urban areas tend to place a much higher value on species conservation than those living in rural areas (Kellert 1996). Legislators from highly urbanised states and districts tend to be more supportive of strengthening the Endangered Species Act in the USA (Mehmood & Zhang 2001). This is especially true where local policies directly promote conservation activities within the urban area such as landscaping for wildlife, green roofs and the cultivation of native plants. There is a psychological benefit for people in the cultivation of local biodiversity (Fuller et al. 2007). This is a crucial point because any definition of 'sustainability' must include the benefit to human welfare as well as biodiversity (McGranahan & Satterthwaite 2003). Conversely, the disconnection from biodiversity in urban environments can have a negative impact on humans as well as conservation, or what Miller (2005) has called the 'extinction of experience'.

These arguments imply that rethinking the way that cities are designed, constructed and maintained could significantly reduce the total and per capita anthropogenic impact on the natural environment, as well as maintain a high quality of life for humans. What follows is a necessarily brief overview of some fundamental ways to rethink cities in this way.

Reducing the urban footprint

Howard Odum (1971) was an early analyst of urban energetics, documenting that modern cities consume 10–100 times more energy (per unit area) than natural ecosystems. His use of input–output models documented the heterotrophic basis of urban habitats and their strong dependence on external sources of matter and energy which are 'metabolised' by the city. In the past few decades, two widely used approaches have evolved that have elaborated on this insight: the ecological footprint and the mass balance (Kaye *et al.* 2006). While neither of these approaches provides a complete accounting of the environmental impacts of cities, they are useful heuristic models for describing general impacts and, most importantly, ways to reduce urban impacts on the environment. Because most of the matter and energy used by cities is imported from distant locations, and because many wastes emitted by cities are then transported to distant locations, the ecological footprint and mass-balance models are useful ways to assess a city's impact on the regional and global environment.

The ecological footprint (EF) of a city is the area of land required to meet its metabolic needs for resource consumption (inputs) and waste assimilation (outputs). Its roots lie in earlier ideas, such as 'ghost acres' and similar concepts developed by Borgstrom (1972) and Ehrlich (1968). Specifically, the EF is the 'total area of productive land and water ecosystems required to produce the resources that the population consumes and assimilate the wastes that the population produces, wherever on Earth that land and water may be located' (Rees 2000, p. 371). Studies of many cities have documented that the EF is orders of magnitude greater than the area actually occupied by a city (Olalla-Tarraga 2006; Wackernagel *et al.* 2006). For example, Rees and Wackernagel (1996) estimate Vancouver's EF at more than 200 times its geographic area. London's EF is estimated to be 125 times its actual size (Jopling & Girardet 1996). All of these studies confirm the comment by Rees (1997) that 'cities as presently conceived are incomplete ecosystems, typically occupying less than 1 per cent of the ecosystem area upon which they draw.'

One of the main attractions of the EF is its simplicity. The EF converts several measures of consumption and waste assimilation into a single unit, the amount of land required to support the consumption and waste. As a result of this simplicity, the EF approach has become widely used and is firmly entrenched in the scientific and popular literature. It is often used by educators and even policy makers to illustrate the easily forgotten but fundamental insight that urban impacts extend far beyond the boundaries of any town or city.

On the other hand, the simplicity of the EF has led to several criticisms. Kaye *et al.* (2006, p. 192), for example, argue that EF studies are 'flawed because they do not incorporate biophysical setting, population size and per capita consumption rate into the footprint calculation. Until these factors are included, footprints depict negative impacts of cities without accounting for the probable efficiency of dense urban living.' Several other criticisms are summarised in Mcmanus and Haughton (2006), mostly echoing the problem of oversimplification resulting in the neglect of several variables, resulting in a number that is an incomplete metric of the overall impact of cities. Examples of neglected parameters include water resources, geography (e.g. biophysical settings noted above), environmental impacts within urban boundaries (such as wetland degradation), and metrics of social and human welfare (Mcmanus & Houghton 2006).

Perhaps the biggest problem from a purely analytical view is that the EF simply does not include many variables that would provide a more thorough assessment of urban impacts on the environment. For instance, the typical EF calculation uses four consumption categories: energy use, the built environment (the land area covered by a settlement and its connection infrastructure), food and forestry products. This is obviously a small subset of all resources, determined by the practical requirements of data gathering. Similarly, EF incorporates only a small subset of urban waste outputs, such as carbon dioxide, to be assimilated by natural ecosystems. These impact variables are then converted into a single number (land area). While this is certainly an interesting number, it obviously compresses many other dimensions of urban impacts in a single value. The International Institute for Environment and Development described the EF analysis, whereby all environmental impacts are aggregated into a simple index, as 'resource reductionism' (Hammond 2000).

Despite the many criticisms of the EF, it has promoted much constructive thinking about how to improve upon it, discussed in the next section. It has also become a way to formulate specific proposals to reduce human impacts (i.e. shrink the urban footprint). Wackernagel *et al.* (2006, p. 107) list these four examples of ways to reduce the EF:

- use resource-efficient technology to reduce the demand on natural capital;
- reduce human consumption while preserving quality of life: for example, reduce the need for fossil fuels by making cities pedestrian-friendly;
- lower the size of the human family in equitable and humane ways, so that total consumption decreases even if per capita demand remains unchanged; and
- invest in natural capital: for example, implement resource extraction methods that increase rather than compromise the land's biological productivity, thereby increasing supply.

The first two of these examples relate directly to urban life. Cities need to be constructed and maintained with more resource-efficient technologies (e.g. buildings with improved energy and water conservation). Also, cities can be redesigned in many ways to reduce consumption of many resources. In addition to less reliance on automobiles to reduce energy consumption, an example would be more area landscaped with native vegetation to reduce energy and water consumption. Moos *et al.* (2006) used the EF as an urban planning tool to show that increasing the density of housing in residential subdivisions can reduce the urban EF. In their words, 'despite the challenges associated with data collection and conversion, it is argued that EF has utility for planners and urban designers because it enables assessment of built form from an environmental consumption point of view' (Moos *et al.* 2006, p. 195).

The EF has also become an excellent metaphor for public education, which has led to specific policy recommendations to reduce the EF of some urban areas. Wackernagel *et al.* (2006) cite several examples, such as California's Sonoma County Footprint project which has inspired every city in the county to sign up for the Climate Saver Initiative of the International Council for Local Environmental Initiatives (ICLEI) – also known as Local Governments for Sustainability.

There have been several attempts to improve the EF by expanding the parameters used in its measurement. For example, Satterthwaite (1997) suggested including criteria to estimate human social welfare, such as health, sanitation and recreational facilities. More commonly, many studies have added biophysical parameters to increase the metrics of human impacts. The amount of land needed to assimilate the nitrogen and phosphorus waste of cities, for instance, is of the order of 400–1000 times the area of the cities (Folke *et al.* 1997).

Reducing urban metabolism

Because of the problems noted above for the EF, many analysts favour the 'metabolic' model as a better method to assess the twin environmental impacts of resource consumption and waste export of cities (Decker *et al.* 2000; Kaye *et al.* 2006). This view examines mass balances which 'quantify inputs and outputs of elements or materials (or these values converted to energy units), and yield information about whether a city is a source (input > output) or sink (input < output) for the material and how the material is transformed by urban activities' (Kaye *et al.* 2006, p. 193).

Mass-balance studies show that cities act as a sink for many substances imported from long distances. For example, building materials such as stone, wood and metals are incorporated into the urban structure and then become stored in landfills within or near the urban area (Decker *et al.* 2000). Indeed, it has been estimated that less than 10% of urban material inputs are exported as goods (Brunner *et al.* 1994). Thus, cities are net sinks for most materials. This is especially true for rapidly growing younger cities, where the ratio of inputs to outputs can be 10:1, whereas the ratio is closer to 1:1 in old cities experiencing little growth (Brunner 2007). At the chemical scale, cities are sinks for many elements such as nitrogen, phosphorus and, sometimes, metals (Kaye *et al.* 2006). A study of Gavle, Sweden, estimated that of the 330 tons of phosphorus per year entering the urban system, only 35% (116 tons) leaves the system, half of it in exported goods and half expelled into the Baltic Sea. The remaining 65% is stored in the city, mainly as sewage sludge in landfills (Nilson 1995).

On the other hand, cities can also act as sources for some substances. These are materials that are imported, often from far away, and then released into the local and regional environment. Carbon, from fossil fuel combustion, is an obvious example. A typical large city consumes of the order of 100 to 1000 petajoules (1 petajoule being 10^{15} joules) of energy per year to operate its transportation, electrical and climate control infrastructures (Decker *et al.* 2000). As most of this energy is produced by fossil fuel combustion, cities are a huge source of carbon dioxide (CO_2). It was estimated, for example, that each million residents of Sydney, Australia, produced about 48 148 tons of carbon dioxide per day in 1990 (Newman & Kenworthy 1999).

'Transformed' inputs are converted to another form and then exported from the urban system as waste (Decker *et al.* 2000). Consequently, cities are typically local and regional sources for many of the transformed materials. Fuels, just noted, are a good example in the transformation of hydrocarbons into CO_2, particulates and many other byproducts that become transported by air and water to localities near and far. Food and water are two other major inputs that become transformed and then largely exported into the environment. New York City, USA, consumes 20 000 tons of food per day, with half being discarded, mostly as organic waste into landfills (O'Meara 1999). Water is notable as a transformed substance (mostly becoming converted into sewage) with the greatest quantity of flux through the urban system. Cities consume an average of roughly 1000 kt of water per day compared to 10 124 kt of fuel in a *year* for Mexico City (Decker *et al.* 2000). Similarly, in 1990 each one million residents of Sydney, Australia, daily consumed over 490 000 tons of water but less than 11 000 tons of fossil fuels (Newman & Kenworthy 1999). Most (over 75%) of the water consumed in Sydney was exported locally as sewage.

As noted above, a key aspect of urban metabolism is that it has environmental impacts at many spatial scales. It destroys habitat in many localities, ranging from near to far away. Among the many examples would be the habitat loss caused by the mining extraction of minerals and timber to supply urban growth and maintenance, and the subsequent habitat loss produced by the export, often by water and air, of polluting waste into natural habitats near and far away from the city. Therefore, reducing the impact of cities on biodiversity requires reducing urban metabolism. Unfortunately for conservation goals, the metabolism of cities has been increasing. The material turnover of a modern city is about one order of magnitude larger than that in an ancient city of the same size (Brunner *et al.* 1994). A review of urban metabolism studies from eight metropolitan regions across five continents conducted since 1965 showed that most cities had increasing per capita (and total) urban metabolism with respect to water, waste water, energy and materials (Kennedy *et al.* 2007).

Reversing the trend toward increasing urban metabolism is clearly a basic way for cities to help achieve global sustainability (Newman 1999). By reducing resource inputs, cities not only can reduce the environmental burdens of resource extraction and non-renewable resource depletion, they will also reduce the environmental burdens created by the export of urban metabolic waste and pollution. There are two (non-exclusive) ways for cities to reduce their metabolic demands for materials: (i) increase the efficiency of use, and (ii) increase the recycling of materials within urban areas. As energy cannot be recycled, reducing the urban metabolic demand for energy must focus on increasing efficiency. Another option will be increasing the use of renewable energy, as long as the sources have relatively low environmental burdens.

Sustainable cities and global biodiversity conservation

Most of the ecological literature on urban biodiversity conservation has tended to focus on ways to preserve biodiversity in localities within and adjacent to a city (McKinney 2002). This literature often provides data that can help mitigate the negative impacts of urbanisation on natural habitats, by preserving or restoring those habitats. The basic goal is a 'win–win' scenario whereby human needs can be met while minimising urban impacts on biodiversity conservation (Miller 2006). A common theme of this literature is the mitigation of local environmental impacts. For example, one can promote the preservation or reintroduction of native plants (and the animals that rely on them) in urban ecosystems such as golf courses (Hammond & Hudson 2007) and residential areas (e.g. Fetridge et al. 2008; Pennington et al. 2008).

Reducing local urban impacts in this way is clearly an extremely important part of an overall strategy to preserve global biodiversity. This is especially true if human settlements are often located in areas with a naturally high richness of native species with high conservation value due to their endemism and phylogenetic uniqueness. Furthermore, many urban areas still contain a substantial number of threatened and other at-risk categories of species, despite being occupied by humans for many years. Examples would include a surprisingly large number of threatened bird species in cities in the UK (Fuller et al. 2009) and plant species in urbanised areas of California, USA (Schwartz et al. 2006).

This work of preserving local biodiversity in and around urban areas is commonly discussed in the ecological literature because there is much that ecologists can do to promote this aspect of global conservation. Ecologists have a direct role in providing data that can help restore and preserve local biological communities and ecosystems.

In contrast to this local emphasis, the approach I have taken here is to emphasise the importance of the sustainable cities concept, and especially the goal of reducing urban metabolic demands for its potential for contributing to

Table 12.1 *Reduction in local and distant habitat alteration and pollution from urban changes.*

Clean (non-fossil-fuel) energy
Distant impact reduction: climate change, acid rain, smog, coal mining, oil extraction and refining
Local impact reduction: smog, carbon monoxide
Reduced and recycled construction materials
Distant impact reduction: deforestation (lumber), oil extraction (plastic), mining impacts (stone, metals)
Local impact reduction: landfill waste

the conservation of *non-local* biological resources. By reducing the demand for vast quantities of natural resources that are extracted and imported from many areas of the planet, there is a consequent reduction in the impacts to biota affected by resource extraction in and pollution of those areas (Table 12.1).

For example, many cities have long exceeded or polluted local freshwater supplies and must rely on water from very distant sources (UN-HABITAT 2003). Dakar, the capital of Senegal, now needs to draw water from a lake 200 km away (White 1992); Mexico City has to supplement local supplies with water drawn from neighbouring basins that has to be pumped over 1000 metres in height and drawn from up to 150 km away (Connolly 1999). Therefore, any reductions in water demands in these (and most other) cities would reduce these environmental impacts on the affected distant biota.

Another example would include the many distant impacts from the extraction and importation of vast amounts of construction materials (e.g. stone, wood and metals). Mining of stone and especially metals involves moving enormous volumes of rock. Most of this is open pit mining which typically involves the complete removal of biological communities in the area. The large overall impact of mining on the biosphere can be inferred from the estimate that humans now move more rock and sediment on the planet than all the natural processes (e.g. water, wind) combined (Meybeck 2004). In addition, the mining and smelting of the ore produces very large amounts of air and water pollution, with high concentrations of heavy metals and other very toxic pollutants (Muezzinoglu 2003). Some of the impacts of mining on the local and regional biota are well documented by ecologists. Examples would include the effects on wildlife of gold mining in the USA and Australia (Donato *et al.* 2007), and copper mining in Poland (Kalisinska *et al.* 2004). Such cases typically document mass mortality of vertebrates, especially waterfowl, caused by acute toxicity of heavy metals and cyanide (used in ore extraction). In addition there is widespread soil contamination leading to severe and long-term impacts on native vegetation (Bech *et al.* 1997). Given that urban areas are the main markets for these materials, often for building construction (e.g. copper) and other uses directly related to urban expansion, any reduction in the demand would reduce the impacts of their extraction on those distant ecosystems.

The export of mahogany *Swietenia macrophylla* from the Amazon basin to global urban markets outside of South America provides an illustration of the benefits that could be derived from reducing urban wood consumption. For each mahogany tree harvested, over 1 square kilometre of forest ground vegetation is removed or damaged, resulting in the cumulative loss of thousands of square kilometres of forest ecosystem (Verissimo *et al.* 1995). Compounding this loss is the growing trend to convert the harvested forests to cattle pasture. The growth of cities in China, combined with logging restriction within the nation, has led to a rapid increase in the importation of forest products from heretofore relatively undisturbed forest ecosystems in more distant localities in Russia, Southeast Asia and Africa (Zhang & Gan 2007).

In addition to water and construction materials (stone, minerals, wood), another major resource input into cities is energy, especially fossil fuels. The extraction of these has severe impacts on the ecosystems where these resources are found. The debate over petroleum drilling in tundra, ocean and other fragile ecosystems has raised awareness of the impacts of oil extraction (Kryuchkov 1993). The local ecological impacts from coal extraction and use are probably even greater. Life cycle assessments that measure the effects of mining, transport, combustion and waste disposal of coal show that all of these have significant negative impacts on local and regional ecosystems in several different localities (Cazplicka-Kolarz *et al.* 2004; Babbitt & Lindner 2008). Most notable in terms of local biotic impacts is the mining of the coal body itself, which typically involves extensive de-vegetation, and air and water pollution impacts on local biota (Elberling *et al.* 2007). Acid rain and other regional air pollution impacts must also be considered as urban impacts on ecosystems that are distant from the point of use. Finally, the increasing global impacts of climate change on the many ecosystems, especially those of high latitudes, also exemplify impacts that are far removed from the urban areas that consume the electricity produced by coal.

Redesigning cities for sustainability

Of course, merely pointing out that cities need to reduce metabolic demands is a relatively easy task compared with applying these concepts. The way that cities are constructed and maintained will need to be redesigned so that they continue to meet human needs but also increase the efficiency of resource use, recycle those resources and rely more on alternative, especially renewable sources. Figure 12.2 summarises this general approach. What follows are some specific examples of ways to implement the approach.

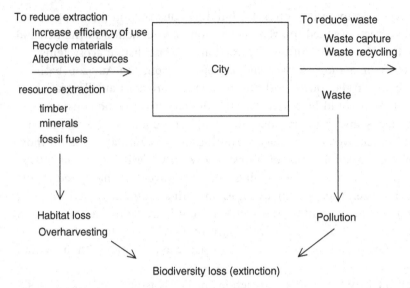

Figure 12.2 Urban resource consumption (extraction) of natural resources and consequent waste disposal are the underlying factors driving many species to extinction. Often these species inhabit areas far removed from the city itself. Cities can reduce these impacts (i.e. become more 'sustainable') by reducing their resource demands (via increased efficiency, recycling, alternative resources) and reducing their waste output.

An obvious method to increase efficiency of resource use in cities is the widespread adoption of more efficient technologies (e.g. compact fluorescent and LCD lighting) by their inhabitants. This citizen-based approach has been the focus of many educational campaigns to promote 'green living', such as the Sonoma County Footprint project noted above (Wackernagel *et al.* 2006). The goal in such cases is to encourage citizens to purchase hybrid cars, purchase organic foods, improve home energy conservation and so on. However, a second approach, at the scale of the city, provides another option for increasing efficiency of resource use: redesigning the city itself (the urban form). It is estimated, for example, that green design in the planning and building phase of urban growth can have a large impact on the operational phase, reducing energy demand by a factor of 5 to 10 (Brunner 2007). Similarly, cities that concentrate people at higher population densities are more efficient in terms of transportation and housing. There is an exponentially inverse relationship, for instance, between per capita energy consumption for transportation and population density in cities (Newman & Kenworthy 1991).

Recycling materials will also be an important way to reduce the urban metabolic demand for consumption of external resources. Future cities will need to rely less on 'linear' metabolic flows and more on 'cyclic' metabolic flows (Doughty & Hammond 2004). This would entail increasing the recycling

of organic and inorganic materials within urban areas, which would simultaneously reduce both consumption of resources and the export of waste and
pollution. Ore mining will be replaced by urban mining, resulting in less
demand for energy and resources and fewer emissions (Brunner 2007).

One problem with the 'sustainable city' concept is that it can foster the idea
that there is a universal solution to the problem of urban impacts on the
environment when, in fact, cities are often quite different in terms of their
natural and social contexts. For example, some cities are currently 'less sustainable' than others. Wealthier cities such as Canberra, Chicago and Los Angeles
have between 6 and 9 times the carbon dioxide emissions per person of the
world's average and 25 or more times that of cities such as Dhaka (Nishioka
et al. 1990). Motor fuel use per capita in cities such as Houston, Detroit and Los
Angeles, with among the world's highest consumption levels, is 100 times or
more that of most cities in low-income nations (Newman & Kenworthy 1999).
Average waste generation per inhabitant in cities can vary more than 20-fold
when comparing urban citizens in high-income nations (who may generate
1000 kg or more of waste per person per year) with those in some of the lowest-
income nations (who may generate less than 50 kg per person per year) (Hardoy
et al. 2001). In Los Angeles, only 10% of solid waste is recycled compared with
over 60% for Tokyo (Decker *et al.* 2000).

In addition, the geographic location of a city can affect its pathway to
sustainability. An emphasis on water conservation is important for cities in
arid regions but may be much less so for cities located in areas with abundant
water supplies, such as many tropical regions. Similarly, the age, or developmental stage, of a city can affect its pathway to sustainability. In young, rapidly
growing cities, the ratio of materials input to materials output can exceed 10:1
whereas this ratio is closer to 1:1 in mature cities (Brunner 2007). An implication is that it may be harder to promote materials recycling in younger cities
because of the large amount of imported primary resources.

Urban sustainability as win–win ecology

Discussions about sustainable development have often invoked sustainability
as a 'win–win–win' scenario for its potential simultaneously to optimise economic, environmental and social goals (Gibbs & Deutz 2005). In the context
here, we can focus on sustaining global biodiversity and especially the preservation of endangered native species in the face of urbanisation. Rosenzweig
(2003) has promoted the use of 'win–win' scenarios to maximise species preservation. This involves establishing and maintaining habitats to conserve species
diversity in places where people live, work or play.

Win–win scenarios are critical for the success of redesigning cities in ways
that reduce their environmental impacts because they also include ways to
meet human economic and social needs. As noted above, most of the ecological

literature on urban nature conservation focuses on such scenarios at local scales. Green roofs, naturalised golf courses and residential landscaping for wildlife are examples (when done properly) of this type of win–win scenario because they meet human aesthetic, recreational and other needs while also promoting biodiversity conservation goals.

A main goal of this chapter has been to extend the spatial scale of these ecological win–win scenarios beyond the urban area itself. A benefit of this approach is to address one of the major criticisms of the win–win scenario, which is that it is often applied in incomplete ways and focuses on small-scale conservation problems (Sorice 2005). For example, urban win–win scenarios are often discussed in terms of designed habitats that maximise local biodiversity, such as optimising the configuration of green spaces (Colding 2007). However, the maximised biodiversity is usually defined simply as species richness, regardless of whether it includes non-native species or if the native species have any conservation value. Conversely, many studies have focused on ways to promote ecological services in urban habitats with the cultivation of designed ecosystems in allotment gardens, cemeteries, city parks (Andersson et al. 2007) and green roofs (Oberndorfer et al. 2007). However, most such proposals do not encourage, or simply omit mentioning, the importance of conserving native species, especially those native species at risk. Indeed, restoration ecology has often included non-native species in its efforts to restore or enhance ecosystem functions (Jones 2003).

By focusing only on the local city ecosystem, efforts to increase urban species richness and urban ecosystem services therefore neglect the critical criterion of global sustainability noted at the beginning of this chapter. The long-term health of the biosphere hinges on preventing the global extinction of endemic native species which are found in the vicinity of urban areas. Because human settlements tend to be found in areas with many native species of high conservation value, proposals to increase urban species richness or ecological services would be incomplete without including rare, endemic and at-risk native species. In a sense, this enhances the 'win–win' scenario of simultaneously meeting 'human needs and ecological function' or 'human needs and local biodiversity' into a 'win–win–win' scenario whereby 'global biodiversity conservation' is added to either of the preceding combinations.

The conservation importance of globally threatened species is not, of course, a novel insight, but I do suggest that it is a neglected dimension in much of the literature on urban ecology. Among the relatively few discussions of urban biodiversity conservation that do address global conservation in urban ecosystems would be the study by Lawson et al. (2008) showing that urbanising habitats, despite years of conservation neglect, still contain many plant populations of high conservation value, and that conservation planning for global biodiversity preservation should allocate relatively more resources to protecting urban-associated plant taxa. Similarly, Turner *et al.* (2005) note that, while residential ecosystems of Halifax, Nova Scotia, are species-rich, the dominance of exotic species reduces the ecological integrity of those communities. Pennington *et al.* (2008) discuss the importance of urban riparian zones for globally threatened migrating Neotropical migrant birds. Finally, Milder *et al.* (2008) propose that residential suburban developments explicitly consider threatened natural resources including rare species and ecological communities in their design, something that traditional 'conservation developments' have tended to ignore.

There are also a few studies that indicate a role for rare or declining native species in restoring ecosystem services. Fiedler *et al.* (2008) discuss evidence that native plants can be as useful in restoration as much as widely recommended non-natives. Winfree *et al.* (2007) show evidence that native bees can provide pollinator functions to replace those of the declining exotic honeybee in the USA.

There is yet one more dimension to this 'win–win–win' scenario. It is possible to conceive of urban conservation initiatives that not only meet human needs, promote local species richness and preserve global biodiversity but in addition reduce other distant urban impacts. An example would be an urban wetland that benefits human needs for recreation, aesthetics and water purification, promotes local species richness, provides habitat for globally threatened avifauna, and also reduces impacts on the overall watershed by reducing urban water demand and the total pollution load on the watershed. Furthermore, as wetlands are net carbon sinks, urban wetlands may have potential for ameliorating anthropogenic impacts on climate change (Erwin 2009).

Again, it is not a novel insight that well-designed urban ecosystems can simultaneously meet several human needs as well as several conservation goals, both local and global. Indeed, it seems likely that discussions of 'sustainable cities' will increasingly converge on this approach for reasons of practicality, efficiency and economics (e.g. Choi *et al.* 2008; Martinez & Lopez-Barrera 2008). My point herein is that the current conservation and urban sustainability literature is highly fragmented in its approach, focusing on just one or two objectives of these potential 'win–win–win–win' scenarios that can potentially achieve several objectives if a larger spatial scale of planning is attempted.

Summary and conclusions

The urbanisation of our planet will clearly continue for many years to come. Efforts to preserve what remains of our biosphere must, consequently, increasingly focus on ways to reduce the impact of cities. Indeed, one could argue that, because cities have the potential to concentrate human populations into relatively small areas as well as reduce per capita resource consumption and per capita pollution via economies of scale, then well-designed (and maintained) cities may be the best opportunity our species has for a sustainable coexistence with the natural biosphere.

However, if this enormous conservation potential for cities is to be attained, conservation planning must extend far beyond the very local scales at which most urban ecology is carried out. We must begin to focus much more attention on the many impacts of cities on distant areas at all spatial scales, ranging from resource consumption of water, minerals, energy and biological products to waste and pollution impacts on the hydrosphere and atmosphere at regional and global scales. 'Sustainable' cities of the future must sustain the biosphere as well as local and regional ecosystems.

This chapter has centred on the ecological footprint and especially the mass-balance (metabolic) approach because they provide well-established metrics for the resource inputs and waste outputs of cities. From the mass-balance view, it seems clear that there are many opportunities to reduce per capita human impacts by finding ways to reduce urban metabolic demands. By reducing per capita resource consumption and waste production, most per capita environmental impacts are simultaneously decreased.

Such a policy of 'source reduction' of the waste stream is commonly discussed in the resource economics and management literature (Fullerton & Wu 1998) but it is not a common theme in urban ecology. One reason is that much of the work needed to reduce urban metabolism is simply outside the realm of what ecologists have traditionally studied: green technologies that increase efficiency of resource use, recycling those resources after use or utilising more renewable resources are of more interest to engineers than ecologists (Montalvo 2008).

There are, however, two crucial reasons that reducing urban metabolism should be of interest to urban ecologists. One is that it is, arguably, the best way of preserving what remains of the most pristine ecosystems on the planet. Urban consumption is the main driver of resource extraction, even in areas that are far removed from the city itself. Conversely, urban waste and pollution is the main driver of climate change and other processes that also affect ecosystems that are far removed from the urban sources that produce them. Another reason that urban metabolism should be of interest to urban ecologists is that ecologists can play a significant role in doing something about the problem. Much ecological literature on cities has focused on the conservation of local native biota, especially via the restoration or preservation of native species habitat. However, ecologists can also help design urban ecosystems that reduce urban impacts at scales beyond the peri-urban fringe, such as artificial wetlands that reduce demand on the regional watershed or providing habitat for globally threatened species. Much of this work has the potential for simultaneously achieving multiple goals ('win–win–win' scenarios) whereby human welfare is improved, economic costs are reduced, and both local and global urban impacts are decreased as well.

References

Andersson, E., Barthel, S. and Ahrne, K. (2007). Measuring social-ecological dynamics behind the generation of ecosystem services. *Ecological Applications*, **17**, 1267–78.

Babbitt, C. W. and Lindner, A. S. (2008). A life cycle comparison of disposal and beneficial use of coal combustion products in Florida. *International Journal of Life Cycle Assessment*, **13**, 555–63.

Bech, J., Poschenrieder, C., Llugany, M. *et al.* (1997). Arsenic and heavy metal contamination of soil and vegetation around a copper mine in Northern Peru. *Science of the Total Environment*, **203**, 83–91.

Bettencourt, L. M., Lobo, J. and Helbing, D. (2007). Growth, innovation, scaling, and the pace of life in cities. *Proceedings of the National Academy of Sciences of the USA*, **104**, 7301–6.

Borgstrom, G. (1972). *The Hungry Planet: The Modern World at the Edge of Famine*, 2nd edn. New York: Macmillan.

Brown, L. R. (2001). *Eco-Economy: Building an Economy for the Earth*. New York: Norton.

Brunner, P. H., Daxbeck, H. and Baccini, P. (1994). Industrial metabolism at the regional and local level: a case-study on a Swiss region. In R. U. Ayres and E. E. Simonis, eds., *Industrial Metabolism: Restructuring for Sustainable Development*. Tokyo: UN University Press, pp. 163–93.

Brunner, P. H. (2007). Reshaping urban metabolism. *Journal of Industrial Ecology*, **11**, 11–13.

Cazplicka-Kolarz, K., Wachowicz, J. and Bojarska-Kraus, M. (2004). A life cycle method for assessment of a colliery's eco-indicator. *International Journal of Life Cycle Assessment*, **9**, 247–53.

Choi, Y., Temperton, V. M., Allen, E. B., Grootjans, A. P. and Torok, K. (2008). Ecological restoration for future sustainability in a changing environment. *Ecoscience*, **15**, 53–64.

Cohen, B. (2006). Urbanization in developing countries: current trends, future projections, and key challenges for sustainability. *Technology in Society*, **28**, 63–80.

Colding, J. (2007). Ecological land-use complementation for building resilience in urban ecosystems. *Landscape and Urban Planning*, **81**, 46–55.

Connolly, P. (1999). Mexico City: our common future? *Environment and Urbanization*, **11**, 53–78.

Czech, B., Krausman, P. R. and Devers, P. K. (2000). Economic associations among causes of species endangerment in the United States. *BioScience*, **50**, 593–601.

Decker, H., Elliott, S., Smith, F. A., Blake, D. R. and Sherwood Rowland, F. A. (2000). Energy and material flow through the urban ecosystem. *Annual Review of Energy and the Environment*, **25**, 685–740.

Devictor, V., Julliard, R., Couvet, D., Lee, A. and Jiguet, F. (2007). Functional homogenization effect of urbanization on bird communities. *Conservation Biology*, **21**, 741–51.

Donato, D. B., Nichols, O. and Possingham, H. (2007). A critical review of the effects of gold cyanide-bearing tailings solutions on wildlife. *Environment International*, **33**, 974–84.

Doughty, M. R. and Hammond, G. P. (2004). Sustainability and the built environment at and beyond the city scale. *Building and Environment*, **39**, 1223–33.

Edderai, D. and Dame, M. (2006). A census of the commercial bushmeat market in Yaounde, Cameroon. *Oryx*, **40**, 472–5.

Ehrlich, P. (1968). *The Population Bomb*. New York: Ballantine.

Elberling, B., Sondergaard, J. and Jensen, L. A. (2007). Arctic vegetation damage by winter-generated coal mining pollution released upon thawing. *Environmental Science and Technology*, **41**, 2407–13.

Erwin, K. L. (2009). Wetlands and global climate change: the role of wetland restoration in a changing world. *Wetlands Ecology and Management*, **17**, 71–84.

Fetridge, E. D., Ascher, J. S. and Langellotto, G. A. (2008). The bee fauna of residential gardens in a suburb of New York City (Hymenoptera: Apoidea). *Annals of the Entomological Society of America*, **101**, 1067–77.

Fiedler, A. K., Landis, D. A. and Wratten, S. D. (2008). Maximizing ecosystem services from conservation biological control: the role of habitat management. *Biological Control*, **45**, 254–71.

Folke, C., Jansson, A., Larsson, J. and Costanza, R. (1997). Ecosystem appropriation by cities. *Ambio*, **26**, 167–72.

Fuller, R. A., Irvine, K. N., Devine-Wright, P., Warren, P. H. and Gaston, K. J. (2007). Psychological benefits of greenspace increase with biodiversity. *Biology Letters*, **3**, 390–4.

Fuller, R. A., Tratalos, J. and Gaston, K. J. (2009). How many birds are there in a city of half a million people? *Diversity and Distributions*, **15**, 328–37.

Fullerton, D. and Wu, W. B. (1998). Policies for green design. *Journal of Environmental Economics and Management*, **36**, 131–48.

Gibbs, D. and Deutz, P. (2005). Implementing industrial ecology? Planning for eco-industrial parks in the USA. *Geoforum*, **36**, 452–64.

Green, R. E., Cornell, S. J., Scharlemann, J. P. W. and Balmford, A. (2005). Farming and the fate of wild nature. *Science*, **307**, 550–5.

Grimm, N. B., Faeth, S. H., Golubiewski, N. E. *et al.* (2008). Global change and the ecology of cities. *Science*, **319**, 756–60.

Hammond, G. P. (2000). Energy, environment and sustainable development: a UK perspective. *Transactions in Chemical Engineering Part B*, **78**, 304–23.

Hammond, R. A. and Hudson, M. D. (2007). Environmental management of UK golf courses for biodiversity – attitudes and actions. *Landscape and Urban Planning*, **83**, 127–36.

Hansen, A. J., Knight, R. L., Marzluff, J. M. *et al.* (2005). Effects of exurban development on biodiversity: patterns, mechanisms, and research needs. *Ecological Applications*, **15**, 1893–905.

Hardoy, J. E., Mitlin, D. and Satterthwaite, D. (2001). *Environmental Problems in an Urbanizing World: Finding Solutions for Cities in Africa, Asia and Latin America*. London: Earthscan.

Jones, T. A. (2003). The restoration gene pool concept: beyond the native versus non-native debate. *Restoration Ecology*, **11**, 281–90.

Jopling, J. and Girardet, H. (1996). *Creating a Sustainable London*. London: Sustainable London Trust.

Kalisinska, E., Salicki, W. and Myslek, P. (2004). Using the Mallard to biomonitor heavy metal contamination of wetlands in north-western Poland. *Science of the Total Environment*, **320**, 145–61.

Kaye, J. P., Groffman, P. M., Grimm, N. B., Baker, V. and Pouyat, R. V. (2006). A distinct urban biogeochemistry? *Trends in Ecology and Evolution*, **21**, 192–9.

Kellert, S. R. (1996). *The Value of Life*. Washington, DC: Island Press.

Kennedy, C., Cuddihy, J. and Engel-Yan, J. (2007). The changing metabolism of cities. *Journal of Industrial Ecology*, **11**, 43–59.

Kryuchkov, V. (1993). Extreme anthropogenic loads and the northern ecosystem condition. *Ecological Applications*, **3**, 622–30.

Lawson, D. M., Lamar, C. K. and Schwartz, M. W. (2008). Quantifying plant population persistence in human-dominated landscapes. *Conservation Biology*, **22**, 922–8.

Lepczyk, C. A., Hammer, R. B. and Stewart, S. I. (2007). Spatiotemporal dynamics of housing growth hotspots in the North Central US from 1940 to 2000. *Landscape Ecology*, **22**, 939–52.

Liu, J., Daily, G. C., Ehrlich, P. R. and Luck, G. W. (2003). Effects of household dynamics on resource consumption and biodiversity. *Nature*, **421**, 530–3.

McGranahan, G. and Satterthwaite, D. (2003). Urban centers: an assessment of sustainability. *Annual Review of Energy and the Environment*, **28**, 243–74.

McKinney, M. L. (2002). Urbanization, biodiversity, and conservation. *BioScience*, **52**, 883–90.

McKinney, M. L. (2006). Urbanization as a major cause of biotic homogenization. *Biological Conservation*, **127**, 247–60.

Mcmanus, P. and Haughton, G. (2006). Planning with ecological footprints: a sympathetic critique of theory and practice. *Environment and Urbanization*, **18**, 113–27.

Martinez, M. L. and Lopez-Barrera, F. (2008). Restoring and designing ecosystems for a crowded planet. *Ecoscience*, **15**, 1–5.

Marzluff, J. M. and Ewing, K. (2001). Restoration of fragmented landscapes for the conservation of birds: a general framework and specific recommendations for urbanizing landscapes. *Restoration Ecology*, **9**, 280–92.

Mehmood, S. R. and Zhang, D. W. (2001). A roll call analysis of the Endangered Species Act amendments. *American Journal of Agricultural Economics*, **83**, 501–12.

Meybeck, M. (2004). The global change of continental aquatic systems: dominant impacts of human activities. *Water Science and Technology*, **49**, 73–83.

Milder, J. C., Lassoie, J. P. and Bedford, B. L. (2008). Conserving biodiversity and ecosystem function through limited development: an empirical evaluation. *Conservation Biology*, **22**, 70–9.

Miller, J. R. (2005). Biodiversity conservation and the extinction of experience. *Trends in Ecology and Evolution*, **20**, 430–4.

Miller, J. R. (2006). Restoration, reconciliation, and reconnecting with nature nearby. *Biological Conservation*, **127**, 356–61.

Mills, G. (2007). Cities as agents of global change. *International Journal of Climatology*, **27**, 1849–57.

Montalvo, C. (2008). General wisdom concerning the factors affecting the adoption of cleaner technologies: a survey 1990–2007. *Journal of Cleaner Production*, **16**, S7–S13.

Moos, M., Whitfield, J., Johnson, L. C. and Andrey, J. (2006). Does design matter? The ecological footprint as a planning tool at the local level. *Journal of Urban Design*, **11**, 195–224.

Muezzinoglu, A. (2003). A review of environmental considerations on gold mining and production. *Critical Reviews in Environmental Science and Technology*, **33**, 45–71.

Newman, P. (1999). Sustainability and cities: extending the metabolism model. *Landscape and Urban Planning*, **44**, 219–26.

Newman, P. and Kenworthy, J. (1991). *Cities and Automobile Dependence: An International Source Book*. Aldershot, UK: Avebury.

Newman, P. and Kenworthy, J. (1999). *Sustainability and Cities: Overcoming Automobile Dependence*. Washington, DC: Island Press.

Nilson, J. (1995). A phosphorus budget for a Swedish municipality. *Journal of Environmental Management*, **45**, 243–53.

Nishioka, S., Noriguchi, Y. and Yamamura, S. (1990). Megalopolis and climate change: the case of Tokyo. In J. McCullock, ed., *Cities and Global Climate Change*. Washington, DC: Climate Institute, pp. 108–33.

Oberndorfer, E., Lundholm, J., Bass, B. *et al.* (2007). Green roofs as urban ecosystems: ecological structures, functions, and services. *BioScience*, **57**, 823–33.

Odum, E. P. (1963). *Ecology*. New York: Holt, Rinehart and Winston.

Odum, H. T. (1971). *Environment, Power, and Society*. New York: Wiley-Interscience.

Olalla-Tarraga, M. A. (2006). A conceptual framework to assess sustainability in urban ecological systems. *International Journal of Sustainable Development and World Ecology*, **13**, 1–15.

O'Meara, M. (1999). Exploring a new vision for cities. In L. R. Brown, ed., *State of the World 1999*. New York: Norton, pp. 133–50.

Pennington, D. N., Hansel, J. and Blair, R. B. (2008). The conservation value of urban riparian areas for landbirds during spring migration: land cover, scale, and vegetation effects. *Biological Conservation*, **141**, 1235–48.

Rees, W. E. (1997). Is 'sustainable city' an oxymoron? *Local Environment*, **2**, 303–10.

Rees, W. E. (2000). Eco-footprint analysis: merits and brickbats. *Ecological Economics*, **32**, 371–4.

Rees, W. E. and Wackernagel, M. (1996). Urban ecological footprints: why cities cannot be sustainable – and why they are a key to sustainability. *Environmental Impact Assessment Review*, **16**, 223–48.

Roach, W. J., Heffernan, J. B., Grimm, N. B. *et al.* (2008). Unintended consequences of urbanization for aquatic ecosystems: a case study from the Arizona desert. *BioScience*, **58**, 715–27.

Robinson, J. M. (2001). The dynamics of avicultural markets. *Environmental Conservation*, **28**, 76–85.

Rosenzweig, M. L. (2003). *Win–Win Ecology: How the Earth's Species Can Survive in the Midst of Human Enterprise*. New York: Oxford University Press.

Sanford, M. P., Manley, P. N. and Murphy, D. D. (2009). Effects of urban development on ant communities: implications for ecosystem services and management. *Conservation Biology*, **23**, 131–41.

Satterthwaite, D. (1997). Sustainable cities or cities that contribute to sustainable development? *Urban Studies*, **34**, 1667–91.

Schwartz, M. W., Thorne, J. H. and Viers, J. H. (2006). Biotic homogenization of the California flora in urban and urbanizing regions. *Biological Conservation*, **127**, 282–91.

Sorice, M. G. (2005). Book review: win–win ecology: how the earth's species can survive in the midst of human enterprise. *Society and Natural Resources*, **18**, 89–91.

Tratalos, J., Fuller, R. A., Warren, P. H., Davies, R. G. and Gaston, K. J. (2007). Urban form, biodiversity potential and ecosystem services. *Landscape and Urban Planning*, **83**, 308–17.

Turner, K., Lefler, L. and Freedman, B. (2005). Plant communities of selected urbanized areas of Halifax, Nova Scotia, Canada. *Landscape and Urban Planning*, **71**, 191–206.

UN-HABITAT (2003). *Water and Sanitation in the World's Cities*. London: Earthscan.

United Nations (2008). *World Urbanization Prospects: The 2007 Revision*. New York: United Nations.

Venter, O., Brodeur, N. N. and Nemiroff, L. (2006). Threats to endangered species in Canada. *BioScience*, **56**, 903–10.

Verissimo, A., Barreto, P., Tarifa, R. and Uhl, C. (1995). Extraction of a high-value natural resource in Amazonia – the case of mahogany. *Forest Ecology and Management*, **72**, 39–60.

von der Lippe, M. and Kowarik, I. (2008). Do cities export biodiversity? Traffic as dispersal vector across urban–rural gradients. *Diversity and Distributions*, **14**, 18–25.

Wackernagel, M., Kitzes, J., Moran, D., Goldfinger, S. and Thomas, M. (2006). The ecological footprint of cities and regions: comparing resource availability with resource demand. *Environment and Urbanization*, **18**, 103–12.

White, R. (1992). The international transfer of urban technology: does the North have anything to offer for the global environmental crisis? *Environment and Urbanization*, **4**, 109–20.

Winfree, R., Williams, N. M., Dushoff, J. and Kremen, C. (2007). Native bees provide insurance against ongoing honey bee losses. *Ecology Letters*, **11**, 1105–13.

Zhang, J. and Gan, J. B. (2007). Who will meet China's import demand for forest products? *World Development*, **55**, 2150–60.

Index

Locators in **bold** refer to major content
Locators in *italic* refer to figures/tables
Locators for headings which also have subheadings refer to general aspects of that topic only.

Printed in the United States
By Bookmasters